Unmanned Aircraft Design

Longbiao Li • Dmytro Tiniakov

Unmanned Aircraft Design

Features of Design Process

Springer

Longbiao Li
College of Civil Aviation
Nanjing University of Aeronautics and Astronautics
Nanjing, Jiangsu, China

Dmytro Tiniakov
College of Aerospace Engineering
Nanjing University of Aeronautics and Astronautics
Nanjing, Jiangsu, China

ISBN 978-981-95-7166-6 ISBN 978-981-95-7167-3 (eBook)
https://doi.org/10.1007/978-981-95-7167-3

This Springer imprint is published by the registered company Springer Nature Singapore Pte Ltd.
The registered company address is: 152 Beach Road, #21-01/04 Gateway East, Singapore 189721, Singapore

Preface

Modern engineering is a continually evolving field, with new technologies emerging on a daily basis to enhance various aspects of human life. Unmanned aerial vehicles (UAVs) represent one such technological advancement.

In the opinion of the engineering community, UAVs will supersede manned aircraft in the foreseeable future. It is likely that there will be a considerable amount of news coverage regarding autopiloted aircraft that have the potential to be utilized in a variety of new and innovative ways. Such applications include private package delivery, public inspection, maintenance of hard-to-access buildings (e.g., power lines, bridges), geophysical research, and others. In the modern era, an autopilot can effectively substitute for a human pilot in the majority of flight modes utilized by commercial airliners. Consequently, a number of companies have approached airworthiness agencies with the objective of modifying existing regulations pertaining to the requisite number of human pilots.

However, it should be noted that the technology utilized by drones is not a novel phenomenon. The use of unmanned flying vehicles has a long history, with evidence of their application dating back to ancient times. What, then, are the novel contributions to this field? What distinguishes the methodologies employed in the fabrication of unmanned aerial vehicles in the past from those utilized in the present? What new requirements and challenges are presented to aviation engineers? The authors of this book will endeavor to address these and other questions.

The objective of this book is to provide readers with a comprehensive overview of the historical development, analysis, and consideration of drones, which have previously been referenced in various contexts. The authors posit that this approach will conserve readers' time, allowing them to devote it to developing novel concepts related to drones. Additionally, the authors will present their own novel ideas about UAVs and analyze them in a manner that may be of interest to researchers.

Humanity continually strives to improve its living environment. To achieve this goal, an engineering discipline within the applied sciences was established. In the modern world, engineering plays a crucial role, driving sustained technological progress.

Aviation engineering holds particular significance within this field. Intensive development, especially in civil aviation, began after World War II. In engineering in general—and aviation in particular—breakthrough ideas periodically emerge. For instance, the advent of gas turbine engines enabled aviation to dominate long-distance passenger transportation, displacing maritime vessels. Later, in the 1960s, the introduction of titanium alloys improved the weight efficiency of civil aviation. Similarly, the use of composite materials contributed to aviation advancements in the late twentieth and early twenty-first centuries.

What new technologies and engineering solutions could become the driving forces of civil aviation in the near future? Currently, two promising directions are emerging: unmanned technologies and so-called green technologies.

This book will focus on the first of these—unmanned technologies. As modern trends demonstrate, unmanned systems offer numerous advantages, not only in aviation but also across various industries. However, this technology also presents several challenges that must be addressed for its successful implementation. This is especially critical in aviation, where any engineering oversight could compromise flight safety.

In just a few short decades, drones have evolved from military tools to ubiquitous devices reshaping industries, hobbies, and even humanitarian efforts. What was once the realm of science fiction is now an everyday reality—UAVs delivering packages, monitoring crops, inspecting infrastructure, and capturing breathtaking aerial footage. The civil drone revolution is here, and its potential is only beginning to unfold.

This book explores the branch of drones beyond the battlefield—their designing, manufacturing, technological advancements, safety issues, regulatory challenges, and transformative applications across diverse fields. From agriculture to filmmaking and disaster response to logistics, drones are proving to be more than just flying cameras; they are powerful tools driving efficiency and innovation and even saving lives.

The design challenges of drones are an important stage of the UAVs' cycling life. At first glance, the design of drones may not appear particularly complex. After all, civil aviation now possesses over a century of experience in aircraft design. However, this impression is deceptive. The defining feature of drones is the absence of an onboard human—a characteristic that might seem to simplify the design process. Yet, paradoxically, it also introduces new challenges for engineers.

Indeed, the lack of a human eliminates many traditional design constraints, such as life-support systems, emergency safety systems, and other similar requirements. Nevertheless, it is crucial to remember that civil aviation remains a branch of transportation logistics. As such, it must ensure operational efficiency within defined constraints. Since some limitations tied to human presence no longer apply, the criteria for evaluating efficiency have also shifted.

This raises critical questions: How must engineering approaches adapt to these changes? What now serves as the primary performance metric for assessing the effectiveness of commercial unmanned aircraft?

In this book, the authors aim to provide well-reasoned answers to these and other pressing questions.

Whether you are an enthusiast, a professional, or simply curious about the drone revolution, this book will guide you through the past, present, and future of civil drones. Fasten your seatbelt—you are about to take off into the exciting world of unmanned flight.

The book's primary contents are as follows:

- The initial chapter provides a comprehensive overview of the historical background, contemporary conditions, and future prospects of UAV technologies. In the absence of a comprehensive understanding of the engineering principles underlying drones and an accurate assessment of the demands of the UAV market, the development of high-efficiency drones will encounter significant challenges.
- The second chapter of the book provides a comprehensive overview of the design process for UAVs. A fundamental aspect of this process entails a comprehensive understanding of the limitations and diverse requirements associated with drones. In the contemporary era, there exists a multitude of long-standing traditions and established principles that must be given due consideration during the process of technological development. UAVs represent a novel branch of engineering. However, its origins are firmly embedded in the domains of aviation, shipbuilding, and space exploration, among others. The objective of this chapter is to explore the application of extant knowledge and skills to novel concepts, with the aim of enhancing efficiency.
- The third chapter of this book delineates the fundamental principles that must be taken into consideration when selecting a power plant for a given class of UAVs. In the modern time, the industry has the capacity to propose a wide range of engines and motors, with several novel types of engines on the horizon. This chapter delineates the various components of power plants utilized for drones.
- The fourth chapter is dedicated to the aerodynamic aspects of UAVs. As previously mentioned, the aerodynamic performance of flying vehicles is a critical factor. The primary objective of this chapter is to elucidate the methodology for evaluating and implementing aerodynamic requirements for a range of UAVs.
- The fifth chapter of this book is devoted to an examination of the structural characteristics of unmanned flying vehicles. The fundamental distinction between conventional aircraft and drones lies in the absence of a human operator on board the former. This specific represents a significant stride forward in the realm of aviation, particularly with regard to the advancement of flying vehicles. The primary objective of this chapter is to deliberate on the safety requirements and the structural integrity of drone frames.
- In Chap. 6, the reader is presented with a comprehensive overview of the various methodologies that can be employed to facilitate the takeoff and landing processes for UAVs. In the modern era, a variety of methodologies have been developed for the configuration of drones. Consequently, engineers have numerous

options for the organization of takeoff and landing procedures. The objective of this chapter is to analyze such methods and determine their application for UAVs.
- The subject of Chap. 7 is the description of the particulars of control systems for UAVs. The absence of a pilot on board has been demonstrated to exert an influence on the control process of drones. This is of particular importance during periods of autonomous flight. The objective of this chapter is to elucidate the fundamental principles of the control process for UAVs.
- A close examination of Chap. 8 reveals a significant correlation with Chap. 7. Navigation constitutes a component of the control process. The contemporary environment is characterized by the presence of satellites and inertial systems. These systems have the capacity to ensure the reliability of navigation for UAVs. The chapter provides a comprehensive overview of the methodologies employed for their implementation.
- The ninth chapter of this book is devoted to an in-depth discussion of the manufacturing process of UAVs. Due to the wide range of sizes in UAVs, spanning from micro- to large-scale dimensions, and the diverse requirements for different purposes, the manufacturing process of UAVs is characterized by its own unique set of characteristics. The chapter is devoted to elucidating the aforementioned subjects.
- The tenth chapter is dedicated to the testing and validation of UAVs. As has been previously outlined, the fundamental characteristics of a drone are predicated on its capacity for autonomous flight, a capability that is made possible by the absence of a pilot on board the flying vehicle. This feature imposes stringent requirements on the safety and reliability of UAVs in general and their constituent components in particular. In such conditions, it is imperative that the testing process yield reliable results that validate the safety indicators of UAVs.
- The final chapter is devoted to an exposition of the process of developing airworthiness regulations, with particular emphasis on the various types of UAVs and the intricacies of regional politics within the aviation safety sector. The UAV industry is in its nascent stages. Presently, aviation regulatory agencies are endeavoring to establish a reliable legal foundation for the future safe development of the drone industry.

The book is intended for a diverse readership, including both general readers and professionals in the field of engineering. The opening chapter provides a historical overview of unmanned aerial vehicles, presenting a wealth of factual information that will be of interest to a diverse audience. The remaining chapters address specific questions, including design approaches, aerodynamic and structural specifics, airworthiness problems, and so forth. The authors posit that the book will prove beneficial for students and engineers engaged in the aviation industry.

Nanjing, Jiangsu, China Longbiao Li
Nanjing, Jiangsu, China Dmytro Tiniakov
8th December 2025

Contents

Chapter 1
UAV Background

1.1 Introduction

As previously noted, unmanned technologies represent a significant development vector in modern civil aviation. While aviation possesses a well-established design tradition, drone engineering introduces several distinctive characteristics. Yet how novel are these features really? Admittedly, intensive research into unmanned aerial systems (UAS) is relatively recent. However, it's important to recognize that drone applications, much like conventional aviation, have their own historical background.

This historical foundation, while not as thoroughly developed as that of manned aircraft, nevertheless exists. A careful examination and analysis of early unmanned flight attempts and their applications may provide valuable insights for addressing contemporary challenges in UAS design, manufacturing, and operational implementation.

Consequently, a systematic study of unmanned aerial vehicle (UAV) history appears both warranted and potentially beneficial. In this chapter, the authors will undertake a brief historical survey of unmanned aviation's developmental trajectory.

1.2 Historical Background

The history of UAV has long-standing roots [1, 2]. As the history of any technology shows, the main driving force of progress is its military application. Nevertheless, a purely civilian application of these technologies always follows the military application. One of the most striking examples of this is aviation. This can also be observed with UAV. The military has achieved new capabilities in its operations through the use of UAVs. Currently, drones are used for various military purposes,

L. Li, D. Tiniakov, *Unmanned Aircraft Design*,
https://doi.org/10.1007/978-981-95-7167-3_1

including tactical and strategic reconnaissance, delivering weapons (bombs, torpedoes, etc.) to specified targets, and managing combat operations.

A brief overview of the historical progression of UAV development is warranted to facilitate comprehension of the technologies employed, their respective efficacies, and the manner in which they have evolved over time.

In ancient China, kites were utilized as drones for military applications, including aerial bombardment [1, 3]:

- During the Song Dynasty, the Chinese scientists had developed a kite known as the "Fiery Raven" that carried an explosive propellant.
- Kites were also utilized for military signaling purposes. A kite painted with specific colors served as a means of transmitting signals. For example, the color red indicated danger, while green signaled an attack.
- According to legend, the strategist Zhang Liang is said to have employed kites to transport several soldiers into an enemy camp. Soldiers were attached to kites.

The history of ancient China shows this. Contemporary drones continue to exhibit the same functions as their historical counterparts.

In the fifth century BC, the mathematician and mechanic Archytas of Tarentum developed an automatic flying machine that mimicked the flight of a pigeon [1]. The device was composed of wood, cylindrical shape, and was hollow on the inside. The structure was composed of two large and two small wings, located on the sides. The vehicle was powered by steam. A perforation was present in the rear part of the vehicle, through which it was connected to a water boiler. Upon the onset of evaporation within the boiler, a corresponding increase in pressure was observed, subsequently prompting the pigeon's takeoff. The flight persisted for several hundred meters, attributable to the force of compressed air within the vehicle hollow.

The aerial bombing of Venice in 1849 represents the earliest documented instance of the utilization of UAVs in history [1, 2]. The event unfolded during the Austrian siege of Venice in the context of the War of 1848–1849. There have been several attempts to use balloons to throw bombs over Venice. The bombs were dropped using clockwork mechanisms on board the vehicles. Many bombs have reached their targets. But, changing wind directions a lot made them less effective. So, these operations were then stopped.

The development of UAVs was significantly accelerated by the invention of the radiotelegraph by Professor A.S. Popov and Guglielmo Marconi in 90th of nineteenth century [1, 4]. Subsequent to this, a significant number of scientists endeavors to employ a wireless communication method for the remote control of vehicles. The demand for wireless signal transmission technology was evident, and by September 1898, Nikola Tesla had already unveiled the first radio-controlled unit to the public [1, 4]. The vessel's operational parameters, such as direction and speed, were modulated in accordance with the radio signals received from the control panel.

The prototype of modern cruise missiles is automatically controlled, unmanned, propeller-driven aircraft. The conception of such devices is believed to have emerged at the close of 1909, manifesting concurrently by engineers in different countries [1, 4]. This concept was predicated on several technological and historical developments, including the invention of the radio, the development of torpedo weapons, the initial experiments on radio control, and the first practical applications of aircraft.

The concept of remotely piloted aircraft emerged during World War I. Efforts were made in Britain and the US to develop "aerial torpedoes" or "flying bombs." Notable examples include Britain's "Aerial Target" (1917) and the American "Kettering Bug" (1918), essentially unmanned biplanes intended to deliver explosives [1, 4]. While these early attempts faced challenges with reliability and precision, they laid the groundwork for future development.

One such project was initiated by the American engineer E. Berliner in 1909 [1, 4]. In fact, the object in question was a winged torpedo that bore a striking resemblance to the Bleriot-11 monoplane, the design of which was conceived by the Frenchman L. Bleriot. It was on this aircraft that Bleriot achieved the distinction of being the first person to successfully navigate the English Channel by air in 1909. The command sequence for the flying torpedo was transmitted via a wire communication system. The utilization of the wing resulted in an augmentation of the torpedo's range and velocity. On June 5, 1917, the United States Patent Office granted a patent to Engineer E. Berliner for his invention.

In 1909, the French engineer R. Lorain began to consider the possibility of using a jet engine on aircraft. In May of that same year, a French engineer conceptualized the development of cruise missiles that would be directed by telemechanical devices. In early 1910, R. Loren published the first draft of a cruise missile in the magazine "Aerophile" ("L'aerophile") [5]. The aircraft was designed with a small wing, a launch weight of 79 kg, and a payload weight of 12 kg. Loren's objective was for his cruise missile to achieve a velocity of 200 km per hour.

In the aftermath of the First World War, German corporations embarked on endeavors to devise a radio-controlled unmanned bomber. One of these projects was Fledermaus, which translates to "the bat." The bomber was distinguished from analogous "aerial torpedo" projects of that historical period by the fact that the device was required to drop a bomb on a target and subsequently return to the airbase, landing with a parachute. The project was executed by several German companies from the autumn of 1917 to November 1918 [1, 4].

In the United States, engineers from Curtiss Aeroplan, in collaboration with Sperry Gyroscope, have dedicated their efforts to the pursuit of automatically controlled flight. The Curtiss N-9 seaplane was selected as the vehicle for the study [6]. By the close of 1917, the N9 aircraft had successfully completed an automatic flight on a predetermined course. During the second test flight of the day, the aircraft deviated from the predetermined course by 12.5%. Despite the low flight accuracy, the leadership of the US Navy issued a technical specification to the companies Curtiss Aeroplan and Sperry Gyroscope for the development of a fixed wing aerial vehicle with a take-off weight of 675 kg, a warhead weight of 450 kg, a flight range of 80 km, and a flight speed of 145 km per hour.

Concurrently, another project involving a flying bomb was underway at the Daiton-Wright Airplane Company [6]. According to the technical specifications provided, the flying vehicle was designed to have a takeoff weight of 280 kg, a warhead weight of 40–80 kg, a flight range of more than 60 km, and a flight speed of 90 km per hour. Consequently, a flying Bug bomb was soon designed. This vehicle was intended for single-use. The precision of the aircraft's trajectory to the designated target was facilitated by an autopilot system that utilized radio commands for navigation. Control in flight along the designated course was executed by E. Sperry's autopilot, which utilized actuators and a rudder to maintain stability and direction. The flight altitude was meticulously regulated by an altimeter, which employed an aneroid barometer to ensure precise measurement. Prior to reaching the target, the wings were undocked, and the vehicle descended towards the target along a semi-ballistic trajectory. During the First World War, the cost of an airplane was approximately $5000, whereas the cost of a Bug flying vehicle was only $575.

The period between 1920 and 1930 was characterized by the refinement and expansion of pre-existing power projects, as well as the development of novel aircraft power plants [1]. The focus shifted towards target drones for anti-aircraft gunner training. Reginald Denny, a British model aircraft enthusiast, played a significant role. His "Radioplane" company produced the OQ-2 [1, 4, 6], a widely used radio-controlled drone during WWII, with nearly 15,000 manufactured for the US Army. The term "drone" is believed to have originated from a model called the DH.82B Queen Bee. Germany also developed the V-1 flying bomb, an early cruise missile.

On June 26, 1928, the German engineer R. Teeling submitted an application for a solid-fuel cruise missile [6]. It is widely considered that R. Teeling was the first to integrate the characteristics of both winged and ballistic missiles into a unified design, thereby developing the inaugural cruise missile. When confronted with identical thrust values and engine operating time, a ballistic missile can attain a considerable altitude during the active segment of its trajectory. At this altitude, the cruise missile demonstrated the capability to traverse a greater distance than a ballistic missile during the passive segment of its trajectory, a feat accomplished through the utilization of aerodynamic gliding techniques. On March 13, 1931, Teeling and Karl Poggensee launched a solid-fuel rocket to an altitude of 1800 m. On April 15, 1931, Teeling unveiled the inaugural postal rocket, which successfully delivered 188 postcards to the public.

In 1933, British engineers developed the first reusable target drone, which was designated the Queen Bee [1, 6]. The prototype for the Queen Bee was the Fairey Queen project. The weapon's design was conceived for deployment from a catapult mounted on warships. The Queen Bee exhibited notable similarities to its predecessor, yet it also demonstrated distinct characteristics. The fuselage was derived from the Moth Major, while the wings were sourced from the Tiger Moth aircraft. The maximum velocity of the aircraft was 175 km per hour at an altitude of 300 m, with a practical ceiling of 4145 m. The Queen Bee UAS is designed to be operable in a remote mode from an operator console situated on another aircraft, a warship, or the ground. The control panel was equipped with a rudimentary rotary dial, enabling the operator to issue commands through the radio channel. The aircraft's operation was dictated by a series of instructions, including "turn left," "turn right," "gain altitude," and other commands. The ignition and throttle were regulated by additional controls. The range of the control channel was 5 km. The production of these aircraft exceeded 400 units, which were deployed to support military operations, persisting in service until at least 1947.

In November 1939, Fritz Gosslau, a doctor-engineer from the German aviation company Argus, presented a project for a "flying bomb" (Fernbombe) [6]. In July 1941, Argus and Fieseler submitted a more detailed design for a vehicle with a range of 250 km, a speed of 450–600 km per hour, a payload capacity of up to one ton, and an estimated circular probable deviation (CPD) of approximately 0.9 km. The Fi.103 aircraft, also called the "Kirschkern," was the prototype for the V-1 cruise missile. The vehicle was simple and relatively inexpensive. The cost of a vehicle was one-fifth the cost of a V-2 ballistic missile and one-thirtieth the cost of a bomber. Additionally, producing one Fi.103 cruise missile took only 280 labor hours. However, the Fi.103 required a catapult for launch. The catapult accelerated the aircraft to 320 km/h, after which the engine gradually increased the speed to 650 km/h. The Fi.103 rocket was designed according to the aerodynamic scheme of a free-roasting midplane with a symmetrical, spindle-shaped fuselage. Steel was widely used in the design to provide the affordability necessary for a single-use aircraft. The cruise missile was designed to fly in a straight line at altitudes between 300 and 2500 m. It had relatively high maneuverability only in the vertical plane and climbed at a speed of 150 m per minute. Due to the lack of ailerons, the cruise missile could only perform flat turns. Low-octane gasoline was used as fuel. The power plant provided a speed of about 650 km/h (0.53 Mach) at the beginning of the flight. As the fuel burned out and the vehicle lightened, the speed increased to 800 km/h (0.65 Mach). The Fi.103 had many disadvantages related to flight control. For example, of the 68 combat launches carried out in June–July 1943, only 28 reached the target. The Fi.103's flight path corresponded to the program laid out during launch preparation. After launch, it could not be altered. After being ejected, the cruise missile reached the required altitude of 300–2000 m and flew straight to the target. After calculating the distance according to the meter, the vehicle shut off the fuel supply and fixed the elevator in the dive position. The Fi.103 had low accuracy (CPD 1–2 km). The designation V-1 (Vergeltungswaffe 1, or "weapon of retribution") did not appear until the summer of 1943. In terms of flight characteristics, the V-1 cruise missile was nearly equivalent to modern aircraft of that time. It flew at high speeds and often at low altitudes. The autonomous control system allowed the cruise missile to be used in any weather and at any time of day. All of this made intercepting the V-1 quite difficult, even with the latest technology of the time, such as radar stations and fighter jets.

From after World War II until the mid-1950s, aviation engineers primarily focused on improving manned aviation by introducing turbojet engines [1, 4, 6]. Projects that had been started before and during the war were finalized for unmanned aircraft. The developments of German engineers were of particular interest. Building on these developments, significant groundwork was laid for cruise missiles in both the US and the USSR. The Cold War spurred significant advancements in UAV technology, primarily for military reconnaissance and intelligence gathering. Drones like the Ryan Model 147 "Lightning Bug" were extensively used by the US during the Vietnam War for surveillance missions, demonstrating their value in reducing risk to human pilots [6].

One example of such a device is the Regulus I projectile aircraft. Development of the Regulus 1 began in 1947, drawing on experience with the Lune projectile aircraft and testing of the German V-1 rocket [6]. The Chance-Vought company led the project. The first flight of the Regulus cruise missile occurred in 1950. From 1952 to 1962, submarines and warships were equipped with Regulus missiles. The Regulus I was decommissioned in the early 1960s. This was due to imperfections of the vehicle; for example, the inertial control system had a CPD of about 4.2 km from the target. They were replaced by the more advanced Regulus II.

American engineers developed an interesting project involving the CIM-10 Bomark unmanned jet aircraft [1, 4, 6]. It was assumed that the aircraft would activate autopilot immediately after launch, climb, and move to the target area while coordinating its flight automatically using the SAGE control system. The homing feature only worked when approaching the target. During the 1952–1958 testing period, approximately 70 missiles were launched from the Cape Canaveral test site. The Bomark was an unmanned interceptor, and it was initially designed for reuse. The unmanned vehicle was supposed to use air-to-air missiles against attacking aircraft and then make a soft landing using a parachute rescue system. However, due to the excessive complexity of this option, it was rejected. Consequently, the developers decided to build a single-use interceptor equipped with a powerful fragmentation or nuclear warhead.

The 1950s witnessed a substantial advancement in UAV technology with the advent of jet engines, enabling enhanced performance characteristics such as increased speed, altitude, and the capacity to more closely resemble actual targets. This decade also marked the tentative steps towards using UAVs for actual intelligence gathering.

The development of the BQM-34A formally commenced in the 1950, initiated under a contractual agreement between the United States Air Force and Teledyne Ryan Aeronautical [6]. The Firebee drone was designed to serve as the paradigm of a jet-powered target drone. The utilization of the system in question was extensive by the United States Air Force, Navy, and Army, with the objective of training anti-aircraft defense crews to counteract fast-moving jet threats. The success of the B-29 led to the development of numerous variants and a prolonged service life, even extending to its use as a reconnaissance platform. Its deployment could be carried out from the ground or from an airborne platform, followed by a parachute recovery. The BQM-34A's production was terminated in 1982; however, in 1986, the manufacturing line was reinitiated to produce BQM-34S targets.

The success of target drones such as the Firebee prompted experimentation in converting them for reconnaissance purposes. These early modifications frequently entailed the installation of cameras and rudimentary telemetry systems.

The Radioplane SD-1 (MQM-57 Falconer) is a noteworthy example of such innovation [1, 4, 6]. Despite its limited deployment for reconnaissance purposes during the 1950s, the SD-1 exemplifies an early initiative by the Army to develop a diminutive, tactical reconnaissance drone. The aircraft in question was a propeller-driven monoplane designed for short-range battlefield surveillance. The MQM-57 remained in service until the mid-1970s, with approximately 1500 drones of all versions having been manufactured.

Research in the field of aviation conducted by engineers within the Soviet Union yielded similar results. A notable example of this design is the Lavochkin La-17 [6]. The target drone in question was jet-powered. Throughout its extensive service life, the La-17 underwent numerous modifications, including the development of a tactical reconnaissance version (La-17R) in the early 1960s. This variant was equipped with cameras and radiation monitoring equipment. The platform remained in service into the 1980s and even early 1990s, making it a very long-lived platform.

The Vietnam War served as a significant theater for the operational assessment of UAVs, showcasing their efficacy in perilous reconnaissance missions, particularly in environments characterized by substantial air defense capabilities. These years witnessed the introduction of reconnaissance drones designed specifically for military applications. The Ryan Model 147 "Lightning Bug" is widely regarded as the most significant drone of the Vietnam War. Ryan Aeronautical modified its target drone with larger wings, more fuel, and various reconnaissance payloads (high-resolution cameras, electronic signal intelligence sensors).

Another example of 60th drones is the Gyrodyne QH-50 DASH (Drone Anti-Submarine Helicopter) [6]. While the QH-50 was principally developed for anti-submarine warfare (ASW) operations by the United States Navy, it also marked a pioneering advancement in the field of remotely piloted helicopters. The vessel was capable of carrying torpedoes and was deployed from destroyers. Subsequently, a select number of QH-50 s were fitted with cameras for reconnaissance as part of the "Snoopy" program, thereby demonstrating the adaptability of rotary-wing UAVs for observation missions.

In the ensuing decade, there will be ongoing endeavors to advance the development and miniaturization of drones. While the rate of significant new UAV deployments may have experienced a slight decline following the Vietnam War, the 1970s witnessed ongoing research and development initiatives, with a focus on enhancing control systems, sensor technology, and the exploration of smaller, more tactical designs.

At that time, an increasing number of companies were expressing interest in smaller, more portable UAVs for the purpose of battlefield surveillance. Projects such as the MOBILE Epervier (Belgium) and the Westland Wasp (UK) exemplify European endeavors to develop tactical reconnaissance drones, frequently equipped with television cameras for real-time observation [6].

The Epervier drone is characterized by its winged configuration, equipped with a turbojet engine, and employed for the purpose of battlefield reconnaissance under both diurnal and nocturnal conditions, in addition to object recognition. The flight was executed in radio-controlled mode within the target approach area and in fully automatic reprogrammable mode within the target area. Subsequent to the mission, the UAV descended by parachute. The system was utilized by the Belgian Armed Forces until the beginning of the twenty-first century.

The Westland Wisp was an experimental unmanned coaxial helicopter developed in the UK during the 1970s by Westland Helicopters [7]. The vehicle was originally designed and developed as a small, remotely piloted aircraft with primary applications in reconnaissance and surveillance. Initial models were equipped with electric motors; subsequent iterations incorporated small combustion engines instead.

The Tupolev Tu-141 Strizh is a Soviet reconnaissance unmanned aerial vehicle [1, 6]. The primary objective of the creation is to procure real-time aerial reconnaissance in conditions. The Tu-123 model served as the basis for the Tu-141. However, the Tu-141 has undergone substantial enhancements and upgrades to ensure optimal efficiency in carrying out reconnaissance missions. The drone was equipped with on-board reconnaissance equipment, including optical and radar systems. The UAV had the capacity to execute a variety of tasks, including artillery guidance, weather modification, and other operations. The Tu-141 was controlled from the ground or from a carrier aircraft that could transmit intelligence information in real time.

In the 1970s, Israel initiated the development of UAVs, thereby establishing the foundation for its subsequent preeminence in the domain of military drone technology [6]. Concurrent with the United States and the Soviet Union's experimentation with large, expensive reconnaissance drones, Israel focused on smaller, more cost-effective systems tailored to its defense needs. The Tadiran Mastiff is an Israeli reconnaissance UAV developed by Tadiran Electronic Systems. The drone's inaugural flight occurred in 1973. The drone is a diminutive tactical UAV that has been designed for surveillance and reconnaissance purposes. It possesses a high degree of endurance and the capacity to transmit data in real time. The drone in question was of the airplane type, equipped with a compact piston engine.

The 1980s witnessed a period of significant transformation for UAVs, largely attributable to the successful operational deployment by Israel during the 1982 Lebanon War [6]. This demonstrated the immense potential of UAVs beyond target practice and high-altitude reconnaissance, paving the way for their integration into modern military doctrine. The advent of GPS also initiated a period of profound transformation in the realm of navigation.

The IAI Scout is a drone that shares a similar conceptual framework with the Mastiff [6]. The Scout was another Israeli tactical reconnaissance UAV that provided real-time video to ground stations. These UAVs played a pivotal role in the suppression of enemy air defenses, prompting a substantial shift in military thinking about UAVs.

Inspired by the Israeli success, the U.S. Navy and Marine Corps rapidly acquired and deployed the AAI RQ-2 Pioneer [1, 4, 6]. This was essentially a licensed-built version of the Israeli IAI Pioneer. The Pioneer was utilized extensively during the Gulf War (1990–1991) for naval gunfire support, battle damage assessment, and even psychological operations (the dissemination of messages). Its deployment can be facilitated from seafaring vessels and subsequent retrieval can be achieved through the use of fishing nets.

During the 1980s, the integration of GPS (Global Positioning System) into drone technology resulted in a paradigm shift in military and reconnaissance operations, thereby enabling precise navigation, autonomous flight, and improved target tracking. While GPS was still in its early stages of development (and was not fully operational until the mid-1990s), experimental and early military drones began incorporating satellite navigation for enhanced accuracy.

As is the case with the RQ-2 Pioneer, the GNAT-750 is a reconnaissance UAV of American origin, marking it as the precursor to the Predator UAV [1, 6]. The aircraft's aerodynamic design incorporates a single-girder fuselage and a propeller mounted at the rear. The tail exhibits an inverted V-shaped configuration. Both of these drones were equipped with the first experimental GPS systems.

The impact of GPS on the development of drones in the 1980s was a seminal development in the field:

- Autonomous Flight. This technology enables drones to operate with reduced reliance on ground control, following pre-programmed GPS waypoints with greater autonomy.
- Precision Strike Coordination. This technology enabled the utilization of drones for the purpose of marking targets for artillery and aircraft.
- Beyond-Line-of-Sight Operations. The advent of drone technology has ushered in a new era of autonomous navigation, enabling these devices to operate far from the control of human operators without compromising their ability to navigate.

Consequently, during the 1990s, GPS became a standard feature in military drones, such as the Predator and Global Hawk models. This development was preceded by the experimental and early operational systems that were introduced during the 1980s, thereby establishing the foundation for the subsequent advancements in drone technology.

The 1990s witnessed the genesis of the sophisticated UAVs of today, marked by the development of platforms with enhanced endurance and capabilities that would delineate the subsequent generation of aerial warfare.

Two examples are provided below: The RQ-1 Predator and the RQ-4 Global Hawk are two UAVs that have been utilized in various military operations [1, 6].

The RQ-1 was developed by General Atomics, and the Predator was revolutionary in its impact. Initially developed for reconnaissance and surveillance missions, the aircraft demonstrated the capacity to operate in hostile environments for extended durations, transmitting real-time video and other sensor data via satellite communication links. While the Predator is most renowned for its post-2001 deployments, it was utilized for intelligence gathering and target identification during the mid-1990s. This drone served to validate the concept of persistent, remote surveillance. Subsequently, it would be equipped with weaponry, undergoing an evolution into the MQ-1.

The development of the RQ-4 Global Hawk commenced in the 1990s, with its operational debut occurring in the early 2000s. The conceptualization of the Global Hawk took place during this decade. The aircraft was developed to function as a high-altitude, long-endurance reconnaissance vehicle, designed to cover extensive areas over protracted missions. This development signified a transition from a predominantly tactical to a more strategic approach in the realm of UAV capabilities.

The 1990s also marked the inception of interest in smaller, hand-launched UAVs, though these would become much more prevalent in the 2000s with advancements in battery and sensor technology [4, 6]. In contrast to the large, costly drones utilized during the Cold War era, these systems were more affordable, portable, and extensively deployed, particularly by the United States and Israel.

One of the earliest examples of such drones was the AeroVironment FQM-151 Pointer, a tactical reconnaissance UAV [6]. The system has been engineered for real-time television or infrared surveillance of terrain and targets. The aircraft possesses wings and has an estimated takeoff weight of approximately 4 kg. The vehicle's propulsion system was powered by an electric motor.

In consideration of the aforementioned observations, it is feasible to assert that the prevailing trends and significant developments that transpired during the period spanning from the 1950s to the 1990s can be characterized as follows:

1. A shift from Target Drones to Reconnaissance: The initial utilization of UAVs was for the simulation of enemy aircraft; however, their primary function has evolved to encompass active intelligence gathering.
2. Increasing Autonomy (limited): While remote control remained the dominant form of control, early attempts at pre-programmed flight paths and basic autonomous functions began to emerge, especially with GPS integration.
3. The development and refinement of sensors has led to a notable reduction in size, a process referred to as sensor miniaturization. Sensor miniaturization has contributed to an improvement in the functionality and efficiency of these devices, paving the way for new and innovative applications in these areas. The capacity to accommodate sophisticated cameras (e.g., still, video, infrared) and electronic intelligence sensors on smaller platforms was imperative.
4. Data Links: Advancements in radio communication, followed by satellite communication, facilitated real-time data transmission and augmented operational ranges.
5. Operational Validation: Conflicts during this period yielded invaluable real-world experience, thereby demonstrating the military utility of UAVs. This, in turn, led to further investment and development in the field.
6. Variety of Propulsion: The evolution of propulsion systems has been driven by the evolving demands of military operations, with successive advancements in engine technology resulting in enhanced performance in terms of speed, altitude, and endurance. This progression, from piston engines to early jets and subsequently to more efficient turboprops, underscores the importance of technological innovation in meeting the specific requirements of military operations.

This period is widely regarded as the foundation for the subsequent widespread adoption and diversification of UAVs that has been observed in the twenty-first century.

1.3 Modern Main Trends

The twenty-first century has seen a proliferation of drone development, transforming these vehicles from specialized military tools to ubiquitous consumer devices and critical instruments across numerous industries. This evolution is driven by advancements in miniaturization, battery technology, computing power, artificial intelligence, and sophisticated sensor systems.

This subsection provides a detailed analysis of the evolution of drone development from the beginning of the twenty-first century to the present.

The early 2000s witnessed a significant advancement in drone technology, propelled by the War on Terror, advancements in satellite communications, and the transition from reconnaissance-only drones to armed UAVs. The United States and Israel dominated the development phase, while Europe and China initiated the expansion of their respective programs.

The US has developed armed UAVs, specifically the MQ-1, which is based on the reconnaissance UAV RQ-1 [4]. In 2001, the inaugural armed Predator successfully launched a Hellfire missile. This marked a significant shift, enabling "hunter-killer" missions where a single platform could find and strike targets. It became a symbol of modern warfare, lauded for its precision and criticized for its controversial use and civilian casualties.

Another trend that has emerged in the domain of mini and tactical drones is the proliferation of "backpack UAVs."

For illustrative purposes, the RQ-11 Raven is a pertinent case. This is a diminutive reconnaissance UAV. The design of the FQM-151 Pointer drone served as the basis for the development of the RQ-11 Raven drone [4]. The initiative was initiated as a component of the US Army's Future Combat Systems program, which aims to augment the technological capabilities of ground troops. The inaugural prototype completed its maiden flight in October 2001. The RQ-11 is utilized for reconnaissance and battlefield surveillance. The device can be launched manually, akin to an aircraft, and does not necessitate a designated landing site. The aircraft's operation can be configured to either automatic mode, facilitated by GPS navigation, or manual remote-control mode.

Furthermore, the research on stealth and high-altitude long-endurance drones was continued.

An example of such a drone is the RQ-4 Global Hawk, an American strategic reconnaissance UAV [1, 4]. The development of the system was conducted by Teledyne Ryan Aeronautical, a subsidiary of Northrop Grumman. The distinguishing characteristic of this drone is its dimensions (wing span is about 40 m) and mass (takeoff mass is about 14.5 tons), which are commensurate with those of an airplane. The flight duration, which is up to 36 h, is another salient feature. In addition to its deployment for military reconnaissance missions, the RQ-4 Global Hawk is utilized for environmental monitoring in support of scientific research initiatives.

In this period, as well, the research efforts were directed towards the study of swarms and experimental drones. The development of such programs has commenced:

- Micro Air Vehicle (MAV). The purpose is to engineer an UAV with a maximum size of 15 cm, which will be utilized for terrain exploration, building detection in urban environments, and other related tasks.
- The Fast Lightweight Autonomy (FLA) is a theoretical framework that aims to optimize the efficiency and autonomy of robotic systems. The objective of this research is to develop algorithms that will enable an autonomous drone to navigate both urban and indoor environments, successfully overcoming obstacles without the intervention of a human operator.
- Aerial Dragnet. The objective of the initiative was to establish a system capable of detecting and tracking small drones at low altitudes, including on rough terrain in urban areas.

One notable example is the Wasp model developed by AeroVironment [4, 6]. The AeroVironment created a miniature UAV that was supposed to perform real-time reconnaissance, surveillance, and targeting tasks. The Wasp variant possesses a wingspan of 33 cm, a weight of 200 g, and is equipped with two miniature video cameras. The AeroVironment Wasp drone is utilized in military operations.

The "Super Swarm Project 2000" constituted foundational programs and technological demonstrations, primarily led by the Defense Advanced Research Projects Agency (DARPA) in the United States [4, 6]. These programs sought to explore and prove the feasibility of multiple unmanned systems working together autonomously. During the early 2000s, there was a significant initiative to develop UAVs capable of both autonomous flight and target engagement. It is imperative to note that the conceptualization of these systems operating in a collaborative manner was a fundamental tenet from the project's inception.

The Boeing X-45 Program was a pivotal program that perfectly encapsulated the "Super Swarm" ambition of the era [4, 6]. The X-45 was an experimental UAV developed for the United States Air Force (USAF). Subsequently, it was incorporated into the Joint Unmanned Combat Air System (J-UCAS) program. The X-45 program sought to demonstrate not only the individual capabilities of UAVs but also the ability of multiple X-45 aircraft to collaborate autonomously. The experiment demonstrated the capacity of these jets to operate in a collaborative manner with minimal human intervention, exhibiting a coordinated response to their combat environment. By 2005, the X-45A prototypes had advanced to the point of executing mock combat missions as an integrated unit, building on the foundation laid in the early 2000s. The software enabled ground operators to direct missions from any location, while the two X-45A aircraft could collaborate through the utilization of algorithms, onboard sensors, and data links. This was the first operational progenitor of the flying UCAV "swarm" concept to be disclosed to the public.

The 2010s decade witnessed a notable increase in the accessibility of drones, resulting in a substantial proliferation of commercial and civilian applications, concurrent with ongoing advancements in military technology. The advent of powerful, small electric motors and advanced flight controllers (often based on smartphone technology) rendered multi-rotor drones a viable technological proposition.

Concurrently, the consumer drone revolution transpired. The DJI Phantom 1, released in 2013, signified the inaugural release of DJI's quadcopter series [8]. This series may have signaled the inception of the consumer drone market. The device's affordability, ease of operation, and incorporation of built-in cameras have collectively contributed to its widespread dissemination, thereby democratizing aerial photography and videography. It is evident that it contributed to the development of a new industry for amateur professionals of various specialties. This technology has been adopted by millions of individuals for personal photography, travel video production, and in early commercial applications such as real estate photography and small-scale inspections.

A little earlier, the Parrot AR (2010) drone was among the earliest and most influential quadrocopter that can be controlled using a smartphone application [2, 8]. It demonstrated the potential of convenient and affordable drone technology, which has the capacity to expand the boundaries of consumer electronics.

DJI has also developed a professional Matrice (for example, the M200 series) of drones. These reliable platforms have been designed for heavy-duty applications and are capable of supporting a variety of professional workloads, such as thermal imaging cameras, lidar sensors, and high-magnification cameras. These vehicles have become indispensable tools in a variety of fields, including infrastructure inspection, geodesy, agriculture, and public safety. The utilization of drones encompasses a wide range of applications, including the inspection of power lines, wind turbines, cell towers, and construction sites. Additionally, these unmanned aerial vehicles are employed in the spraying of crops and the provision of assistance in search and rescue operations.

Concurrently, there was a resurgence of interest in drones capable of vertical takeoff and landing. This enhancement can be attributed to the advancements in flight control systems, battery technology, and a more efficient "hybrid VTOL" design, which contributes to an extended service life and enhanced speeds. These vehicles are equipped with both fixed wings, which facilitate efficient flight, and multiple rotors, which enable vertical lift.

The following types of VTOLs are presented:

1. The utilization of dedicated vertical lift propellers is a common practice, followed by a transition to fixed-wing flight, often characterized by the aircraft "standing" on their tails during take-off and landing. This approach is exemplified by certain military reconnaissance drones.
2. The tilt-rotor/tilt-wing configuration is a specialized aircraft design that incorporates both rotor and wing mechanisms, allowing for vertical takeoff and landing capabilities. The propulsion system, or wing, is designed to rotate, thereby providing both vertical and horizontal thrust. This concept has been influenced by the V-22 Osprey, a manned aircraft that has inspired numerous UAV designs, leading to the development of smaller, uncrewed versions.

A significant number of contemporary commercial inspection, mapping, and surveillance drones are now equipped with hybrid VTOL designs (e.g., SenseFly eBee series RTK/PPK, various military reconnaissance platforms).

The development of VTOL UAVs has undergone a series of significant transitions, evolving from an initial adaptation of basic helicopter principles to a state of fundamental transformation initiated by the multirotor revolution [7]. This transformation has led to the emergence of efficient, hybrid platforms that are adaptable to a wide range of applications across consumer, commercial, and military sectors.

The current decade is characterized by an increasing trend of autonomy, the integration of advanced artificial intelligence (AI), the emergence of drone swarms, and the profound impact of inexpensive, off-the-shelf drones in conflicts [9]. Military operations worldwide have been a substantial catalyst for rapid innovation and the widespread adoption of various drone types.

The highest level of adjustment was applied to AI-powered autonomy and obstacle avoidance drones. For instance, companies such as Skydio (X2/X4 models) have been at the forefront of developing vision-based autonomous flight systems [10]. These drones possess the capability to navigate complex environments, avoid obstacles in real-time, and track subjects without the need for constant human input. This reduction in workload has been shown to enhance safety and enable operations in challenging areas, such as indoor inspections and dense forests. These devices find application in a variety of scenarios, including search and rescue operations by public safety agencies, law enforcement applications for situational awareness, and autonomous infrastructure inspection by enterprise clients.

The employment of "kamikaze" or loitering munitions, including but not limited to Switchblade, Shahed-136, and FPV drones, signifies a novel approach within the domain of military drones [11]. Although the concept had existed previously, it was not until the 2020s that it gained widespread and devastating use, thus becoming a defining feature of the era. These UAVs, engineered to autonomously navigate to their targets and execute strikes with precision, offer a cost-effective alternative to traditional aerial bombing campaigns. These systems vary in size, with some being small enough to be carried on backpacks (e.g., Switchblade) and others being larger and able to travel over longer distances (e.g., Shahed-136). The adaptation of hobbyist FPV (First Person View) racing drones for combat has been particularly impactful, providing a low-cost, precise strike capability.

The advent of delivery drones signifies a paradigm shift in the realm of logistics, proffering a more expeditious, economical, and environmentally sustainable alternative to conventional delivery methodologies [12]. While China has emerged as a global leader in the realm of commercial drone delivery, other countries, including the United States, various European nations, and Rwanda, have also made significant strides in this field.

The development of delivery drones commenced in the preceding decade. In approximately 2013, Amazon, Google, and SF Express initiated their research endeavors. However, it is important to note that these drones underwent a significant surge in development only after the year 2020. Presently, this technology remains under active development. This approach has the potential to significantly mitigate logistical losses.

Key trends defining twenty-first-century drone development:

1. The process of miniaturization, characterized by the reduction of size of vehicles or components, and cost reduction, the decrease in the financial expenditure required for a product or service, are of particular concern in this field. The advent of smartphone components and the advent of mass production have facilitated the dissemination of drones to a broader market segment.
2. The advent of increased autonomy and AI has profoundly impacted various facets of aviation. From GPS-guided flight to advanced computer vision, machine learning, and decision-making capabilities, there has been a notable reduction in the reliance on human pilots.
3. The integration of sensors is a critical component of the system. UAVs have undergone a series of technological advancements, resulting in the development of increasingly sophisticated payloads. These sophisticated payloads include high-resolution cameras, thermal imagers, LiDAR systems, multispectral sensors, and chemical/radiation detectors.
4. The term "Beyond Visual Line of Sight" (BVLOS) is employed to denote the condition of operating a drone in a manner that is not constrained by the physical limitations of visual perception. The advent of regulatory and technological advancements has rendered drones capable of autonomous flight over extended distances, a pivotal development for applications such as delivery services, large-scale inspection operations, and strategic military operations.
5. Multi-Purpose Platforms: The utilization of UAVs has evolved beyond the realm of surveillance; they are employed in a diverse array of applications, including military strikes, logistics support, communications relay, electronic warfare, and beyond.
6. Swarming and Collaborative Operations: The capacity for multiple drones to operate in conjunction, whether autonomously or with minimal human oversight, has the potential to transform military tactics and efficiency.
7. Counter-drone measures are a critical component of contemporary security strategies. As drones become increasingly prevalent, the necessity to protect against their misuse has also increased, resulting in a competition among countries to develop advanced counter-unmanned aerial system (C-UAS) technologies.
8. Ethical and Regulatory Challenges: The accelerated rate of technological development has outpaced the capacity for regulatory oversight, thereby giving rise to intricate inquiries concerning privacy, accountability, and the ethical implications of autonomous weapons systems.

The twenty-first century has been characterized by the proliferation of UAVs, which have profoundly impacted various aspects of warfare, commerce, and daily life. This transformation is particularly remarkable when considering the limited imagination of the preceding decades regarding the potential of such technology. The rate of innovation is accelerating, driven by both military necessity and commercial opportunity.

1.4 UAV Vs Traditional Aircraft

Traditional aircraft and UAVs represent two distinct approaches to aviation, each with unique advantages and disadvantages [1, 4]. The following is a comparative analysis:

1. The most obvious and fundamental difference is the presence of a human pilot on board.

   The design of traditional manned aircraft is such that they are intended to be operated by one or more human pilots who are physically present in the cockpit [3]. The pilot, who is the aircraft's primary operator, is responsible for the aircraft's control and the making of real-time decisions. These decisions are based on sensory input and instrument readings. The following are illustrative examples: the Boeing 747/Airbus A380 are commercial airliners, the Lockheed Martin F-22 Raptor/Chengdu J20 Mighty Dragon are fighter jets, the Cessna 172/Piper Archer are general aviation aircraft, and the Sikorsky UH-60 Black Hawk/Mil Mi-8 Hip are helicopters.

   The UAVs/Drones are capable of operating without a human pilot on board vehicles are capable of operating in two distinct modes: either through remote control by an operator from a ground station, or through autonomous flight based on pre-programmed flight plans and onboard intelligence. The following are illustrative examples: The following drones are notable examples of the diverse range of UAVs in current use: the DJI Mavic series [8] (which is designed for commercial use), the MQ-9 Reaper [1] (a military surveillance and attack drone), the Global Hawk (a high-altitude reconnaissance drone), and various small quadcopters that are employed for photography and inspection.
2. Safety and risk to human life [2, 13].

   For traditional manned aircraft: the inherent risk to human life (i.e., pilot, crew, and passengers) in the event of an accident or combat scenario is a primary concern. Extensive safety protocols and training have been implemented to mitigate this risk.

   For UAVs/Drones: it is important to note that no human life is directly at risk in the aircraft itself. This feature enables the utilization of the vehicle in perilous or hazardous environments, such as combat zones, disaster areas, and chemical spills, without endangering the personnel.
3. Cost [2, 3].

   For traditional manned aircraft: in general, the acquisition, operation, and maintenance of the aforementioned system are accompanied by significantly higher financial expenditures. These expenses encompass high fuel costs, extensive maintenance schedules, and the substantial expense of pilot training and salaries.

   For UAVs/Drones: this option is often considerably more cost-effective. The acquisition cost of smaller UAVs is typically in the hundreds of dollars, whereas even advanced military drones generally have a lower price tag than their manned counterparts. Operating costs are reduced due to decreased fuel consumption and reduced personnel requirements.

4. Operational flexibility and deployment [2, 3].

   For traditional manned aircraft: they necessitate a comprehensive array of infrastructure, including but not limited to runways, air traffic control systems, and ground support facilities. Moreover, they are subject to a meticulous pre-flight procedure. The deployment process may be characterized by a reduction in velocity and an augmentation in intricacy.

   For UAVs/Drones: it is imperative to offer greater operational flexibility. A significant number of these vehicles can be deployed and retrieved from a variety of locations that do not necessitate the use of runways. These vehicles can be deployed expeditiously and are well-suited for repetitive or persistent tasks. It is also possible to deploy they in a swarm configuration.

5. Endurance and range [2, 3].

   For traditional manned aircraft: the endurance of an aircraft is constrained by two primary factors: pilot fatigue and onboard resources, such as fuel and oxygen. While certain individuals possess the capacity to sustain flight for extended periods, limitations in human physiology and psychology serve as constraints on this capability.

   For UAVs/Drones: the vehicle's endurance capabilities are notable, with the primary constraints being fuel/battery capacity and mechanical reliability. A select number of HALE UAVs possess the capability to remain airborne for extended periods, often spanning several days. These vehicles persistent surveillance or monitoring.

6. Payload capacity and size [2, 3].

   For traditional manned aircraft: the capacity of these vehicles to transport substantially larger and heavier loads (including passengers, cargo, and armaments) is attributable to their dimensions and power output.

   For UAVs/Drones: the payload capacity of drones exhibits a wide range of variability. Small cameras affixed to hobby drones serve as a lower end of the spectrum. Substantial sensor packages and weaponry, on the other hand, are characteristic of military drones. However, their capacity is generally less than that of large traditional aircraft. The absence of a cockpit allows for more compact designs for a given payload.

7. Control and Autonomy [2, 3].

   For traditional manned aircraft: the aircraft are piloted directly by humans, who maintain continuous, real-time control.

   For UAVs/Drones: drones can be remotely controlled by human operators located thousands of miles away or operated autonomously using pre-programmed missions and AI. The level of autonomy exhibited by these systems is continuously increasing, enabling the execution of complex tasks without direct human intervention for extended periods.

8. Applications [1, 2, 3].

   For traditional manned aircraft: aircraft have a wide range of applications, including commercial passenger and cargo transport, military combat and transport, general aviation, search and rescue, aerial firefighting, and executive travel. Exhibiting superior performance in high-threat environments that demand rapid movement, dexterity, and substantial offensive capabilities (e.g., air-to-air combat).

For UAVs/Drones: the utilization of drones is undergoing a rapid expansion, encompassing a wide range of applications.

- Military. These include surveillance, reconnaissance, precision strikes, target designation, electronic warfare, and border patrol.
- Commercial. These include infrastructure inspection (e.g., pipelines, power lines, bridges), agriculture (e.g., crop monitoring, spraying), mapping and surveying, delivery services, search and rescue, and disaster response.
- Scientific. These include the scope of environmental monitoring encompasses the observation and analysis of ecological systems, with a focus on the atmosphere, which is the scientific study of the composition, behavior, and interactions of gasses in the atmosphere.
- Recreational. These include the hobby flying, drone racing, videography and photography.

9. Vulnerabilities [2, 8].

   For traditional manned aircraft: an aircraft is vulnerable to anti-aircraft fire, missile attacks, and mechanical failure.

   For UAVs/Drones: due to their reliance on digital communication and navigation, they are vulnerable to a variety of threats, including cyberattacks, jamming, electronic warfare, and GPS spoofing. Their constrained defensive capabilities render them vulnerable in situations involving the contestation of airspace.

10. Regulation and airspace integration [2, 14].

    For traditional manned aircraft: aircraft are subject to a series of well-established and stringent aviation regulations (e.g., the Federal Aviation Administration in the United States, the European Aviation Safety Agency in Europe) and are integrated into controlled airspace.

    For UAVs/Drones: the rapid proliferation of these vehicles has precipitated the development of evolving regulations. The integration of UAVs into shared airspace with traditional aircraft represents a significant ongoing challenge, primarily due to safety concerns and differing operational requirements.

    The utilization of traditional aircraft remains paramount for tasks that demand high velocity, substantial payload capacity, human judgment in intricate and evolving scenarios, and direct human presence (e.g., commercial passenger transport).

    Conversely, UAVs are profoundly impacting various industries and military operations by virtue of their cost-effectiveness, reduced risk to human life, flexibility, and the capacity to execute persistent or hazardous tasks. Despite their current limitations in terms of speed, agility, and heavy-lift capabilities when compared to certain traditional aircraft, their rapid technological advancement suggests a future in which they will play an even more significant role, potentially complementing or even replacing manned aircraft in specific niches. The selection of either a conventional aircraft or a UAV is contingent upon the particulars of the mission, financial constraints, and the acceptable level of risk.

1.5 Conclusion

In this chapter, we have conducted a comprehensive review of the historical development of UAVs, spanning from antiquity to the contemporary era. A review of the predominant trends in the development of unmanned vehicles was also conducted. A comparative analysis has been conducted to elucidate the fundamental distinctions between conventional aircraft and UAVs. In light of the aforementioned evidence, it can be concluded that the UAV sector is undergoing rapid development within the contemporary aviation industry. Concurrently, numerous challenges persist for drone designers. Consequently, the investigation and description of the drone design process represent an obligatory task, which will persist for numerous years.

References

1. Li L, Kretov A, Tiniakov D. Aeronautics—Characteristics and Emerging Technologies. Chapter: UAVs and Hydrogen Technologies are the Future of Civil Aviation [Internet]. IntechOpen; 2025. https://doi.org/10.5772/intechopen.1008352.
2. Marshall D, Barnhart R, Shappee E, Most M. Introduction to unmanned aircraft systems. New York, NY: CRC Press; 2016.
3. Tiniakov D, Li L. Airworthiness engineering basis. Datian Aviation Beijing Beixue Publishing House; 2020.
4. Prisacariu V. The history and the evolution of UAVs from the beginning till the 70s. J Def Resour Manag. 2017;8(1):181–9.
5. Loren R. La Torpille Aerienne. J Aerophile. 1909, 15 October:463. [Internet]. https://gallica.bnf.fr/ark:/12148/bpt6k65639070/f86.item. Accessed 15 Sept 2025
6. Pavlushenko M, Evstafyev G, Makarenko I. Unmanned aerial vehicles: history, distribution and future development. Moscow: Pir Center; 2004.
7. Moiseev V. Unmanned aerial vehicles: history and nowadays classification. Kazan: Shkola; 2022.
8. Custers B. Drones the future of drone use. T.M.C. Asser Press; 2016. https://doi.org/10.1007/978-94-6265-132-6_1.
9. Shinde S, More J, Chaudhari C. Mastering drone technology with AI. India: BPB Publications; 2025.
10. Skydio. Skydio 2X. USA: Skydio. [Internet]. https://pages.skydio.com/rs/784-TUF-591/images/skydio-x2e-datasheet-x2-pg.pdf. Accessed 15 Sept 2025
11. Nikolsky M. Shahed-136 and other UAVs of Iran. Moscow: Yauza; 2023.
12. Xu J. Design perspectives on delivery drones. Santa Monica, CA: Rand Corporation. 2017. [Internet]. https://www.rand.org/content/dam/rand/pubs/research_reports/RR1700/RR1718z2/RAND_RR1718z2.pdf. Accessed 15 Sept 2025
13. Sun Y, Li L, Tiniakov D. Reliability engineering. Beijing: Springer/Science Press; 2024. https://doi.org/10.1007/978-981-99-5978-5.
14. Tarr A, Tarr J-A, Thompson M, Ellis J. Drone law and policy global development, risks, regulation and insurance. New York, NY: Routledge; 2022.

Chapter 2
Design Process

2.1 Introduction

The development of a new aircraft/UAV model is contingent upon the formulation of a conceptual framework. In this context, the term “aircraft concept” is defined as a set of data that determines the external forms and internal design of the UAV [1, 2]. However, the development of a promising UAV concept is not an isolated process. The employment of drones in simulating various scenarios within challenging UAV operational environments facilitates an assessment of the aircraft’s capabilities in accomplishing designated tasks [3, 4]. These models utilize the structural, mass-inertial, and aerodynamic performance of the drone, data about its control system, and the environmental conditions in which the flight is conducted. Conventionally, the design of aircraft and UAVs is initiated with the formulation of a mission requirements package.

Contemporary UAVs are equipped with advanced control systems that are capable of performing a multitude of tasks. These systems can be described as intelligent, in the sense that they are capable of processing a vast array of data and utilizing it to complete a given task. This data may encompass information from digital terrain maps, which are utilized for the planning of flight routes, or it may comprise instructions from an autopilot system, which is engineered to maintain the vehicle’s position and trajectory. Furthermore, these systems frequently utilize decision support tools to assist in selecting the most suitable maneuver to overcome obstacles.

It would be premature to designate modern drones control systems as systems with elements of artificial intelligence (AI) [5]. Indeed, for such systems, the primary consideration should be the absence of human involvement in decision-making processes. This necessitates a high degree of system autonomy, a capability that UAVs are only beginning to demonstrate. This is particularly salient in the context of representing drones as carriers of artificial intelligence.

L. Li, D. Tiniakov, *Unmanned Aircraft Design*,
https://doi.org/10.1007/978-981-95-7167-3_2

By systematically adjusting the technical properties of the vehicle, it is possible to create a variety of models that can be used to create highly efficient UAVs capable of performing specific tasks with optimal efficiency.

2.2 Basic Concept of Designing Process

In this section, the focus will be on the intricacies of the UAV design process.

2.2.1 Initial Data and Limitations

The initial data utilized in the design process are as follows:

The customer's specified tasks for this aircraft, operating conditions, and the specifics of drone usage are also taken into account. Furthermore, limitations resulting from the design methods employed must be considered.

The limitations on the design results, in conjunction with the specified tasks for this UAV, are as follows [1, 4]:

1. The operating conditions of the UAV.
2. The conditions of manufacture of the drone in production.
3. The Airworthiness Regulations.

Limitations due to design methods include the following:

1. Design deadlines.
2. The possibilities of available computing hardware and software.
3. The capabilities of the available test equipment.

To illustrate this point, the conditions of use of UAVs can be considered [1].

The Atmosphere A table of the International Standard Atmosphere (ISA) offers data on a range of physical characteristics of the atmosphere, including temperature, pressure, specific gravity, air density, the ratio of air density at a given altitude to air density at sea level, the speed of sound in the air, and other characteristics depending on the altitude above sea level.

The characteristic atmospheric conditions are employed in the design of drones in accordance with the customer's specifications (see Table 2.1). The following conditions must be met:

- medium duty conditions;
- design conditions utilized for the simulation of take-off and landing characteristics;
- potential conditions for specific UAV tasks.

Table 2.1 The basic data set comprises information on the characteristic conditions of the atmosphere

Atmosphere characteristics	Units	Medium duty conditions	Design conditions	Potential conditions (for example, Stratosphere)
Altitude	m	0	0	≥11,000
Temperature	°C	+15	+30	−56.5
Pressure	Pa	101,325	973.35	–
Density	kg/m^3	1.225	1.12	–
Sound speed	m/s	340.3	350	295

Table 2.2 The basic data set for the climatic conditions

Climatic zone	Conditions	Temperature (°C)		Specific humidity (%)
		max	min	
Desert zone	Dry heat, intense solar radiation, sand dust	Atmosphere at +60 Ground at +75	In night at −10	5
Tropical zone	Humid heat, high specific humidity, mold growth, detrimental effect of insects	Atmosphere at +40 Ground at +70	In night at +25	100
Sea zone	Wet salty dust	+50	−40	100
Arctic zone	Low temperature, ice dust	−25	at −70	–
High altitude zone	Low temperature, low pressure, condensation of water vapor caused by rapid change in temperature	–	at −90	–

The Extreme Climatic Conditions In the process of designing a UAV, it is essential to consider the following factors, as outlined in Table 2.2.

In tropical conditions, non-metallic materials are susceptible to mold and insect infestation. This is a particular concern when insulating electrical wiring, as damage to the insulation can result in short circuits and failure of equipment and control systems.

In conditions of elevated temperature and humidity, the target level is set at 10% relative humidity in outdoor environments where the temperature exceeds 40 °C. This is the primary design condition for all cooling systems.

The Wind In order to calculate the take-off and landing characteristics, it is necessary to assume a crosswind speed of 15 m per second.

The rolling loads are calculated at an indicator speed corresponding to a vertical gust of wind with a speed of 15 m per second.

As can be observed, the conditions in question exhibit a considerable degree of variability. It is not feasible to consider all of these factors when designing a single UAV. A logical approach would be to assess the drone's capabilities in the majority of potential operational environments. Consequently, a new variant may be designed on the foundation of the existing project, in accordance with the specified conditions.

2.2.2 General Design Process of the UAV

The current state of the industry allows for a well-developed and established process for aircraft design, comprising a set of defined laws and stages [6, 7]. This development is largely attributable to the advancements in aviation that have taken place over the past century. It is evident that the characteristics of UAVs have a significant impact on the design process. This is primarily attributable to two factors: the absence of a human aboard the aircraft and the dimensions and operational capabilities of existing drones. The take-off weight and dimensions of contemporary large-scale drones are analogous to those of analogous general aviation aircraft [3]. Consequently, the design process for such UAVs is more straightforward than that of traditional aircraft.

Nevertheless, should UAVs eventually possess capabilities analogous to those of contemporary civil aviation, such as the transportation of substantial cargo or even passengers, the design process for such UAVs would then be structured in a manner analogous to that of transport category aircraft.

The preliminary design phase is illustrated in Fig. 2.1 [2].

As demonstrated in Fig. 2.1, the preliminary phases of the design process are intricately linked with statistical analysis. Each company gathers statistical data for analysis from a multitude of potential sources. The integration of a company's proprietary manufacturing expertise in the domain of aircraft or drones has been shown to significantly accelerate the design process, ultimately resulting in a more reliable final design outcome. In instances where a company lacks in-house expertise in drone manufacturing, it is imperative that the enterprise acquires statistical data for design purposes from publicly accessible sources. A notable trend among young aviation engineers is the development of projects related to unmanned taxis, drones for parcel delivery, and related applications.

While experience in the design of aircraft or UAVs may enhance the probability of success, it is not a definitive guarantee. The success of a new project is contingent upon a multitude of factors that are not directly related to the design process.

Consequently, statistical analysis allows for the selection of a general concept for a promising UAV, based on the collective experience of researchers [3]. This approach serves to mitigate the potential for erroneous decision-making.

The statistical data provide insight into the current state of development of UAVs. They offer information on:

(a) the types of UAVs and the tasks they perform;
(b) the potential flight performance of UAVs;
(c) the operational and service data of UAVs;
(d) the equipment, structural materials, and power plants used in UAVs;

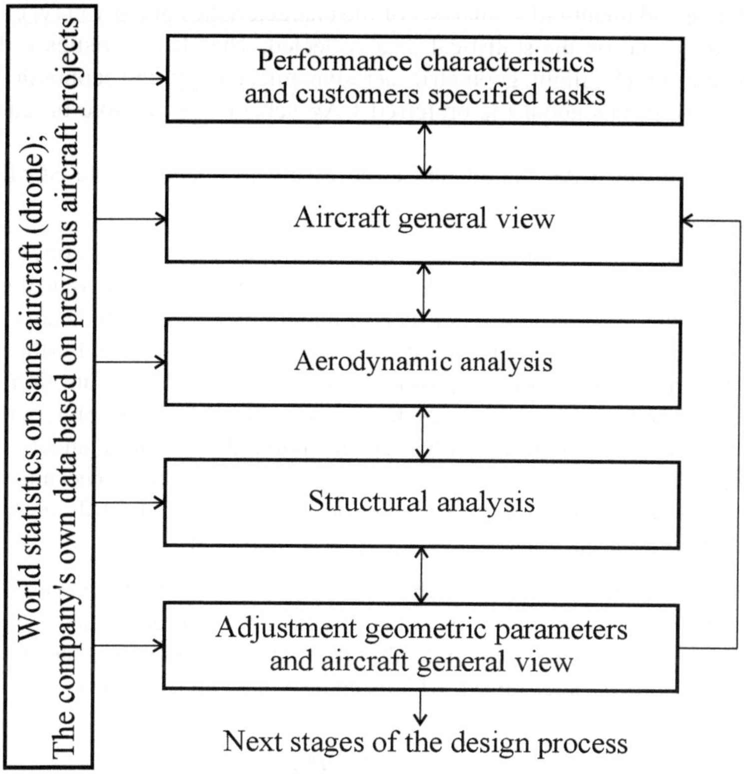

Fig. 2.1 The preliminary design phase of an aircraft (UAV)

(e) the types of structures employed in UAV design;
(f) the potential costs of UAV project and its transport efficiency.

The utilization of statistical data must be approached with discernment and employed judiciously. A comprehensive evaluation necessitates the consideration of the following factors:

1. It is imperative to ascertain the altitude and mode of engine operation at which maximum speed, cruise speed, rate of climb, altitude ceiling, and other pertinent data are indicated.
2. Furthermore, it is imperative to ascertain the altitude and loading option at which the range and duration of the flight are indicated.
3. Moreover, it is imperative to ascertain the conditions that permit takeoff and landing. Furthermore, the following data should be considered: the required takeoff distance or area for takeoff and landing, takeoff and landing speeds, etc.
4. Additionally, it is imperative to ascertain the components encompassed within the full load and payload.

A qualitative and quantitative analysis of the characteristics of a given type of UAV is necessary based on the statistical data collected. The characteristics to be analyzed include weight, flight, geometric, aerodynamic, energy, and economic factors. Drawing conclusions about the preferred UAV schemes is possible based on this analysis.

A review of the UAV parameters is necessary to conduct a statistical analysis [1–8].

The Speed Within the domain of aircraft, the minimum speed is conventionally delineated as the lowest flight speed at which reliable, stable, and controlled flight can be sustained. A lower minimum speed value for a given UAV indicates a higher level of operational safety, particularly during takeoff and landing. Furthermore, a shorter runway length is a prerequisite for the operation of such an aircraft. The value of the minimum speed is a determining factor in the safety and cost-effectiveness of the operation. To illustrate this point, the definition under consideration highlights the high efficiency and safety of multicopters. The multi-copter scheme has emerged as a prominent solution for the operation of flying taxis and parcel delivery drones.

In the context of heavy transport drones, an increase in flight speed is more often associated with cruising speed, the value of which has a direct impact on the efficiency of the operation. Consequently, the cost of transportation is inversely proportional to the value of block speed, which can be expressed as follows:

$$V_{\text{block}} = \frac{L \cdot V_{\text{cruise}}}{L + \Delta t \cdot V_{\text{cruise}}}, \tag{2.1}$$

here L represents the distance between the takeoff and landing points, expressed in kilometers; V_{cruise} denotes the cruising flight speed, expressed in kilometers per hour; Δt signifies the time spent on starting the power plant, on taxiing before takeoff and after landing (if necessary), on takeoff and climb, on maneuvering after takeoff and before landing, and on descent and landing.

The Flight Range An increase in flight range, accompanied by maximum commercial load and maximum fuel reserve (or augmented battery power), has been demonstrated to result in a concomitant increase in flight speed, as illustrated by Eq. (2.1). The value of L is directly proportional to the cost of transportation and, consequently, the cost-effectiveness of UAV operation. This is due to the fact that cost is a primary determinant of efficiency.

The Rate of Climb The rate of climb, or RoC, is a critical factor in the performance of emergency drones. In the context of civilian drones, an augmentation in RoC (particularly in the event of an engine failure) contributes to enhanced operational safety, particularly during take-off.

The Minimum Turn Radius This performance metric is critical, as it quantifies a UAV's maneuverability. In the context of heavy drones, this characteristic enhances operational reliability and safety, facilitates the achievement of desired flight paths, and bolsters the safety of takeoff and landing procedures. Conversely, in the case of light and high-maneuverable drones, it enhances their operational efficiency.

The Payload Weight The payload weight of commercial drones is specified in absolute form (in kilograms or tons). In addition to the payload weight, the size of the cargo (i.e., its dimensions) and its nature (e.g., commodity cargo or in containers, whether specific environmental conditions are required in the cargo compartments) are critically important for such UAVs.

Irrespective of the purpose of a UAV, the greater the commercial load, all other things being equal, the greater the useful return of such a drone, the greater its economic efficiency.

The Equipment (Hardware and Software) The incorporation of contemporary equipment into the UAV guarantees autonomous flight along a designated trajectory. It prompts the UAV operator, if applicable, to select the most rational flight pattern, thereby enhancing autonomous control of the drone in challenging meteorological conditions during both diurnal and nocturnal periods. The implementation of automatic landing protocols has been demonstrated to contribute to increased flight safety and enhanced passenger comfort. This feature will become increasingly important in the context of passenger unmanned systems in the future. Consequently, these measures contribute to a high operational efficiency.

The Takeoff and Landing Performance The following is a list of takeoff and landing characteristics: $V_{\text{t-o}}$—takeoff speed; V_{ap}—approach speed; V_{land}—landing speed; $L_{\text{t-o land}}$—takeoff or landing run; or $S_{\text{t-o land}}$—area required to ensure safe takeoff and landing. It is imperative to reduce all these UAV characteristics to ensure the convenience, reliability, and safety of its takeoff and landing. The optimization of these characteristics is imperative for ensuring the optimal performance of UAVs:

1. the ability to base the drone on limited and poorly prepared sites;
2. makes it possible to locate takeoff sites closer to destinations;
3. reduce takeoff site construction costs;
4. increase operational efficiency;
5. increase flight regularity;
6. reduce weather restrictions.

The Aerodynamic Requirements The UAV must possess the following characteristics:

- The external shape of the parts and their mutual arrangement, surface quality, and dimensions are designed to ensure high flight characteristics with minimal energy expenditure. In essence, it is imperative that the drone possess a minimum drag coefficient (C_D) value, a maximum lift coefficient (C_L) value, and a lift-to-drag factor (K) value.

- Consequently, the UAV must be designed to ensure minimal energy expenditure while maintaining optimal flight characteristics. Achieving this objective demands minimizing specific values linked with the drone's mass, thrust, and fuel needs, especially the takeoff mass m_{to}, the mass of the power plant m_{pp}, and the mass of the fuel m_{fuel} (or batteries m_{bat}).
- Flight stability and controllability must be guaranteed.
- Specified takeoff and landing characteristics must be provided.
- It is imperative that the frame temperature does not increase significantly due to aerodynamic heating, particularly during extended flights at high supersonic speeds.

The Strength, Rigidity and Minimum Mass The requirements of strength, rigidity, and minimum mass are inextricably linked. The specified strength and rigidity must be met by the minimum extra mass. In essence, the objective is to ensure that the mass of the drone's design is kept to a minimum.

The following properties are requisite for the structural integrity of the frame:

- The absence of destructive stresses in the frame under the action of loads is less than the destructive.
- The absence of fatigue damage and destruction under the action of repeated and dynamic loads is a crucial consideration.
- The absence of unacceptable deformations under the action of operational loads and high temperatures is imperative.
- The structural elements exhibit sufficient general and local stiffness, thereby excluding the possibility of violation of aerodynamics and flight stability. This, in turn, ensures the absence of vibrations.

The requirements for sufficient strength and rigidity, which are essential for meeting strength standards, are unconditional. Failure to meet these requirements will result in the inability of the design to perform its functions under all calculated flight modes.

The Minimum Mass The following measures are taken to ensure that the minimum mass of the frame is obtained:

- The loads for all flight modes are correctly determined.
- Rational carry-load schemes for units, components, and parts are selected.
- Rational choices are made regarding structural materials.
- The technological infrastructure necessary for the production of parts and the assembly of units and assemblies is in place.
- The development of parts, components, and assemblies is meticulous and competent, characterized by a culture of design and production.

The Reliability UAVs are intricate systems, the effectiveness of which is contingent upon their reliability. That is to say, the capacity of the system to execute tasks assigned to it without failure for the requisite duration under stipulated conditions.

The following measures have been identified as effective strategies to enhance reliability:

- The implementation of established operating modes is to be executed unconditionally.
- It is imperative that periodic inspections and regulatory maintenance procedures be conducted in accordance with established schedules.
- The execution of current and overhaul procedures is to be carried out in a timely manner.

Service Life In the context of the aviation industry, there is a pronounced emphasis on the longevity of products. Assuming all other requirements for drones remain constant, this expectation is particularly pronounced. This is particularly salient for the most inexpensive drones, where maintaining affordability is imperative. The objective of maximizing the service life of UAVs is to ensure optimal economic efficiency.

The service life is defined as the duration of the UAV's operation, measured in flight hours or the permissible number of landings, under a pre-established level of reliability in the conditions of flights and maintenance.

Requirements of Survivability Survivability is defined as the capacity of a drone to sustain flight and landing operations in the event of damage, whether intentional or unintentional, with the objective of safeguarding cargo and the external medium.

Manufactural Requirements The following manufacturing requirements have been established:

- Minimal complexity;
- Minimal costs for materials and energy;
- Reduced preparation and production time;
- Reduced expenses.

Operational Requirements These items are classified into two categories: those that pertain to the UAV as a transportation vehicle and those that pertain to it as an object of maintenance. The operational requirements for the aircraft as a transportation vehicle primarily comprise a series of cargo requirements (in the case of passenger function, too; passengers are a specific type of cargo). The drone's maintenance requirements encompass a comprehensive array of issues, collectively termed "maintainability."

The Ability of the UAV and Its Systems to Modify The capacity for modification of UAV systems is a pivotal element in the development of novel and sophisticated tasks for these vehicles. The augmentation in the range of tasks executed by drones is directly associated with the proliferation of drone samples. However, the expansion of the UAV family has led to the adoption of technology for drone usage. A more rational approach would be to create new versions of existing models that incorporate new functionality. This approach has been successfully implemented not only in the aviation sector but also in other industries. However, for the successful implementation of this approach, it is essential to incorporate the possibility of modification into the drone's design from the pre-commercial stage of its development.

2.2.3 The Selection of a Drone Configuration

When selecting the configuration of the UAV, it is imperative to consider all its operational performance.

Contemporary drone configurations encompass the following types (Fig. 2.2) [3, 4]: Firstly, fixed-wing UAVs possess the capability to fly without the necessity of rotary engines. Secondly, flapping-wing drones have been developed that feature movable bionic wings. Thirdly, rotary-wing helicopters are distinguished by their two-rotor configuration. Fourthly, multicopters are classified as such due to their three or more rotors. The fifth distinguishing characteristic of airships is their capacity to traverse the atmosphere independently of terrestrial infrastructure, such as the use of propellers or rotors. Finally, combined schemes integrate elements of both fixed-wing and rotary-wing aircraft, thereby offering a multifaceted solution to aerial operations.

In order to facilitate a comprehensive analysis, it is imperative to consider the various factors that contribute to the efficiency of UAVs.

1. The range and duration of flight for airship-type drones are maximized due to their high energy efficiency, while these parameters are slightly diminished for

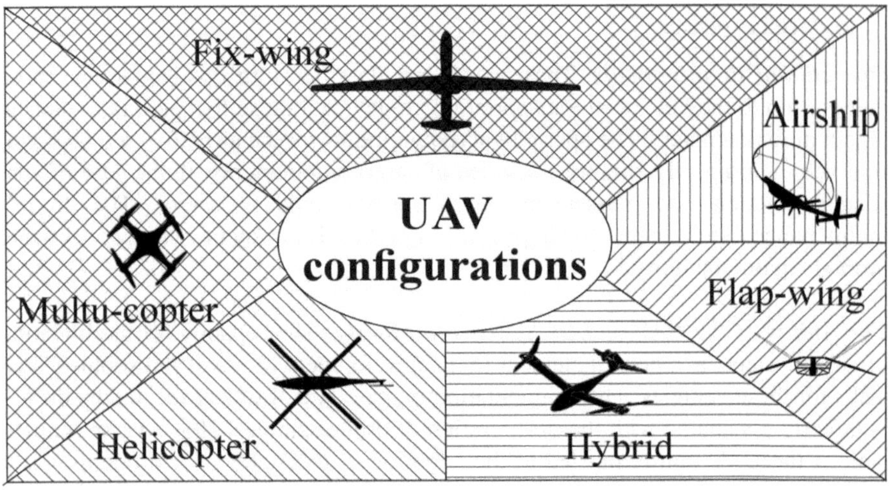

Fig. 2.2 Drones' brief configurations

fixed-wing drones. In contrast, rotary-wing UAVs (copters) and ornithopters demonstrate the least favorable performance metrics.
2. Secondly, rotary-wing drones, which include multi-copters and helicopter-type drones, exhibit the greatest maneuverability. A noteworthy maneuver within this category is the ability to execute a hover, which involves maintaining a stationary position in the air. Conversely, airplane-type drones possess adequate maneuverability but lack the capacity to execute a hover. Airship-type drones demonstrate the least maneuverability, a consequence of the presence of a gas shell, which engenders substantial aerodynamic drag. Nonetheless, drones of this category have the capacity to execute hover maneuvers, similar to the capabilities exhibited by multi-copters.
3. With regard to the ease of takeoff and landing, rotary-wing aircraft and airship-type drones are the optimal choice. Airplane-type drones demonstrate suboptimal performance metrics, as they require the utilization of a runway (even a brief one) or a launch pad (an additional equipment requirement).
4. Drones equipped with wings have been shown to exhibit exceptional stability and balance. Conversely, the stability of copter-type drones is suboptimal, a quality that is consistent with their high degree of maneuverability.
5. In terms of reliability, the optimal drone configuration is that of an airplane. This phenomenon can be attributed to the fact that it possesses the least number of moving parts. Nevertheless, the power plant and control surfaces are of particular concern due to their vulnerability. Consequently, the reliability of copter-type drones is the least optimal, and the presence of additional propellers contributes to a further reduction in reliability.
6. UAVs that incorporate an airship configuration are distinguished by their remarkable survivability capabilities. The gas shell guarantees autonomous emergency descent and landing, even in the event of a failure of all onboard systems. Conversely, drones equipped with wings possess the capability to glide in the event of a power plant failure, thereby ensuring a safe landing under normal conditions and adequate aerodynamic perfection. Conversely, the survivability conditions for copter-type drones are the least favorable. In the event of engine failure, these UAVs are unable to execute an uncontrolled descent, which would otherwise lead to a catastrophic outcome. In order to mitigate the aforementioned risks, it is imperative to incorporate an auxiliary emergency system, such as a parachute system, to ensure the safety of the drone and its cargo.
7. Within the parameters established for the placement and ground transportation of drones, it is evident that copters, under optimal conditions, demonstrate the most compact dimensions. The development of additional solutions for the folding of drone components is a conceivable prospect; however, this would result in an increase in the structure's weight.

Therefore, in consideration of the aforementioned conditions, the following examples can be posited:

1. In scenarios where maximum speed and range are paramount, fixed-wing drones emerge as the optimal solution.

2. For missions necessitating reliable and secure cargo transportation over extended distances, the utilization of airship-based drones is particularly advantageous.
3. In scenarios that necessitate high maneuverability, rotary-wing drones have demonstrated a marked superiority in executing the designated task.

The determination of the optimal drone configuration for a given application necessitates further research, given the variety of conditions that may be present.

2.2.4 Fixed-Wing Drone Design Process

The design process for a fixed-wing drone is analogous to that of an airplane [1–3, 6, 7]. This approach signifies the most advanced design method currently available.

The preliminary stage of the process invariably commences with a meticulous delineation of the mission requirements. After this preliminary stage, statistical material is collected on UAVs with similar or identical characteristics. The statistical database currently contains several thousand airplane schemes of the aforementioned type. It is imperative to note that the number of studied schemes that were developed during the process of designing existing aircraft is significantly larger. Given the extensive range of variants, it is feasible to select a number of suitable alternatives that align with the required characteristics.

At the second stage, the final conceptual appearance of the drone must be approved. The external appearance and geometric parameters of the drone are contingent upon the requirements delineated in the mission requirements, the principles of aerodynamics, and the required degree of structural strength.

The determination of weight parameters is a prerequisite for the design of any aircraft. The selection of a power plant is contingent upon these weight parameters.

The configuration of a drone airframe is invariably initiated with the wing. The wing is the primary component of the design. The perfection of the wing is instrumental in ensuring the high efficiency of the future UAV, encompassing efficiency in production and operation.

The development of the wing exerts a substantial influence on all subsequent stages of design. The geometric parameters of the wing, including the dimensions and location of the tail unit, the fuselage layout, the integration of the power plant, and the configuration of the landing gear, are determined by these design choices.

The ensuing discourse will meticulously explore the complexities inherent in the design of a drone equipped with a fixed wing configuration.

Wing Geometric Parameters Selection It is evident that the wing parameters are contingent upon the mass of the UAV. The subsequent section of this chapter will address the weight limitations for drones [1–3, 6, 7].

The wing area is thus determined by the following equation:

$$S_w = \frac{m_{to} \cdot g}{10 p_0}, \tag{2.2}$$

where the take-off mass of the drone, m_{to}, is expressed in kilograms. The gravitational acceleration, g, is measured in meters per second squared. The specific wing loading, p_0, is defined as the force per unit area, in decanewtons per meter squared. The concept of specific wing loading can be regarded as a metric of conceptual significance, which can be assessed through the utilization of statistical data during the preliminary phase of design. The adjustment can be made during the subsequent design stages. For traditional aircraft, this metric ranges from 100 to 800 decanewtons per meter squared.

The wing span is determined by the equation:

$$L_w = \sqrt{AR \cdot S_w}, \tag{2.3}$$

where *AR* is the aspect ratio of the wing, it is a unitless parameter. The parameter of *AR* is of significant importance. It determines the aerodynamic efficiency of any aircraft.

For subsonic speeds, it ranges from 7 to 12. Subsequently, from the perspective of aerodynamic perfection, the superior option is evident. Nevertheless, the elevated value of this metric exerts a deleterious effect on the strength and stiffness of the wing, consequently impacting its mass. The prevailing contemporary trend is the high valuation of *AR*. The negative impact can be reflected by the application of composite materials or a reduction in specific wing loading.

In the context of supersonic flight, the aerodynamic perfection of the aircraft is less consequential than its strength and stiffness, particularly at high speeds. Consequently, for these velocities, *AR* should be maintained within the range of 2–5.

Subsequently, the parameters of the wing are the chords: the root, the board, and the tip. However, in the absence of limitations pertaining to the internal space of the wing or specific conditions associated with it, it is possible to initially prioritize aerodynamic considerations, followed by the consideration of manufacturing conditions.

In the context of aerodynamics, it is imperative to consider the velocity of flight. It is not feasible to calculate variations of chords if the flight speed is less than 300 km per hour. In this instance, manufacturing conditions have been identified as the fundamental basis. In the event that the velocity of an aircraft exceeds 300 km per hour, it is incumbent upon the aircraft designer to undertake a comprehensive analysis to substantiate the design solution. In general, the wing should be tapered. The grade of tapered wing is contingent upon the speed level. In order to achieve transonic velocity, it is necessary that the wing be not only tapered, but also possess a specific swept angle. The following recommendations are provided for the wing taper ratio, contingent upon the drone's flight speed:

- For drones with low airspeed and about 300 km per hour, $TR = 1$.

- For subsonic drones, *TR* is in the range of 0.5–0.2.
- For supersonic drones, *TR* is less than 0.2.

On the basis of a curated selection of *TR*, it becomes feasible to determine the root and tip chords of the win:

$$b_{\text{root}} = \frac{S}{L} \cdot \frac{2TR^{-1}}{TR+1}, \tag{2.4}$$

$$b_{\text{tip}} = b_{\text{root}} \cdot TR. \tag{2.5}$$

The adjustment of wing parameters is an ongoing process that occurs throughout the entirety of the UAV design. In the initial stages, these adjustments can exhibit significant variability. Subsequent to the validation of the aerodynamic performance within the stipulated structural constraints, the wing geometric parameters are preserved and remain constant for the remaining stages of the design process.

Fuselage Geometric Parameters Selection In the case of a conventional aircraft, the fuselage occupies a significant position. It is imperative for the arrangement of cargo, equipment, and the various components of aircraft systems [1–3, 6, 7]. The particular requirements for the fuselage are contingent upon the nature of the cargo being transported. The passenger-specific requirements are notably more stringent. In all other instances, requirements are contingent on additional conditions, though these conditions tend to be less stringent.

In the contemporary period, there is an absence of authentic drones that are designed for the purpose of passenger transportation over long distances. The term "flying taxi" refers to an UAV that operates within a short range and does not necessitate the implementation of specialized pressurization techniques within the fuselage [1]. Consequently, the requirements for the drones' fuselage are less stringent.

The geometric parameters of the fuselage are contingent upon the cargo and aerodynamic specifications.

The diameter of the fuselage is determined by the overall size of the cargo; however, in the absence of cargo within the fuselage, the diameter is influenced by the structural specifics, including the attachment fittings of the wing and tail unit, the landing gear, and the arrangement of the power plant.

The length of the fuselage is traditionally dictated by aerodynamic requirements. Consequently, it is the responsibility of the designer to determine the drone's configuration in advance. In the case of a tail unit, the lever must be taken into account. For drones that feature a flying-wing configuration, the length of the fuselage is contingent on the wing chord, the length of the engine, and other factors.

In the preliminary phase of the design process, the length of the fuselage can be determined through statistical analysis.

$$L_{\text{f}} = AR_{\text{f}} \cdot D_{\text{f}}, \tag{2.6}$$

where the length-to-diameter factor (also known as the aspect ratio) of the fuselage is denoted by AR_f.

At this stage, the factor can be based on aerodynamic reasons, with consideration given to structural limits (see Table 2.3).

Tail Unit Geometric Parameters Selection The tail unit is an integral component of the drone design, incorporating a fixed-wing configuration [1–3, 6, 7]. In the event that the drone is configured without a tail (see Chap. 1), this section can be bypassed. The primary functions of the tail unit are the provision of stability, balance, and controllability.

In order to ensure the stability of an UAV, it is essential to provide equivalent wing aerodynamic moment and horizontal surface aerodynamic moment about the drone's gravity center (see Fig. 2.3).

As illustrated in the Fig. 2.3, the drone's equilibrium is maintained through the control of the HS's pitch moment.

$$M_{\text{wpitch}} = L_{\text{a w}} \cdot x_{\text{w}} = L_{\text{a HS}} \cdot l_{\text{HS}} = M_{\text{HSpitch}} \tag{2.7}$$

The efficacy of this balancing process is contingent on the arm and the HS area, requiring a balanced consideration by the designer between these two components. Achieving an increased arm length can be accomplished by reducing the HS area, although this concurrently increases the length of the UAV. Conversely, a decrease in the arm of the HS results in an increase in the required HS area, which causes an increase in its mass.

Table 2.3 The fuselage aspect ratio statistical metric

Fuselage aspect ratio	Subsonic speed (M < 0.6)	Subsonic speed (M = 0.7–0.9)	Supersonic speed
AR_f	6–9	8–13	10–23

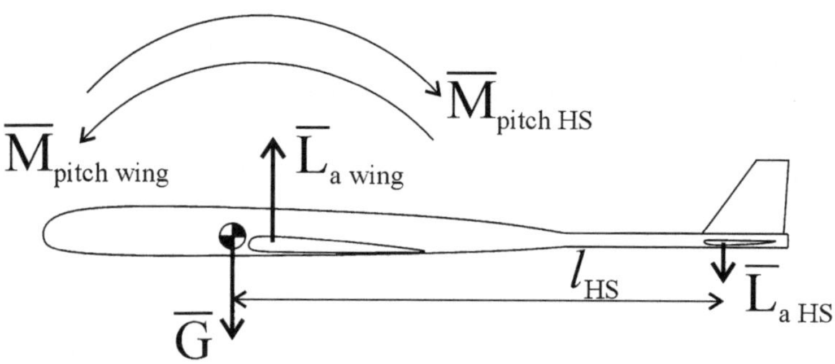

Fig. 2.3 Schematic representation of wing and horizontal surface load locations

$$M_{\mathrm{HS}} = L_{\mathrm{aHS}} \cdot l_{\mathrm{HS}} = C_{\mathrm{LHS}} \cdot \frac{\rho \cdot V^2}{2} \cdot S_{\mathrm{HS}} \cdot l_{\mathrm{HS}}, \tag{2.8}$$

where $C_{\mathrm{L\,HS}}$ is the lift coefficient of the HS; ρ is the density of the air; V is the flying speed; S_{HS} is the area of the HS.

The horizontal-tail volume ratio VR_{HS} of the HS:

$$VR_{\mathrm{HS}} = \frac{S_{\mathrm{HS}} \cdot l_{\mathrm{HS}}}{S_{\mathrm{w}} \cdot b_{\mathrm{MAC\,w}}}. \tag{2.9}$$

According to the statistical materials, the range for VH_{HS} is from 0.35 to 0.55. The values decrease in proportion to the maneuverability of the UAVs. UAVs that exhibit superior maneuverability are assigned higher values, while those with inferior maneuverability are assigned lower values. Consequently, the calculation of the HS's potential area becomes feasible.

$$S_{\mathrm{HS}} = \frac{VR_{\mathrm{HS}} \cdot S_{\mathrm{w}} \cdot b_{\mathrm{MAC\,w}}}{l_{\mathrm{HS}}}. \tag{2.10}$$

A similar rationale should be employed in the case of vertical tail. In this scenario, the primary determining factor is the controllability of the horizontal plane (yaw plane).

$$M_{\mathrm{VSyaw}} = L_{\mathrm{a\ VS}} \cdot l_{\mathrm{VS}} = !_{\mathrm{L\ VS}} \cdot \frac{\rho \cdot V^2}{2} \cdot S_{\mathrm{VS}} \cdot l_{\mathrm{VS}} \tag{2.11}$$

The vertical-tail volume ratio VR_{VS} of the VS:

$$VR_{\mathrm{HS}} = \frac{S_{\mathrm{VS}} \cdot l_{\mathrm{VS}}}{S_{\mathrm{w}} \cdot L_{\mathrm{w}}}. \tag{2.12}$$

As indicated by statistical materials, the volume ratio of VR_{VS} ranges from 0.04 to 0.06 for drones with a straight wing and from 0.06 to 0.15 for high-speed drones with swept and delta wings.

Subsequently, the determination of the VS's potential area becomes feasible.

$$S_{\mathrm{VS}} = \frac{VR_{\mathrm{VS}} \cdot S_{\mathrm{w}} \cdot L_{\mathrm{w}}}{l_{\mathrm{VS}}}. \tag{2.13}$$

Another salient factor that must be considered in the design of the tail is the fact that a scheme with a combined tail, such as a V-shaped or Y-shaped configuration (see Fig. 2.4), is particularly well suited for UAVs. This approach is also widely employed in contemporary fixed-wing drones.

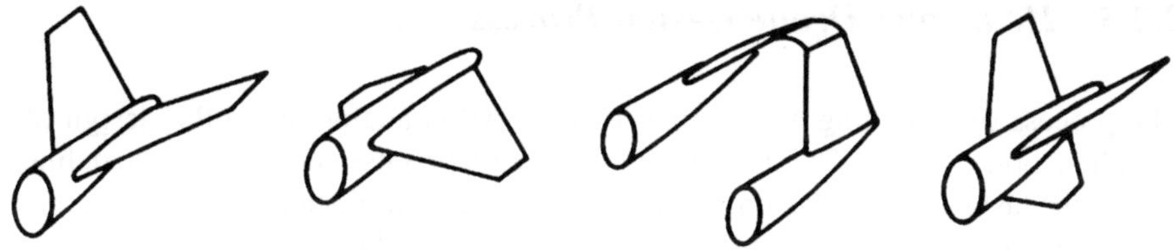

Fig. 2.4 Possible variants of a tail unit for drones

Table 2.4 Statistical data about a tail unit for an UAV

Tail unit parameter		Metric of the parameter
Specific area of the horizontal surface		0.15–0.3
Specific area of the vertical surface		0.1–0.2
The arm of a tail unit	Straight wing	$3.5 \cdot b_{MAC}$
	Swept angle of the wing 10–30°	$(2.5–3.5) \cdot b_{MAC}$
	Swept angle of the wing 30–60°	$(2.0–2.5) \cdot b_{MAC}$
	Delta wing	$(1.0–1.5) \cdot b_{MAC}$
	Canard wing	$(1.0–1.5) \cdot b_{MAC}$

Due to their smaller size compared to conventional transport category aircraft, UAVs are capable of meeting all the requirements for drone maneuverability and stability. Additionally, the weight of the airframe is reduced automatically in these cases, and the production process is less expensive.

For such a scheme, all the described design approaches are mandatory, taking into account the fact that for such a tail it is necessary to consider combinations between cases both in the pitch plane and in the yaw plane.

In the preliminary design stage, collected statistical materials can be utilized to ascertain the arm and tail area. Furthermore, the implementation of general aviation statistical data can be initiated at the onset of the design process (see Table 2.4).

In this particular instance, the specific area of the surface is defined as the ratio of the surface area to the wing area

$$\overline{S}_{HS(VS)} = \frac{S_{HS(VS)}}{S_w}. \tag{2.14}$$

The equation is predicated on statistical data pertaining to extant UAVs, where the dimensions of the tail surfaces and wings are known.

The determination of the remaining dimensions of the tail unit can be achieved through the utilization of the aforementioned approach, which was employed in the calculation of the wing's dimensions. It is imperative to acknowledge that the aspect ratio, taper ratio, and chords manifest analogous dependencies.

Other Design Components The landing subsystems and powerplant systems encompass the propeller, a subject to be addressed in subsequent chapters.

2.2.5 Helicopter Drone Design Process

The process of designing a helicopter-based UAV is analogous to the design of a light helicopter [9–12]. The initiation of the design process necessitates the delineation of the tasks for the mission, a task that can be accomplished through the collection and analysis of statistical materials. The initial calculation requires the selection of several key variables, including the mission tasks, the payload, the possible takeoff weight, and the flight speed. The initiation of the design process is contingent upon the availability of critical data points, in addition to other information derived from the statistical materials and the mission task.

Main Rotor Geometric Parameters Selection The main rotor is engineered to generate the thrust force necessary to propel the drone through the atmosphere, in addition to enabling longitudinal and transverse control [9–12]. The main rotor is composed of blades that generate lifting force and a sleeve that connects the blades to the main gearbox shaft.

The geometric parameters of the main rotor are defined as follows: radius R_{mr}, measured from the axis of rotation to the extremities of the blades; disk area S_{mr}, equal to $\neq R_{\mathrm{mr}}^2$; the number of blades $n_{\mathrm{b\ mr}}$; and filling coefficient σ_{mr}, equal to the ratio of the calculated area of all blades to the disk area.

The radius R_{mr} of the main rotor of a single-rotor UAV is calculated using the following equation:

$$R_{\mathrm{mr}} = \sqrt{\frac{m_{to} \cdot g}{\pi \cdot p}}, \tag{2.15}$$

where p is the specific load on the disk area. The specific load on the area swept by the rotor is a pivotal parameter that influences the flight characteristics of the drone with the helicopter scheme. The necessary data can be obtained through a thorough analysis of the statistical materials currently available for drones. An increase in the specific load on the disk area leads to an increase in the power consumed by the rotor and the velocity of the ejected air flow. For UAVs, this metric ranges from 100 to 300 decanewtons per meter squared.

The characteristics of the main rotor are found to be significantly affected by the solidity ratio, denoted by σ. The ratio is determined by the ratio of the total area of the blades to the disk area, as expressed by the following equation:

$$\sigma_{\mathrm{mr}} = \frac{n_{\mathrm{b}\ mr} \cdot b_{0.7}}{\pi \cdot R_{\mathrm{mr}}}, \tag{2.16}$$

where $b_{0.7}$ is the chord of the blade at the 0.7-specific blade radius.

The maximum efficiency of the main rotor is attained at the optimal solidity ratio, which is determined by the number of rotor blades. The recommended solidity ratios, contingent upon the number of rotor blades, are enumerated in Table 2.5.

The width of the blades (the chord of the cross-section) b and the specific aspect ratio AR_{mr} of the blades are calculated using the accepted solidity ratio σ and the selected number of blades n_b.

$$b_{mr} = \frac{\pi \cdot R_{mr} \cdot \sigma_{mr}}{n_{b\ mr}}; \tag{2.17}$$

$$AR_{mr} = \frac{R_{mr}}{b_{mr}}. \tag{2.18}$$

The optimal aspect ratio of the main rotor blades is 18. In instances where the calculated aspect ratio significantly deviates from the optimal value, it becomes imperative to modify the accepted number of blades or adjust their chord.

Tail Rotor Geometric Parameters Selection It is recommended that the design of a tail rotor be executed through the utilization of methodologies analogous to those employed in the design of the main rotor of a helicopter [9–12].

The tail rotor is mounted on the tail boom of the helicopter with the objective of positioning it outside the rotation zone and, if possible, outside the intense aerodynamic influence of the main rotor. In this instance, the configuration of the tail rotor's shoulder assumes particular significance.

The offset of the tail rotor can be determined as follows:

$$l_{tr} = R_{mr} + R_{tr} + \delta, \tag{2.19}$$

where δ is the space represents the separation between the radii of the propellers, which is equivalent to

$$\delta = 0.15 - 0.2\,\text{m}. \tag{2.20}$$

$$R_{tr} = (0.16 - 0.2) \cdot R_{mr}. \tag{2.21}$$

It has been observed that the tail boom of small drones may be absent due to the simplification and weight reduction of the airframe. Furthermore, the tail rotor thrust may be comparatively minimal.

The quantity of tail rotor blades can be determined through the following relationship:

Table 2.5 The recommended solidity ratios

Blades number	2	3	4
Solidity ratio	0.04–0.06	0.05–0.07	0.06–0.08

$$n_{\mathrm{b\,tr}} = \left[\left(\frac{2}{3} \cdot n_{\mathrm{b\,mr}} \right) \right], \tag{2.22}$$

where $[(\cdot)]$ is an integer part of a number.

The disk area of the tail rotor can be determined by the following relationship:

$$\sigma_{\mathrm{tr}} = (1.7 - 2.3) \cdot \sigma_{\mathrm{mr}}. \tag{2.23}$$

Fuselage Geometric Parameters Selection The dimensions of the fuselage are determined based on the required volumes for equipment (including the engine), fuel or batteries, and cargo placement [9–12]. Subsequent to determining the required volume, it is necessary to choose the optimal shape of the fuselage in terms of creating the least aerodynamic drag.

Empirical evidence indicates that the aspect ratio (length-to-diameter ratio) of the fuselage is a critical factor in minimizing drag. Studies have shown that achieving an aspect ratio of 2.5 typically results in optimal performance in terms of drag reduction.

The rational shape of the fuselage is thus a conventional ellipsoid. The absence of a cabin on board to accommodate the crew eliminates the necessity for the provision of the requisite viewing angles, thereby obviating the need to complicate the nose of the fuselage as would be the case in a traditional helicopter.

2.2.6 Multicopter Drone Design Process

A multicopter is a drone distinguished by its low aerodynamic perfection [5, 13–18]. This drone does not achieve high flight speeds in comparison with fixed-wing drones; consequently, the requirements for its aerodynamic perfection are less stringent. Concurrently, the aerodynamic requirements for its propellers are analogous to those for helicopter-type propellers of drones.

Given the aforementioned factors, the multicopter design process can be distilled into the selection of the requisite number of propellers, their diameter, and their arrangement scheme (Fig. 2.4). The dimensions of the multicopter frame are determined by the diameters of the propellers and their arrangement scheme.

A multicopter is a complex flying vehicle, the designing and optimization of which parameters must be carried out in a complex manner. In essence, it is not feasible to optimize the propellers, electric motors, batteries, and airframe independently.

Numbers of Propeller Selection A single-rotor and twin-rotor copter is reduced to a helicopter UAV scheme [5, 13–18].

The copter, equipped with three propellers (Fig. 2.5a), exhibits good control in the direction of motion, as well as in pitch and roll.

The quadrocopter (Fig. 2.5b), a four-rotor copter, has been determined to be the most rational number of rotors. Control is carried out by:

In direction of motion: total 4 rotors;
In roll: the gap in thrust between the two left and two right rotors;
In pitch: the gap in thrust between the two front and two rear rotors;
In yaw: the gap in thrust between two rotors on a diagonal and two rotors on the other diagonal.

In essence, the only aspect necessary to control the quadrocopter is the variation in the rotational velocity of particular rotors.

In addition to the quadcopter, there are hexa- (Fig. 2.5c) and octocopters (Fig. 2.5d), which have six and eight rotors, respectively. These are not novel features; they simply have duplicated rotors. These features are not novel; they simply incorporate duplicated rotors. This configuration is advantageous in certain circumstances, such as when higher thrust or enhanced survivability is required.

Diameter of Propeller Determination Geometric restrictions on the diameter of the propeller are imposed on multicopters due to the size of the drone [5, 13–18]. It is imperative that propellers do not adhere to each other, and the interaction of closely spaced swirling flows from neighboring propellers is also to be avoided. It is imperative that there be adequate clearance between the components, equivalent to the diameter of a solitary propeller. In essence, the augmentation of the propeller diameter, with the objective of enhancing thrust, is constrained to a specific threshold.

In the preliminary stage, the diameter of the propeller for a multicopter can be determined for the hover mode. This determination is made by calculating the required static thrust, which must equal the drone's weight. The thrust for a single propeller can be calculated by dividing the vehicle's weight by the number of propellers.

$$P_{1\text{prop}} = \frac{m_{\text{to}} \cdot g}{n_{\text{prop}}}, \tag{2.24}$$

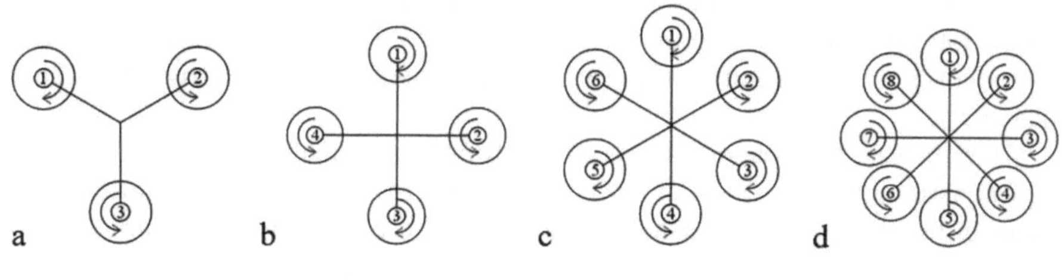

Fig. 2.5 Multicopter rotor schemes: (**a**) tricopter; (**b**) quadcopter; (**c**) hexacopter; (**d**) octocopter

where m_{to} is the takeoff mass of the drone; n_{prop} is the quantity of propellers.

The required power for the single propeller can be calculated using the following equation:

$$P_{req} = \frac{1}{D_{prop}} \cdot \frac{C_P}{C_T^{3/2}} \cdot \frac{1}{\rho^{1/2}} T_{req}^{3/2}, \tag{2.25}$$

where D_{prop} is the smallest possible diameter of a single propeller; C_P is the propeller power coefficient; C_T is the propeller thrust coefficient; ρ is the atmosphere density; T_{req} is the required thrust.

The take-off weight of the drone is contingent upon the mass of the power plant, with greater power requirements resulting in greater take-off weight. This process is iterative in nature, necessitating the initial determination of the power of an existing engine and subsequent calculations based on the available power. The diameter of a single propeller can then be determined by the following equation:

$$D_{prop} = \frac{1}{P_{avail}} \cdot \frac{C_P}{C_T^{3/2}} \cdot \frac{1}{\rho^{1/2}} T_{req}^{3/2}. \tag{2.26}$$

Determination of the Airframe Dimensions The dimensions of the multicopter frame are determined by the diameters of the propellers and their arrangement scheme [5, 13–18]. It is imperative that there be adequate clearance between the components, equivalent to the diameter of a solitary propeller. The dimensions of the airframe can be definitively established only subsequent to the determination of the number and diameter of the propellers.

2.2.7 Hybrid Drone Design Process

Hybrid UAVs are typically composed of a combination of fixed wings and propellers [19, 20]. Such configurations should offer a range and flight time typical of fixed-wing drones, as well as vertical takeoff and landing capabilities akin to those of helicopter or copter drones. The configuration may vary, encompassing turned wings, total propulsion, or propeller components. Alternatively, the configuration may involve a vertical wing during takeoff and landing, followed by a horizontal plane turn.

Nevertheless, the design process for such drones is analogous to the design of their constituent schemes. In essence, it is imperative to identify the most critical stages of the flight and establish the parameters of the wing and propellers.

2.3 Weight Limits

The determination of the mass of the UAV commences only subsequent to the selection of the drone configuration, as the mass of its components is contingent on their external shapes, sizes, and their co-locations [4, 12, 13, 19]. Consequently, the external shapes of the wing or propellers, and other structural components, affect the values of the coefficients of drag and lift. The necessary engine power, and thus the mass of the power plant and fuel or batteries, is contingent on the value of these coefficients. The geometric characteristics of the wing, propeller, tail, and other components directly affect their mass.

The design process entails the identification of the most efficient UAV structure. In this case, calculations of the drone's mass with various schemes should be carried out on a competitive basis according to a predetermined criterion for evaluating the vehicle as a whole. Consequently, a multivariate search for a UAV scheme that will meet the specified requirements is envisaged.

2.3.1 The Determination of Common Dependence for the Takeoff Mass

A statistical method for determining the mass of a UAV is considered, in which the minimum take-off weight is set as an optimality criterion while ensuring the basic mission requirements [1, 3]. This method is predicated on the analysis of statistical material and is available for the preliminary design stage. It is imperative to acknowledge that the aircraft design process is an iterative process. In subsequent design stages, when more precise parameters of the drone are known, it will be necessary to use more precise dependencies.

The takeoff mass of the UAV is the sum of its main functional components:

$$m_{\text{to}} = \sum_{i=1}^{4} m_i = m_{\text{str}} + m_{\text{p.p}} + m_{\text{fuel.s}} + m_{\text{targ}}, \quad (2.27)$$

where m_{str} is the mass of the airframe of the UAV. The calculation involves the aggregation of all constituent components into a unified entity (e.g., the wing, fuselage, tail unit, etc.). This process involves the secure placement of the payload within the airframe. $m_{\text{p.p}}$ represents the mass of the engines and powerplant subsystems that are responsible for thrust creation. This component encompasses engines (motors), nacelles, attachment units, and other components associated with the power plant. A statistical analysis reveals a direct correlation between the power plant system's mass and the mass of the engines (motors):

$$m_{\text{p.p}} = k_{\text{p.p}} \bullet m_{\text{eng}}, \quad (2.28)$$

where m_{eng} denotes the mass of the engines (motors). In the preliminary stage, the factor $k_{p.p}$ is equivalent to 1.1.

$m_{fuel.sys}$ signifies the mass of the system tasked with supplying fuel to the engine during flight. This system includes the fuel (or the batteries) and the elements that are necessary for its arrangement and for the transfer of the fuel or the electric circuit. m_{targ} is the mass of all parts provided for the main purpose of the UAV. In order to conduct a thorough analysis of this component, it is imperative to consider the following elements: the payload, the crew, the payload equipment, and the equipment necessary for reliable and safe flight.

This preliminary phase in the design process involves the delineation of the fundamental mission objective, a responsibility shared by the client and the designer. They engage in collaborative efforts to determine the approaches and instruments necessary to accomplish the mission target. Consequently, the target mass m_{targ} is determined prior to this stage.

The remaining three components of the Eq. (2.27) are influenced by the takeoff mass m_{to}. To advance to the subsequent stage of determining the total mass, it is possible to utilize the specific values of these masses:

$$\bar{m}_i = \frac{m_i}{m_{to}} \tag{2.29}$$

Subsequently, the aforementioned Eq. (2.27) can be expressed in the following form:

$$1 = \frac{m_{str}}{m_{to}} + \frac{m_{p.p}}{m_{to}} + \frac{m_{fuel.sys}}{m_{to}} + \frac{m_{targ}}{m_{to}} = \bar{m}_{str} + \bar{m}_{p.p} + \bar{m}_{fuel.sys} + \frac{m_{targ}}{m_{to}} \tag{2.30}$$

A statistical analysis of existing drones can facilitate the estimation of such specific masses ($\bar{m}_{str}, \bar{m}_{p.p}, \bar{m}_{fuel.sys}$) for a wide range of vehicles. This analysis provides a basis for determining the takeoff mass of the UAV during the preliminary stage of design:

$$m_{to} = \frac{m_{targ}}{1 - \left(\bar{m}_{str} + \bar{m}_{p.p} + \bar{m}_{fuel.sys}\right)}. \tag{2.31}$$

This suggests that establishing a target mass is sufficient to initiate the process of evaluating the potential takeoff mass of the UAV during the preliminary stages of the design process. This approach permits a comparative analysis of diverse schemes, thereby enabling the assessment of their mass efficiency.

2.3.2 Specific Masses for the Fixed Wing Scheme

The fixed-wing configuration has received the most extensive investigation, and as such, designers are at liberty to utilize statistical data from light aircraft, with the exception of equipment utilized for human survival [1–8]. Furthermore, the type of power plant must be taken into consideration. In the case of engines that utilize traditional fuels, reliance on aircraft statistics is a viable option. Conversely, in the case of an electric power plant, it is imperative to acknowledge that electric motors are typically less weighty than conventional fuel engines. Conversely, electric batteries have been observed to exhibit a higher mass compared to conventional fuel.

The utilization of cryogenic fuels, including natural gas and hydrogen, is a viable option. However, further research is necessary to fully explore the conditions of these fuels.

The statistical data regarding the potential distribution of specific masses among the components of the Eq. (2.31) is provided in Table 2.6.

2.3.3 Specific Masses for the Helicopter Scheme

In the context of UAVs operating in a helicopter configuration, the mass of the structure and the mass of the power plant exhibit significant variations, contingent on the implemented scheme [9–12]. A notable distinction emerges when comparing a single main rotor scheme to a coaxial scheme. Consequently, Eq. (2.31) assumes a distinct formulation:

$$m_{\text{to}} = \frac{m_{\text{targ}}}{k_{\text{lf}} - \overline{m}_{\text{fuel.sys}}}, \tag{2.32}$$

where k_{lf} denotes the load factor of the drone at maximum capacity. It is determined by equation

$$k_{\text{lf}} = 1 - \overline{m}_{\text{empt}}, \tag{2.33}$$

Table 2.6 Statistical values of specific masses for fixed-wing UAVs

UAV type	Specific structural mass, $\overline{m}_{\text{str}}$	Specific power plant mass, $\overline{m}_{\text{p.p}}$	Specific fuel system mass, $\overline{m}_{\text{fuel.sys}}$
Subsonic light	0.30–0.32	0.12–0.14	0.20–0.24
Subsonic heavy	0.28–0.30	0.10–0.12	0.29–0.33
Supersonic	0.28–0.32	0.18–0.22	0.28–0.33
With electrical power plant	0.25–0.50	0.02–0.05	0.15–0.45 (batteries mass)

where $\overline{m}_{\text{empt}}$ is the specific mass of an empty drone, that is to say, the mass of the drone devoid of fuel and cargo. In the event that a drone is designed for a particular task and structured accordingly—for instance, an electrical drone utilized for the purpose of visual inspection—the empty mass would refer to the mass of the drone devoid of batteries and the camera.

According to statistical materials, the value of k_{lf} typically ranges from 0.3 to 0.6.

The calculation of the specific mass of the fuel system for a traditional fuel-powered UAV is determined by the following equation:

$$\overline{m}_{\text{fuel.sys}} = 1.1 \cdot \left(\overline{q}_{\text{fuel}} \cdot L + 0.33 \cdot \overline{Q}_{\text{h}} \right), \tag{2.34}$$

where $\overline{q}_{\text{fuel}}$ is the specific fuel consumption per kilometer, $\overline{q}_{\text{fuel}} = (0.025\text{–}0.03)\cdot 10^{-3}$; $\overline{Q}_{\text{h}}$ is the specific fuel consumption per hour, $\overline{q}_{\text{fuel}} = (0.059\text{–}0.063)\cdot 10^{-3}$; L is the mission flight range.

The specific mass of the battery system for the electrical-powered UAV ranges from 0.4 to 0.8. Research has demonstrated a direct correlation between the mass of the battery system and the flight time of the drone.

2.3.4 Specific Masses for the Multi-Copter Scheme

Presently, there is an absence of explicit, readily accessible, and standardized statistics concerning the precise mass of multicopter components, akin to those available for conventional aircraft and helicopters.

Why are there no statistics available on this topic, or are they difficult to find? There are several reasons (it is also interesting, because it is an integral part of the design process):

1. The youth and rapid development of the industry. Multicopters have been commercially available for approximately 10–15 years. Technological advancements, particularly in the domains of batteries, motors, and materials, are occurring at a rapid pace, rendering outdated statistics obsolete.
2. The commercial secret. Manufacturers of serial drones (e.g., DJI) meticulously safeguard the internal structure and mass summaries of their products as a trade secret.
3. The strong dependence on the purpose. The composition and mass of the components vary significantly depending on the task, for example:

   – The racing drone is composed of a minimum number of components, including a frame, a power plant, a battery, and a simplified control system.
   – The cinematic drone is characterized by its robust frame, which includes a substantial frame and a vibration-dampening system. It is also equipped with advanced suspension mechanisms and a high-performance camera system.

- The delivery drone is equipped with a high-performance propulsion system, a robust frame, and a specialized cargo system.
- A flying taxi must be equipped with safety systems, life support, and reliable avionics. This radically changes the mass balance of the vehicle.

The design articles concerning multicopters or eVTOL frequently contain mass diagrams. For instance, in [5, 13–20], the mass fraction of the frame is approximately 30%, the power plant (including propellers) is around 25%, the power system (batteries) is about 35%, and the payload is approximately 10%.

Approximate specific masses for various categories of drones are provided in the Table 2.7. However, it must be noted that these values are entirely arbitrary.

2.4 Systems Content Selection

UAVs are distinguished from conventional aircraft in that they do not have a human pilot on board [1–3, 6, 7]. Presently, the developmental stage of UAVs does not encompass the integration of passengers. These projects are currently in the development, testing, or certification phase. This feature enables the exclusion of human life support systems from the equipment. Nevertheless, for the purpose of passenger transportation, the provision of such equipment for drones is imperative.

In all other respects, the composition of systems and onboard equipment for UAVs and traditional aircraft is identical:

1. Power plant subsystems;
2. Flight control system;
3. Electrical system;
4. Hydraulic system and (or) Pneumatic system;
5. Environmental control system (for the specific equipment);
6. Emergency systems.

Table 2.7 Approximate specific masses for different categories of drones (very approximate)

Category of drone	Mass range (kg)	Frame	Powerplant (motors, props, governors)	Battery	Payload	Other (electronic components, wire, connectors, etc.)
Racing drones	0.5–0.8	0.2–0.25	0.25–0.3	0.25–0.35	–	0.1–0.15
Personal drones	1–8	0.15–0.2	0.2–0.25	0.3–0.4	0.05–0.1 (camera)	0.1–0.15
Cinematic drones	4–10	0.1–0.15	0.15–0.2	0.4–0.5	0.1–0.2 (gimbaled camera)	0.05–0.1
Delivery drones (heavy)	25+	0.2–0.3	0.2–0.3	0.25–0.4	0.1–0.2	0.05–0.1

It is evident that such systems are intricate, comprising numerous additional subdivisions and components.

2.4.1 Powerplant Subsystems

The engine is the primary component of the power plant and the focus of Chap. 3. However, to ensure optimal operation, additional subsystems are necessary, including propellers, air intakes, and a fuel system.

Propeller The propeller and propeller gearbox for the fixed-wing scheme and the transmission for the helicopter scheme must generate thrust or lift [2, 4].

The number of propellers required for the fixed-wing scheme depends on the number and type of engines used. In the context of drones, configurations employing push-pull designs are sometimes observed. In these cases, the number of propellers doubles.

The number of propellers in the helicopter scheme depends on the configuration type, as previously discussed. Additionally, configurations that use twin propellers exist. This configuration is advantageous for generating higher thrust. In this case, the number of propellers is doubled.

Energy Storage and Transmission System for Power Plant This system is common for electrical power plants and will be elaborated upon subsequently [2, 4]. In the case of a traditional power plant, this system is a conventional fuel system.

The fuel system supplies the engines with fuel under specific conditions and ensures the continuous availability of fuel for a flying vehicle. Additionally, it is imperative that it ensure convenient refueling in both ground and flight conditions.

Currently, there are several projects underway that aim to replace oil-based fuels with gas-based fuels, such as natural gas or hydrogen. This transition introduces a new level of intricacy to fuel systems because these fuels operate at low temperatures. These fuels are classified as cryogenic due to their nature.

Lubrication System The lubrication system provides friction surface lubrication and removes heat and wear from the friction zones of engines [2, 4]. This system is similar to those found in conventional aircraft in terms of functionality and operational characteristics.

Engine Mounts The engine mounts are critical components that facilitate the attachment of the engine to the airframe [2, 4]. Additionally, the system's functionality and operational features bear a resemblance to those of conventional aircraft systems. However, this is not the case for electrical power plants. The propeller's operation manifests the lowest oscillation, a consequence that merits further investigation. This particular case demonstrates a relatively uncomplicated structural design.

Air Ingestion System The air intake system is composed of two components: the air intake and its duct [2, 4]. Its function is to prepare and transfer the required quantity of air into the engine under predetermined conditions.

It is evident that this subsystem is indispensable solely for conventional power plants, rather than for electric ones. Nevertheless, protective covers are installed around the propellers. Design methodologies employed for air intakes can be applied to these components.

Exhaust System The exhaust system plays a pivotal role in the formation of the jet stream subsequent to the engine's operation, a process that facilitates the dissipation of the combustion products [2, 4]. It is evident that this subsystem is indispensable solely for conventional power plants, rather than for electric ones.

The exhaust system of conventional aircraft is equipped with a thrust reversal system, a component that is not necessary for light UAVs. The implementation of this system does not necessitate the use of heavy UAVs when operated on a runway designated for conventional aircraft. In scenarios where heavy drones are to be operated in challenging conditions, such as the absence of a prepared runway, the utilization of thrust reversal becomes a viable consideration. However, it is imperative to acknowledge that this approach invariably introduces an additional mass and complexity, necessitating the safe turning of reverse channel gases.

Furthermore, the provision of a controllable nozzle for maneuverable UAVs, analogous to those employed on the F-35 and Su-35, is a viable option. This modification has the potential to substantially augment the operational capabilities of the drone.

Powerplant Control System The power plant control system facilitates the transmission of control signals from the autopilot to the designated unit [1, 2, 4]. In the case of a drone, it is to be incorporated into the drone's common automatic control system.

Fire Protection System The primary function of the fire protection system is to facilitate the transmission of an alarm signal to the ground service personnel in the event of a fire [1, 2, 4]. The system is also equipped with automated fire suppression mechanisms designed to contain and extinguish fires.

It is reasonable to conclude that such a complex system should be implemented for expensive drones, the rescue of which is an important task, or to prevent an incident on the ground in the event of an emergency landing of the drone. The employment of a rudimentary fire alarm system constitutes a viable solution for drones, distinguished by their economical and lightweight nature.

Deicing System It is imperative that the system be capable of both preventing and detecting the formation of ice on the air intake and the blades' edges [1, 2, 4]. It is imperative to note that the aforementioned functions are equally applicable to the fuel system.

It is evident that the implementation of such a system necessitates a comprehensive understanding of the operational parameters of UAVs. It is evident that circumstances involving low-flying drones and regions with a warm climate do not necessitate the implementation of such a system.

2.4.2 Flight Control System

To ensure the effective operation of the UAV, it is imperative to delineate the requisite flight trajectory [1, 4]. The system that enables the UAV to navigate the designated trajectory is designated as the control system. This system comprises a variety of mechanical, hydraulic, electrical, and computerized components, among others, that function collectively to enable the UAV to achieve its desired flight path.

The primary control system encompasses the control of the elevators, rudder, ailerons of the fixed-wing drones or rotors of the copters. The secondary control system is responsible for the operation of the landing gear, engines, systems, equipment, and other components.

The control system can be classified into various types, including mechanical, electrical, hydraulic, and fly-by-wire systems.

It stands to reason that the control system of drones equipped with electrical power plants would also be of an electrical nature. However, in the case of heavy drones, alternative options can be considered. For instance, the autonomous hydraulic system can be employed for the landing gear.

It is evident that the components of the mechanical system invariably function as links between the control system and the effector unit.

Contemporary drones can be equipped with either an autonomous or a semi-autonomous control system.

An autonomous control system is defined as a system that ensures the completion of a mission without human-pilot interaction, according to a pre-programmed plan. Concurrently, elements of artificial intelligence can be integrated to address unanticipated circumstances.

A semi-autonomous control system is a system in which a human pilot can control a drone from the ground. These drones may also possess automated functions, such as the capability to execute an emergency landing in the event of an on-board UAV accident or to execute a flight program under standard conditions.

The degree of mission autonomy is contingent upon the complexity of the mission itself, as well as the complexity of the drone. It is evident that as the complexity of a mission increases, the complexity of the drone and its software required to execute it also rises. Consequently, the manufacturing and maintenance costs of such drones increase proportionately.

It is only natural that the financial resources allocated to the development of UAVs represent a minor proportion of the total budget designated for this endeavor. A considerable portion of the budget is allocated to the development of software that is commensurate with the objectives of the mission.

2.4.3 Electrical System

Electricity is a form of energy that is employed on an UAV board [1, 4]. The preponderance of equipment on an aircraft is powered by electricity. This phenomenon is particularly evident in the case of drones that are equipped with an electrical power plant.

The employment of electric power in drones confers numerous distinct advantages:

1. High efficiency. In the case of a brushless motor, for instance, it has been demonstrated that the motor can reach 95% efficiency.
2. Low mass. Furthermore, these drones possess a lower mass compared to internal combustion or gas-turbine engines, a feature that contributes to enhanced agility and maneuverability. Moreover, the absence of a fuel supply eliminates the need for a fuel tank, thereby expanding the design possibilities of the drone.
3. Good stability. The stability exhibited by these drones is a consequence of their design, which eliminates the need for fuel tanks and relies exclusively on batteries for power. During the object's flight, it exhibits neither fuel consumption nor any displacement of its gravitational center, which would be associated with such consumption. This finding suggests that the aircraft maintains a stable position of the collocated gravity center and aerodynamic center.
4. Low explosion possibility. Furthermore, the absence of a fuel supply negates the risk of explosion, a critical safety consideration. The safety of the electric motor and its power supply system (battery) is a significant advantage over similar systems with internal combustion or gas-turbine engines, as they are less prone to explosion.
5. Low thermal visibility. Furthermore, the absence of significant thermal radiation contributes to an enhancement in their safety profile. Consequently, thermal imaging radars are less effective at detecting drones equipped with electric motors.
6. Low environmental impact. The operation of these drones results in no harmful emissions into the atmosphere due to the combustion of fuel.

The integration of electricity sources on flying vehicle boards coincided with the emergence of the first aircraft. For instance, the Ilya Muromets aircraft, developed by Igor Sikorsky, was equipped with a 500-watt power source.

In contemporary aircraft, both batteries and generators function as sources of electrical energy. In the case of light drones, the application of batteries is the primary concern. In all other scenarios, both batteries and generators are necessary. During flight, generators recharge the batteries and supply all necessary systems. These generators have the capacity to obtain power from piston or jet engines, thereby ensuring the generation of electricity in any condition.

The integration of electrical engines and conventional engines through the generators is designated as a "hybrid power plant." This approach is a viable method for mitigating the environmental impact of aviation.

2.4.4 Hydraulic and Pneumatic Systems

The advent of hydraulic systems in the field of aeronautics coincided with the development of retractable landing gear in the early 1930s [1, 4, 6, 7]. Since that time, an increasing number of tasks have been performed by means of the hydraulic system. Hydraulic power has emerged as an efficient medium for high-energy consumption in drones.

The hydrodynamic system is composed of vane machines, specifically a centrifugal pump and a hydraulic turbine. The operational wheels of these components are situated in close proximity to each other, with one of the centrifugal pump's wheels aligned with the turbine wheel. The rotation of the pump wheel, driven by the input shaft, results in the movement of fluid onto the turbine, thereby transferring power to it. This configuration facilitates the transfer of power, thus circumventing the constraints imposed by rigid mechanisms.

The contemporary turbofan engine functions as an advanced gas generator, thereby facilitating the utilization of engine bleed air in numerous aircraft systems. The utilization of this energy is predominantly influenced by two primary factors: the provision of heating and the provision of motive power. Bleed air, which is extracted from the engine compressor, undergoes a series of processes prior to its utilization for various functions. These processes include cooling and pressure reduction/regulation.

In the engine, high pressure bleed air functions as the motive power—colloquially referred to as "muscle power"—for numerous valves implicated in the bleed air extraction process. Bleed air is also utilized for the purpose of anti-ice protection, accomplished by heating the engine intake cowling. Moreover, bleed air functions as the propulsion source for the engine thrust reversers. Pneumatic systems are engineered for the purpose of energy transfer from a power source to pneumatic (hydraulic) actuators and their associated control mechanisms.

A comparison of pneumatic and hydraulic systems reveals a high degree of similarity between the two.

It is imperative to note that both applications necessitate additional space and mass within the drone. Consequently, the necessity of these components must be substantiated through rigorous testing procedures. In instances where the UAV possesses substantial mass and adequate internal volume, the implementation of hydraulic or pneumatic actuators is rational, as they demand less electrical energy compared to electromechanical actuators.

2.4.5 Emergency Systems

The design, manufacturing, and service engineers of UAVs allocate substantial effort to ensuring the safety of these vehicles [1, 4]. Notwithstanding, the potential for accidents remains a persistent concern. In such circumstances, the potential for damage to the drone itself, as well as the risk of harm to people on the ground, is a salient concern. It is impractical to consider and mitigate every potential hazard during the design process. However, the potential for such occurrences can be assessed and mitigated to a certain extent through systematic forecasting.

In the event that the probability of such an occurrence is deemed to be relatively high, and the consequences of the incident in question are deemed to be adequately grave, the drone structure is to be equipped with specialized emergency systems, the purpose of which is to enhance its survivability.

The primary function of emergency equipment is to ensure the safety of the UAV and its associated equipment, as well as to facilitate the rescue of people and buildings on the ground.

The classification of emergency equipment can be categorized as follows:

1. Anti-fire subsystem;
2. Emergency landing subsystem.

The necessity of the anti-fire subsystem has been previously discussed.

The emergency landing subsystem is typically a software and hardware platform that is hosted on an autonomous computer on board a drone. The system's primary function is to oversee the operation of various drone systems and to make decisions offline regarding the urgent landing of the drone in the event of system malfunctions or the continuation of flight in the event of any problems.

The system contains a database in which information regarding potential landing sites is stored in the event of a drone failure. The decision to execute an emergency landing is contingent upon the drone's geographical location, the time of day, the type of terrain, and an array of additional parameters. In the event that a problem arises that renders continued flight impossible, the system autonomously selects a suitable location and executes a soft landing.

The capacity of a drone to execute a soft landing is contingent upon its specific design and operational characteristics. In the case of a fixed-wing UAV, the soft-landing process can be facilitated by gliding. Conversely, if the drone is of the copter type, its soft landing can be facilitated by the safe rotation of its operable propellers. In the event that these options prove insufficient, an emergency parachute system can be integrated into the drone.

2.5 Environmental Problems

Presently, the field of aviation is undergoing a period of development that is unparalleled in its significance [2, 3, 6, 7]. Nevertheless, the aviation sector is perpetually confronted with novel challenges. A pressing contemporary concern pertains to the mitigation of its environmental impact. In this regard, the International Civil Aviation Organization (ICAO) has established a set of guidelines aimed at achieving a 50% reduction in pollution levels across the global civil aviation sector by the year 2050. This objective is ambitious, and achieving it will require substantial effort. There are several approaches that could be taken to achieve these objectives.

The initial alteration pertains to the nature of power plants, wherein corporations are currently investigating the viability of employing electric, hybrid, and hydrogen power plants [3]. However, the implementation of these advancements in the domain of civil aviation is constrained by the dimensional constraints and mass limitations of aircraft. Consequently, it is advantageous to formulate potential design solutions for UAVs, given their smaller size and reduced manufacturing costs.

The second approach pertains to the automation of civil vessel traffic [3]. The implementation of autopilots will eliminate the human factor in traffic control, thereby reducing the associated environmental impacts. In this context, the development of potential solutions using drones is advantageous. This approach is expected to facilitate the accumulation of statistical data and mitigate the risk of human casualties in emergency situations.

Therefore, it is evident that the UAV is an optimal instrument for the assessment of novel technologies in the domain of civil aviation.

The employment of automatic flight control systems for UAVs has emerged as the prevailing trend in this field. The implementation of drone transportation along pre-designated routes and an established timetable ensures optimal efficiency in the delivery of cargo. This domain is undergoing rapid advancements across numerous nations.

The employment of electric propulsion systems for light drones has become obligatory. The merits of electric power plants are manifold. They include the capacity to rapidly replace batteries, enhanced fire safety, reduced weight, minimized noise levels, and zero emissions.

In the context of heavy UAVs, the utilization of pure electric propulsion systems is constrained by the limited weight output capacity of batteries. Consequently, the adoption of power plants that rely on the combustion process emerges as a more rational approach for heavy UAVs. These power plants can be classified as either hybrid or hydrogen-based.

In the context of hybrid power plants, the generation of electricity is facilitated by conventional piston or jet engines. This approach has been demonstrated to reduce fuel consumption by up to 20%, thereby leading to a corresponding decrease in emissions.

The operation of power plants that utilize hydrogen technology is not, in principle, associated with the release of harmful emissions. However, the unique storage requirements and the substantial complexity of the power plant design pose significant challenges. Furthermore, hydrogen demonstrates a reduced weight efficiency in comparison to conventional fuels. This phenomenon results in an increase in the necessary volume for fuel storage.

In addition to the detrimental effects of harmful emissions, noise levels must be considered in regard to environmental and human health concerns. In this context, electrical installations represent the only viable solution. Electric motors, in particular, are characterized by their minimal noise output. The predominant source of noise in this context is the aerodynamic noise resulting from the interaction between the aircraft body and its propellers and the surrounding airflow.

Given the aforementioned points, it can be concluded that the utilization of UAVs equipped with electric propulsion systems constitutes a priority undertaking for the domain of civil aviation.

2.6 Conclusion

In this chapter, a number of issues pertinent to the design of unmanned aerial vehicles (UAVs) were discussed. The initial step in the design of any flying vehicle is a thorough analysis of its potential limitations. The overarching structure of the aircraft design process was reviewed, accompanied by elucidations concerning the features of the UAV. For certain exemplary drone configurations, the prevailing principles of design and mass distribution were delineated. In addition, approaches were analyzed to determine the composition of UAV systems.

References

1. Tiniakov D, Li L. Airworthiness engineering basis. Datian Aviation Beijing Beixue Publishing House; 2020.
2. Tiniakov D, Li L, Su Y. Airworthiness Design of Civil Aircraft Systems and Structures. Datian Aviation Beijing Beixue Publishing House; 2018.
3. Li L, Kretov A, Tiniakov D. Aeronautics—characteristics and emerging technologies. In: Chapter: UAVs and hydrogen technologies are the future of civil aviation [Internet]. IntechOpen; 2025. https://doi.org/10.5772/intechopen.1008352.
4. Marshall D, Barnhart R, Shappee E, Most M. Introduction to unmanned aircraft systems. New York: CRC Press; 2016.
5. Shinde S, More J, Chaudhari C. Mastering drone technology with AI. India: BPB Publications; 2025.
6. Torenbeek E. Advanced aircraft design: conceptual design, analysis, and optimization of subsonic civil airplanes. Chichester, UK: Wiley; 2013.
7. Raymer DP. Aircraft design: a conceptual approach. 6th ed. Reston, VA: AIAA; 2018.
8. Sun Y, Li L, Tiniakov D. Reliability Engineering. Beijing: Springer/Science Press; 2024. https://doi.org/10.1007/978-981-99-5978-5.
9. Krivtsov V, Karpov Y, Losev L. The helicopters design. Kharkov: KhAI; 2003.

10. AMC Pamphlet. Engineering Design Handbook. Helicopter Engineering. Part One. Preliminary Design. Headquarters, U.S. Army Materiel Command; 1974.
11. NASA. Helicopter calculation and design. Washington: NASA; 1968.
12. Moiseev V., Moiseeva L. Methods of preliminary designing unmanned helicopters. Kazan: Shkola. 2022.
13. Xu J. Design perspectives on delivery drones. Santa Monica, CA: Rand corporation; 2017. [Internet]. https://www.rand.org/content/dam/rand/pubs/research_reports/RR1700/RR1718z2/RAND_RR1718z2.pdf. Accessed 15 Sept 2025
14. Mueller M., Lee S., D'Andrea R. Design and control of drones. The annual review of control, robotics, and autonomous systems, 5. [Internet]. https://www.annualreviews.org/. Accessed 15 Sept 2025
15. Magnussen Ø, Ottestad M, Hovland G. Multicopter design optimization and validation. J Model Identific Control. 2015;36(2):67–79.
16. Pollet F, Delbecq S, Budinger M, Moschetta J-M, Liscouët J. A common framework for the design optimization of fixed-wing, multicopter and VTOL UAV configurations. In: Stockholm: 33rd Congress of the International Council of the Aeronautical Science; 2022. p. 1–18.
17. Allaka G, Anasuya B, Yamini C, Vaidehi N, Ramana Y. Modelling and analysis of multicopter frame and propeller. Int J Res Eng Technol. 2013;2(4):481–3.
18. Oh S, Kim M, Kim H, Lim D, Yee K, Kim D. The solution development for performance analysis and optimal design of multicopter-type small drones. In: Greece, Athens, 2020 International Conference on Unmanned Aircraft Systems (ICUAS); 2020.
19. Tyan M, Nguyen N, Kim S, Lee J-W. Comprehensive preliminary sizing/resizing method for a fixed wing—VTOL electric UAV. Aerosp Sci Technol. 2017;71:30–41. https://doi.org/10.1016/j.ast.2017.09.008.
20. Ozdemir U, Aktas Y, Vuruskan A, et al. Design of a commercial hybrid VTOL UAV system. J Intell Robot Syst. 2014;74:371–93. https://doi.org/10.1007/s10846-013-9900-0.

Chapter 3
Propulsion System

3.1 Introduction

The intended purpose, size, and aerodynamic layout of an UAV largely determine its flight and tactical characteristics, such as range and flight duration, altitude and speed range, and so on. This is why special attention is paid to the choice of engine when designing an UAV. Absence of this component with the required technical parameters significantly reduces the efficiency of the designed drone and raises questions about the feasibility of creating this unmanned complex.

Traditionally, combustion engines (reciprocating and gas turbine) have been used for flying vehicles [1, 2]. Currently, these engines are primarily used for large and medium-sized drones. Electric motors and gasoline-electric hybrid engines are usually used in mini drone designs with a maximum takeoff weight of less than 25 kg [3]. The electric motors of mini UAVs provide cost reduction, elimination of a thermal signature, reduction of a sound signature, and an absence of exhaust gases.

Modern electric batteries have a relatively low energy density, which limits the flight time of vehicles to an average of about 60 min [4, 5]. Therefore, combining batteries and fuel cells in hybrid power plants seems to be a promising approach.

Microjet and microturbine motors are becoming more common. Designers and operators are closely monitoring the development of microturbine engines. However, this type of propulsion does not meet the requirements of small UAVs in terms of flight time.

The efficiency of the UAV propulsion system is contingent upon the configuration of the power plant:

- The efficiency of the batteries is measured at more than 70%.
- The efficiency of the fuel cell is approximately 45%.
- The efficiency of the internal combustion engine is approximately 40%.

L. Li, D. Tiniakov, *Unmanned Aircraft Design*,
https://doi.org/10.1007/978-981-95-7167-3_3

Consequently, the conclusion naturally suggests itself: the exclusive utilization of battery-powered propellers. Consequently, over 95% of contemporary commercial UAV designs are powered by lithium polymer (Li-Po) or lithium-ion (Li-Ion) batteries.

However, the energy source and the type of propulsion that provide flight characteristics are directly dependent on the entire system. Consequently, the incorporation of batteries into the system does not result in an augmentation of flight time or payload capacity. The augmentation of the time duration is contingent upon the enhancement of energy storage quality. In summary, the objective is to enhance the specific gravity (watt per kilogram) and volumetric specific energy (watts per decimeter cube) of the batteries.

The most optimal values for the specific gravity of power plants are derived on the basis of hydrogen fuel [6]. In comparison to liquid hydrocarbon mixtures, such as kerosene, gasoline, and liquefied gases, these values are significantly superior.

This finding lends further credence to the notion that among the diverse approaches for storing electrical energy, fuel cells are regarded as the optimal solution. Figure 3.1 presents a graphical representation of the specific energy (electrical) of various storage technologies, including fuel cells.

The authors posit that a promising trajectory for the advancement of contemporary UAV power plants lies in the implementation of hybrid systems comprising batteries and fuel cells [1]. This is typically a combination of high-specific energy batteries and hydrogen fuel cells. The storage of hydrogen fuel on board an unmanned aerial vehicle can be accomplished through the use of pressurized cylinders or in the form of chemical compounds.

A more complex closed fuel system is also present. In this variant, hydrogen is produced directly on board the drone by an electrolyzer that uses electrical energy from photovoltaic cells. However, it should be noted that the solution remains in the experimental stage of development.

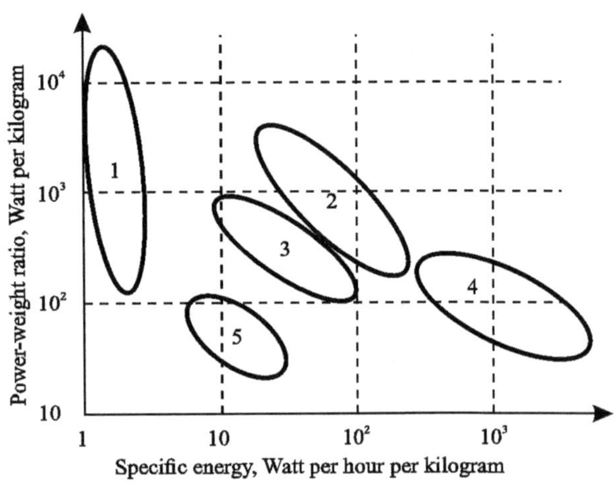

Fig. 3.1 Specific energy graph [5]: 1—Ionistors; 2—Lithium-ion; 3—Nickel metal hydride; 4—Fuel cell; 5—Lead

The predominant energy sources employed in drones are Li-Po and Li-ion batteries [4, 5]. Consequently, there is a high probability of the imminent availability of more advanced types of batteries with higher energy density. For instance, extensive research is underway on lithium-thionyl chloride batteries (Li-$SOCl_2$), which possess twice the specific gravity of Li-Po batteries.

Despite the exorbitant costs associated with these designs, technological advancements have led to the promise of more affordable pricing in the future. We can also provide an example of lithium-air or lithium-sulfur batteries (Li-S), which are expected to provide an even greater energy density, increased by approximately 7–10 times.

The following are some features of the batteries used for modern drones:

1. Li-Po batteries are a type of battery that use lithium as the primary component in their composition. This particular type is favored for its high energy density and lightness. However, these devices are susceptible to overcharging and mechanical damage, necessitating meticulous handling and specialized charging procedures.
2. Li-ion batteries are particularly well-suited for larger drones, offering enhanced capacity and durability. While they exhibit reduced sensitivity to damage, they nevertheless necessitate meticulous handling.
3. Nickel-metal hydride (NiMH) batteries are a type of rechargeable battery that uses a nickel-metal hydride positive-electrode and a hydrogen-absorbing negative-electrode. While their utilization in drones is less prevalent, they offer distinct advantages, including cost-effectiveness and resilience to overcharging. The energy density of these batteries is lower than that of Li-Po and Li-ion batteries, which limits their application.
4. Fuel cells are an alternative power source. They are employed in specialized drones for long-term missions or in research drones.
5. Solid-state batteries represent a significant advancement in the field of energy storage technology. These batteries are based on graphene. Solid-state batteries offer enhanced safety and reduced energy consumption in comparison to conventional lithium batteries.

As demonstrated above, the contemporary UAV industry boasts a plethora of options for propulsion systems. The selection of the power plant type and the configuration of the UAV are influenced by numerous parameters (see Chap. 2). In this chapter, an exploration will be conducted into the features of the most applicable power plants for drones.

3.2 Basic Concepts of Powerplants

In order to comprehend the particulars of employing disparate power plants for UAVs, a comprehensive review of the salient characteristics of power plants suitable for drones is imperative.

Modern aircraft power plants are essential for two primary functions: the generation of thrust for flight and the provision of auxiliary services [1, 2]. The latter include the generation of electricity, the operation of pneumatic and hydraulic systems, and other critical functions.

In accordance with the principle of generating energy for the propulsion of aircraft, power plants can be categorized into the following:

1. Based on the combustion process
2. Electrical basis
3. Hybrid approach

A more detailed classification of power plants for UAVs is shown in Fig. 3.2. In subsequent analyses, the focus will be limited to a select number of possible power plant types.

The first group encompasses reciprocating, gas turbine, ramjet/scramjet, and rocket engines. The second group encompasses electric power plants that utilize a variety of energy sources, including batteries or fuel cells. The third group of power plants incorporates systems that combine two or more energy sources. Typically, the secondary source is traditional combustion engines, which are utilized to generate electrical energy.

A consideration of the principles of operation of such power plants is in order.

3.2.1 *The Principle of Operation of Engines that Is Based on the Process of Combustion*

The operation of any type of combustion engine is predicated on the occurrence of a specific thermodynamic process. The most prevalent engine types for UAVs are piston and gas turbine engines. A comparison of the thermodynamic principles

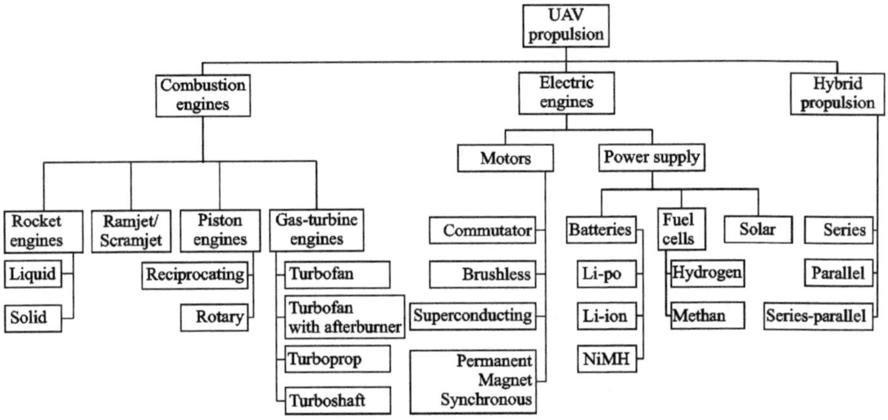

Fig. 3.2 General classification of UAVs propulsion systems [1, 2]

underlying the operation of these engines will provide a more comprehensive understanding of their operation.

For an internal combustion engine (petroleum), the Otto thermodynamic cycle is the closest match (see Fig. 3.3a) [7]. The operation of a gas turbine engine (e.g., a turbojet) is analogous to the Brayton cycle (see Fig. 3.3b) [7].

Now, let's take a closer look and conduct a comparative analysis.

3.2.1.1 Internal Combustion Engine (Piston, Petrol)

The Otto cycle was developed as a theoretical model to explain the operation of the earliest internal combustion engines.

1. The intake stroke. The cylinder is filled with a fresh mixture.
2. The compression stroke. The piston is responsible for the compression of the fuel-air mixture. Concurrently, the temperature and pressure are increasing. It is important to note that no heat exchange (idealization) occurs. This phenomenon is analogous to adiabatic compression, as illustrated in Fig. 3.3a (1–2).
3. Operating stroke. The mixture is ignited by a spark. The rapid combustion process occurs so rapidly that the volume of the gas does not have sufficient time to undergo a phase change (the piston is at the top dead center). This corresponds

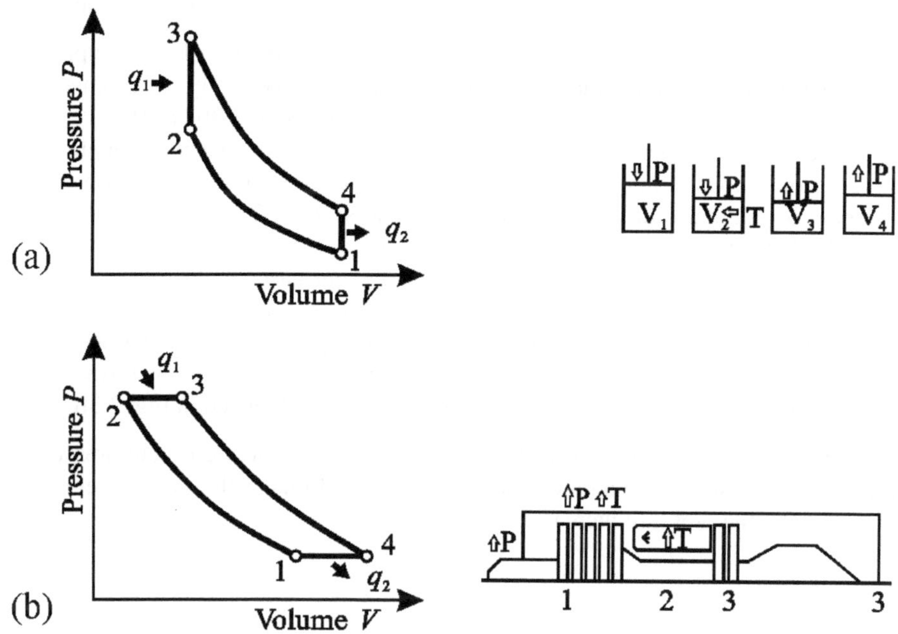

Fig. 3.3 The Thermodynamic cycles diagrams for different engines type: (**a**) The diagram of the Otto cycle and the reciprocating engine; (**b**) The diagram of the Brayton cycle and the gas-turbine engine

to the isochoric heat supply depicted in Fig. 3.3a (2–3). The incandescent gases exert a force on the piston, causing it to move in a backward direction and thereby performing useful work. This phenomenon is analogous to adiabatic expansion, as illustrated in Fig. 3.3a (3–4).

4. The release stroke. The opening of the exhaust valve results in a rapid decrease in pressure, maintaining a relatively constant volume. The exhaust gases are subsequently discharged. This phenomenon is analogous to isochoric heat dissipation, as illustrated in Fig. 3.3a (4–1).

The Otto cycle is characterized by the provision of heat at a constant volume, constituting an isochoric process. The efficiency is contingent upon the compression ratio:

$$\eta = 1 - \varepsilon^{(1-\gamma)},$$

where η is the theoretical efficiency of the Otto cycle, ε is the compression ratio, and γ is the adiabatic index.

3.2.1.2 Gas Turbine Engine

For a gas turbine engine, such as a turbojet or turboprop, the Brighton cycle is the most analogous. This is a fundamental cycle for all gas turbine-based engines.

1. The compressor. The air is compressed by the rotary blades of the compressor. Concurrently, the temperature and pressure are increasing. This phenomenon is analogous to adiabatic compression, as illustrated in Fig. 3.3b (1–2).
2. The combustion chamber. The fuel undergoes combustion at a constant pressure. The continuous influx of air and the continuous egress of combustion products serve to maintain the internal pressure of the system. This corresponds to the isobaric heat supply depicted in Fig. 3.3b (2–3).
3. The turbine and exhaust nozzle. The incandescent gases undergo expansion in two distinct phases. Initially, they expand within the turbine, thereby imparting a portion of their energy to the compressor. Subsequently, they expand within the jet nozzle, where their energy is converted into a high-speed jet (thrust). This phenomenon corresponds to the adiabatic expansion depicted in Fig. 3.3b (3–4).

The Brayton cycle is characterized by the provision of heat at constant pressure, a process that is isobaric in nature. The effectiveness of this cycle is contingent upon the extent of the pressure increase:

$$\eta = 1 - \pi^{[(1-\gamma)/\gamma]},$$

where η is the theoretical efficiency of the Brayton cycle, π is the rate of pressure increase in the compressor, and γ is the adiabatic index.

The Otto and Brighton cycles represent practical yet idealized models that enable engineers to calculate parameters and evaluate engine efficiency. While these

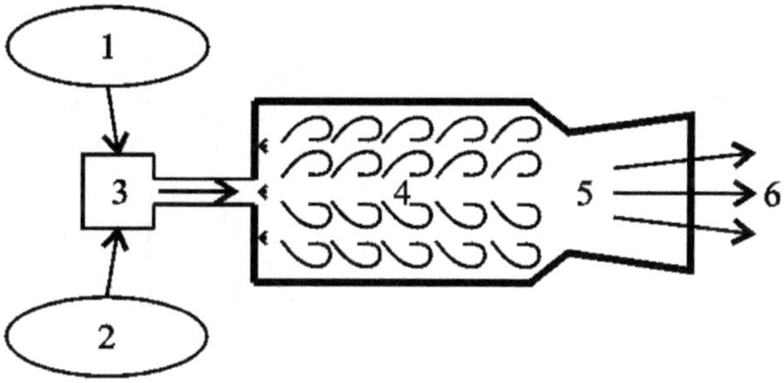

Fig. 3.4 Principle of a rocket engine operation: 1—Fuel tank; 2—Oxidizer tank; 3—Mixing chamber; 4—Combustion chamber; 5—Engine nozzle; 6—Jet stream

engines deviate from the theoretical Carnot cycle, they serve as the foundational models for the two primary categories of heat engines.

3.2.1.3 Rocket Engine

The Brayton thermodynamic cycle is the closest thing to a rocket engine [7]. However, a notable distinction emerges when examining the rocket engine. Contrary to the atmospheric air utilized by the former, the rocket engine employs a closed system, albeit with respect to the vehicle itself. Conversely, it is considered open with respect to the working fluid.

The operation of a rocket engine is predicated on the conversion of the chemical energy of the fuel into the kinetic energy of the gases emitted.

The principle of operation of a rocket engine (Fig. 3.4):

1. The fuel is then mixed with the oxidizer in the combustion chamber, resulting in an exothermic oxidation reaction. This process involves a significant release of thermal energy, leading to the combustion of the fuel and the subsequent release of heat.
2. The subsequent gases are subjected to elevated temperatures, which causes them to expand rapidly and achieve supersonic speeds within the engine nozzle, thereby forming a jet stream.
3. Upon the jet's egress from the nozzle, in accordance with the law of conservation of momentum, a jet thrust is generated, propelling the rocket in the opposite direction to that of the jet stream.

3.2.2 The Principle of Operation of Electrical Engines

The operation of electric motors in UAVs is governed by distinct principles, contingent on the motor's classification as either commutator (Fig. 3.5a) or brushless (Fig. 3.5b) [8]. These engines provide thrust and drive for flight; however, they possess distinct design and control specifics.

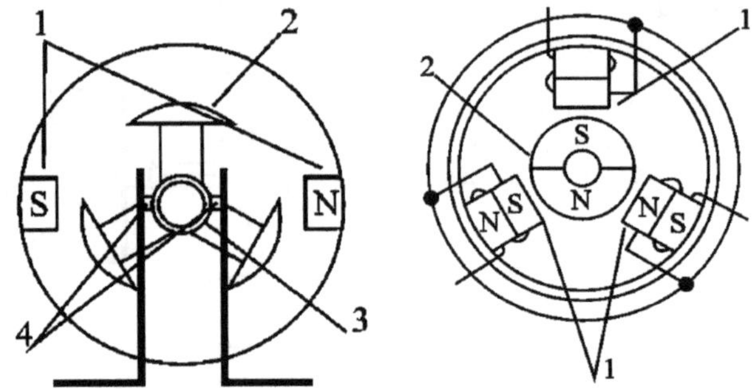

Fig. 3.5 Electrical motors: (**a**) The commutator motor; (**b**) The brushless motor. 1—Stator; 2—Rotor; 3—Commutator; 4—Electric brushes

The commutator engines are equipped with brushes as an integral component of their design. The principle of operation is as follows: the entry of an electric current into the armature windings gives rise to the creation of an electromagnetic field therein. The device interacts with the magnets of the stator, thereby activating the armature and inducing rotation. Subsequently, the current traverses the collector and contacts the subsequent winding. The shaft of the electric motor is in constant rotation with the armature when an electrical current is applied to it.

The following characteristics are exhibited by the commutator engines:

- The direction of rotation is contingent upon the polarity of the supply voltage, while the frequency of rotation is determined by the voltage value.
- Collector engines are regarded as unreliable due to their proximity to moving components, their substantial dimensions, and their weight.

Brushless engines are motors that do not contain brushes as part of their structural design. The principle of operation entails the application of an electric current to the stator windings, thereby inducing the creation of a magnetic field. The direction and intensity of this field can be modulated using the controller.

The following characteristics are exhibited by the brushless engines:

- The permanent magnets affixed to the rotor interact with the magnetic field of the stator, thereby generating a torque that prompts the rotor to rotate.
- The controller determines the timing and method for applying current to the stator windings. The system utilizes sensors, including Hall sensors, to ascertain the position of the rotor and synchronize the current supply. This enables the maintenance of efficient rotation.
- In certain systems, the utilization of feedback plays a pivotal role in the regulation of speed and torque, thereby facilitating the engine's operation with enhanced efficiency and precision.

The energy required to operate the engines is derived from batteries or fuel cells.

Electric batteries for UAVs are essential components for the operation of engines, electronics, and other drone systems. The selection of an appropriate battery has a direct impact on the range, operational duration, and overall efficiency of the UAV.

The following battery types are available for drones [4, 5, 8]:

- LiPo batteries are distinguished by their high energy density and light weight, rendering them well-suited for brief but high-intensity flights.
- Li-ion batteries are utilized in larger UAVs — offer superior capacity and durability.
- NiMH batteries are not as prevalent in UAVs as LiPo and Li-ion batteries due to their comparatively lower energy density.

The principle of operation is as follows: In lithium polymer batteries, energy is stored in the form of ions that move through the electrolyte from the anode to the cathode. In the context of lithium-ion batteries, energy conversion occurs through an electrochemical reaction, thereby generating electrical energy. For instance, in hydrogen fuel cells, the process of direct conversion of chemical energy (hydrogen and oxygen) into electrical energy is exhibited.

It is imperative to note that the battery must conform to the technical specifications of the drone to which it is intended to be attached. This includes considerations such as size, weight, capacity, voltage, and permissible current loads. In order to enhance both capacitance and current output level, the cells are connected in a parallel configuration. To facilitate voltage summation, the cells are arranged in a series configuration.

Charging stations are utilized for UAVs. The following is a list of charging technologies:

- Contact methods refer to the physical connection of the drone to the charger via contacts. These systems offer high charging efficiency; however, they necessitate precise positioning of the drone on the platform.
- Non-contact methods rely on electromagnetic induction to facilitate energy transfer. These devices offer a certain degree of convenience due to their ability to function without the need for precise positioning. However, their efficacy is limited, and there is a risk of electromagnetic interference.

The following are some requirements for electric batteries and charging systems for UAVs:

- Compatibility is a critical consideration in the design of charging systems. The charger must provide the correct voltage and current to ensure that the battery is charged safely and efficiently.
- Integration with drone control systems—a significant number of contemporary drones are equipped with automatic return-to-base charging functions, which necessitate a UAV communication system from the charging station.
- Fuel cells for UAVs are driven by chemical reactions that convert fuel and oxygen from the ambient air into electrical energy [6, 9]. This enables the drones to receive energy continuously, thereby increasing their flight time.

The following types of fuel cells are utilized in UAVs:

1. Hydrogen fuel cells. These batteries offer higher energy density and faster refueling time compared to traditional battery systems.
2. Methane fuel cells. These systems are well-suited for long-term missions, extreme conditions, and tasks that require high autonomy.

It is important to note that combustion does not occur within the fuel cell; instead, hydrogen oxidation is facilitated through an electrochemical mechanism. The reaction of hydrogen atoms with oxygen atoms results in the formation of water. During this process, the released electrons flowing through the external circuit create an electric current.

The following specifics are presented by the fuel cells:

1. The fuel cell itself is not a store of energy; rather, it derives its energy from external sources, specifically the fuel and oxidizer that serve as its primary reactants.
2. The utilization of a catalyst is frequently employed to expedite the reactions occurring on the electrodes.
3. The dried membrane exhibits high electrical resistance, resulting in significant losses and heat generation. Consequently, the gases involved in the chemical reaction are moistened.

Fuel cells have the potential to be incorporated into hybrid systems, for instance, in conjunction with a battery. In this configuration, the battery is utilized during periods of peak demand, such as takeoff and climb, while the fuel cell functions as the primary energy source during flight and refueling operations.

The following are some areas of development for the use of fuel cells for UAVs:

- The development of systems that ensure proper functionality at elevated altitudes in conditions characterized by low temperatures and reduced oxygen levels is of paramount importance.
- The creation of closed-loop fuel systems that generate hydrogen directly on board the drone is of particular interest. This process involves the use of an electrolyzer, which utilizes electrical energy from photovoltaic cells to produce hydrogen. However, it should be noted that this solution remains in the experimental stage of development.

3.2.3 The Principle of Operation of Hybrid Power Plants

Hybrid power plants for UAVs integrate two or more energy sources [6, 10]. Typically, the second source is traditional combustion engines. However, recent advancements have emerged in the field of fuel cells, with ongoing research focusing on their integration as a secondary energy source. The employment of hybrid propulsion systems is gaining prevalence in the domain of drone technology, owing to their enhanced efficiency, adaptability, and environmental sustainability.

The main types of hybrid power systems for UAVs include:

1. Series hybrid circuit (Fig. 3.6a);
2. Parallel circuit (Fig. 3.6b);
3. Series-parallel circuit.

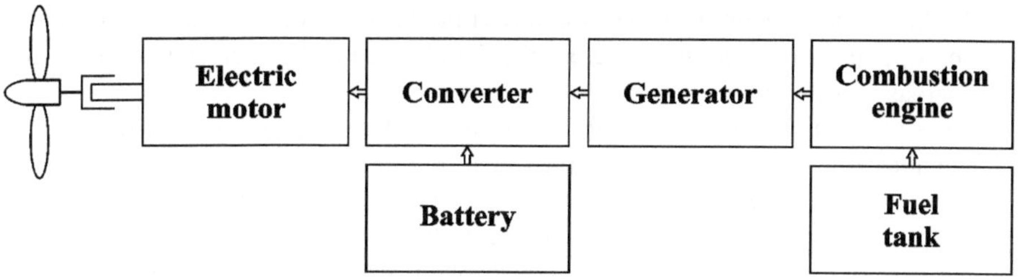

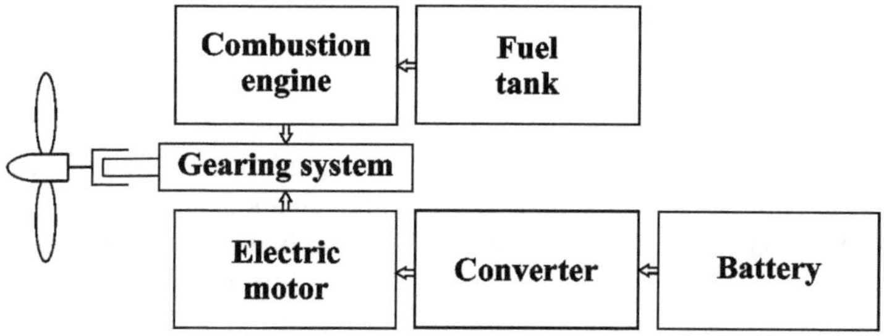

Fig. 3.6 Hybrid power plant schemes: (**a**) The Serial; (**b**) The parallel

In the case of a serial scheme, the second source functions as a generator, thereby charging the batteries, while the electric motors drive the propellers. This configuration ensures optimal functionality for the secondary source, particularly in the context of combustion engines, while concurrently reducing fuel consumption and enabling flight operations to be conducted exclusively on electric power. However, such a scheme entails a loss of energy during the double conversion of energy from mechanical to electrical and then back to mechanical.

A distinguishing feature of a parallel hybrid scheme is the capability of both the secondary source and the electric motor to directly drive the propellers through a mechanical gearing system. This configuration enables high efficiency and the potential for energy recovery (for example, during descents). However, this configuration necessitates a complex gearing system, which complicates the synchronization of engine operation from two distinct sources.

In the context of a series-parallel scheme, a synthesis of the aforementioned approaches is employed. This feature enables the operator to transition between various modes of power plant operation. This solution offers a high degree of flexibility and enhanced efficiency for a variety of flight modes and conditions. Conversely, the implementation of such a scheme is accompanied by a considerable degree of complexity and costs.

In general, hybrid power plants provide several advantages over traditional power sources:

1. Increased range and flight time. The combustion engine serves as a backup energy source, while batteries provide supplemental power, thereby extending the overall flight duration.

2. Reduced fuel consumption. Hybrid power plants have been developed to enhance the operation of combustion engines, thereby reducing fuel consumption and increasing efficiency.
3. Improved reliability. In the event of a failure in one power source, the other is capable of assuming control, thereby ensuring uninterrupted power provision.
4. Noise reduction. The utilization of electric propulsion facilitates the attainment of quieter flight, thereby reducing noise pollution and emissions.
5. Flexibility of use. This particular power plant is well-suited for a wide range of long-duration missions, including patrolling, reconnaissance, and cargo transportation.

The prospects for the development of hybrid power plants are substantial. For instance, there are ongoing developments in the field of hydrogen fuel cells. Another potential avenue for enhancement involves the advancement of battery technology, such as the broader implementation of solid-state batteries or supercapacitors.

The integration of hybrid propulsion systems into UAVs enhances their operational versatility and economic efficiency, a feature that is particularly salient for commercial applications. It is reasonable to hypothesize that their market share will continue to expand in the future.

3.3 Powerplant Selection for UAVs

The selection of an appropriate engine [1, 3, 10] for a particular UAV is primarily influenced by the following factors: drone performance, structural requirements, size constraints, engine weight, engine propulsive efficiency, drone stability and controllability, stealth, manufacturing cost, operating cost, maintainability, installation requirements, integration, noise and vibration, manufacturability, disposability, safety, etc.

A thorough examination of some of these criteria and their particulars is warranted.

3.3.1 Drone Performance

Initially, a comprehensive analysis must be conducted to ascertain the potential applications of diverse engine types in various flight modes and conditions.

One of the most critical factors in selecting an engine type pertains to the flight performance of the UAV. The two parameters that exert the most significant influence on the selection of an engine are the absolute ceiling and the maximum speed [1, 3]. The operational range of the UAV is a pivotal factor in determining these parameters.

In general, a propeller's efficiency is sufficient up to a speed of approximately Mach 0.6. At higher speeds, jet or rocket engines are capable of providing the necessary thrust.

For propeller-based power plants (piston, turboprop, electric), an increase in flight altitude is associated with a decrease in propeller efficiency due to a decrease in air density. For such power plants, the maximum altitude is limited to approximately 10 km.

In the context of combustion engines, the maximum operating altitude is constrained by the amount of oxygen necessary to ensure the efficient burning process. The altitude of these components varies according to the engine type, whether it is a piston engine, a gas turbine engine, or a ramjet engine.

Additionally, the altitude of flight exerts a substantial impact on the operation of electric motors. The provision of cooling is imperative for such engines. As the altitude increases, the complexity and cost of the cooling system concomitantly increase. Additionally, there is a possibility of complications arising from inadequate insulation of the engine's operational components at elevated altitudes.

The approximate speed and altitude limits for different types of engines are illustrated in Fig. 3.7 [1].

3.3.2 Engine Propulsive Efficiency

The efficiency of a power plant is contingent upon its capacity to generate the required power/thrust for diverse flight conditions [1, 10, 11]. As illustrated in Fig. 3.8, the efficiency of power generation and thrust for various engine types can

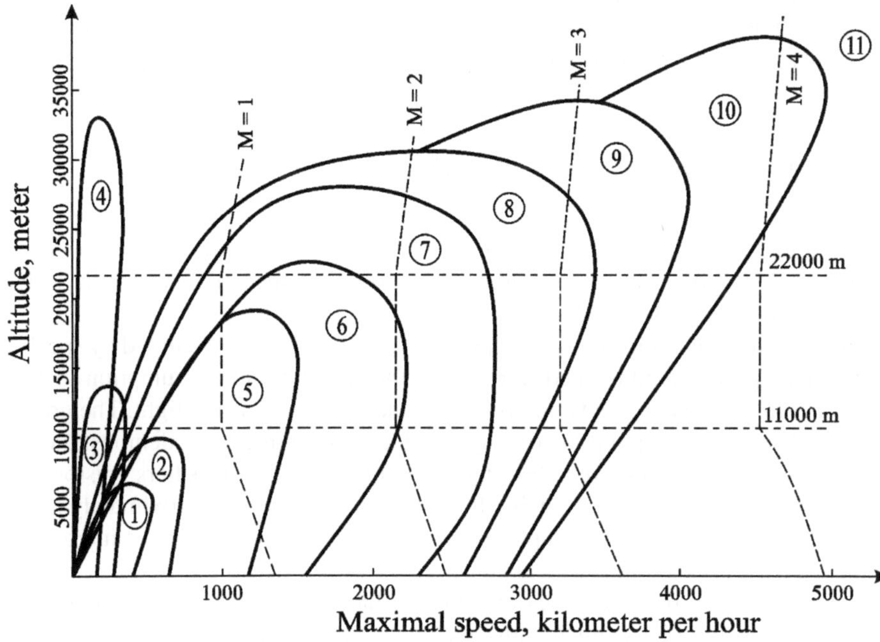

Fig. 3.7 A comparison of the operational limits for various engines: 1—piston; 2—turboprop; 3—electric; 4—solar; 5—turbofan; 6—turbofan with afterburner; 7—turbojet; 8—turbojet with afterburner; 9—combined turbo-ramjet; 10—ramjet/scramjet; 11—rocket

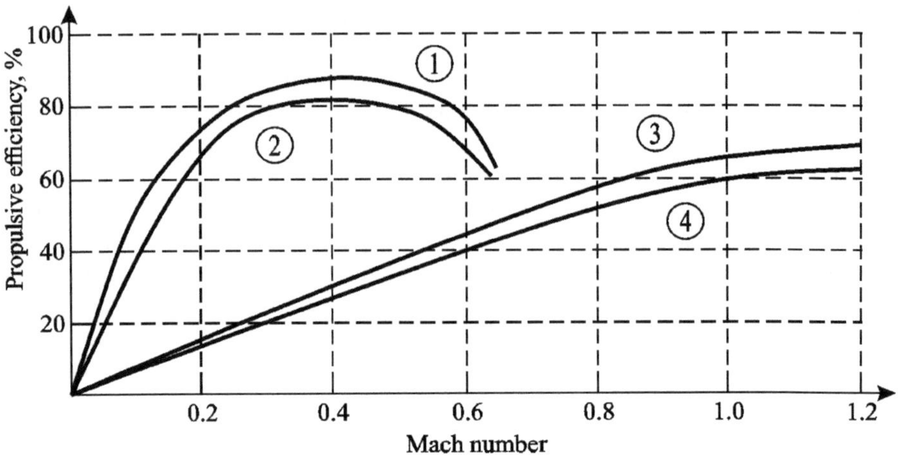

Fig. 3.8 A comparison between the propulsive efficiency of various engines: 1—Propeller; 2—Turboprop, piston, electric, solar; 3—Turbofan (BPR = 5); 4—Turbojet

vary significantly. The graphs provided are for cruising flight mode at a specific thrust level.

As demonstrated in Fig. 3.8, propellers, encompassing reciprocating, turboprop, electric, and solar-powered engines, exhibit the optimal efficiency levels. It has been demonstrated that turbofan engines exhibit a higher degree of efficiency in comparison to turbojet engines. This phenomenon can be attributed to the presence of a secondary channel within the engine.

3.3.3 Engine Fuel/Energy Efficiency

The efficiency of the engine, particularly in the context of a drone employed for commercial purposes, can be evaluated by assessing the efficiency with which energy is utilized [1, 3, 10, 11].

In the context of combustion engines, this indicator can be substituted by specific fuel consumption. As illustrated in Fig. 3.9, the specific fuel consumption of four distinct combustion engines varies with the Mach number during flight. The graphs presented herein correspond to cruising flight at a given thrust. As demonstrated in the accompanying figure, vehicles engines equipped with propellers (piston and turboprop) exhibit the lowest specific fuel consumption, up to approximately Mach 0.4. On average, the fuel consumption of a turboprop engine tends to be marginally higher in comparison to that of a piston engine.

Furthermore, turbofan engines exhibit a reduced specific fuel consumption in comparison to turbojet engines, attributable to the incorporation of a secondary bypass channel. Conversely, an increase in the bypass ratio has been shown to result in a decrease in specific fuel consumption. Consequently, the turbofan engine is the

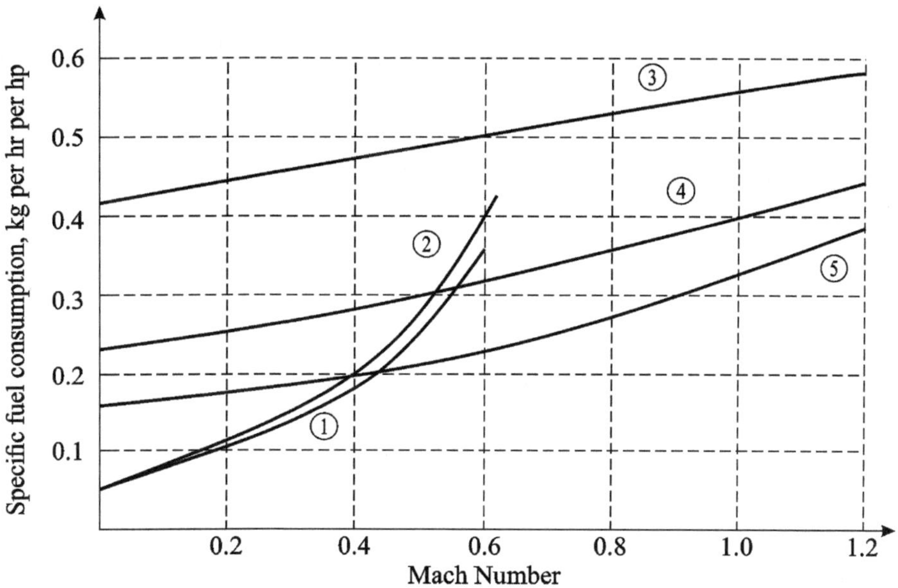

Fig. 3.9 Comparison of specific fuel consumption for combustion engine: 1—piston; 2—turboprop; 3—turbojet; 4—turbofan with BPR = 1; 5—turbofan with BPR = 5

most economical option. Furthermore, kerosene, a more economical alternative to gasoline, is utilized as a fuel source. Turbofan engines have been demonstrated to be the most economical in terms of specific fuel consumption at high subsonic speeds.

When it comes to electric motors, it is important to compare possible losses during energy transfer from the source to the propeller. The same can be done for traditional combustion engines. To estimate losses, the efficiency of the chain from the energy source to the propeller will be used.

According to the values shown in Fig. 3.10 [10], the total efficiency of a combustion-powered system is approximately 39%, whereas typical battery-powered propulsion systems have a total efficiency of 73%. As can be seen, electric power plants have lower losses. Therefore, subject to the limitations described above, such power plants are preferable.

A critical attribute of any aircraft is its flight range, which is defined as the distance a vehicle can travel while in flight. A direct relationship exists between payload, flight range, and fuel reserve (see Fig. 3.11).

In the case of flying vehicle equipped with a conventional fuel-powered propulsion system at maximum payload capacity, it is customary for fuel tanks to be partially drained due to restrictions imposed on takeoff weight. This condition delineates the leftmost vertical segment of the graph between points A and B. In order to achieve a longer flight range, it is necessary to reduce the payload. Concurrently, a more substantial fuel supply can be furnished. Consequently, the graph segment from B to C will exhibit an incline. Point C in the graph represents a ferry flight. Such a flight is considered a service case. The flying vehicle possesses

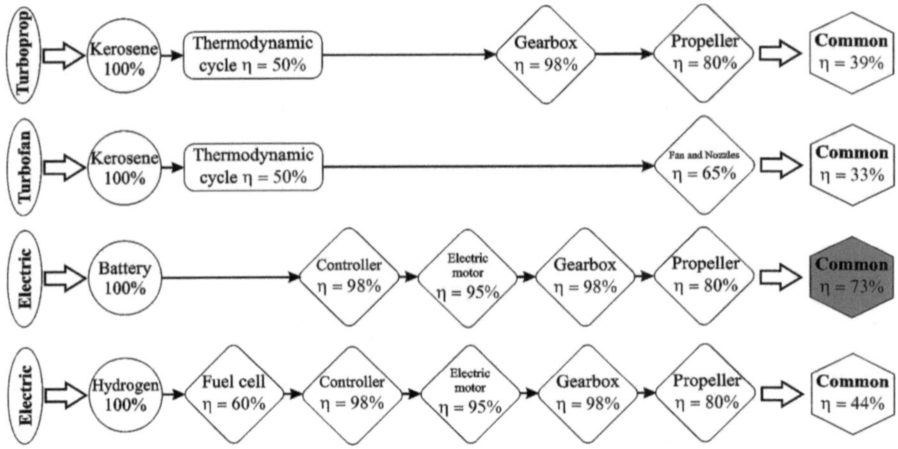

Fig. 3.10 Typical conversion schemes with the typical component efficiencies in a power plant chain

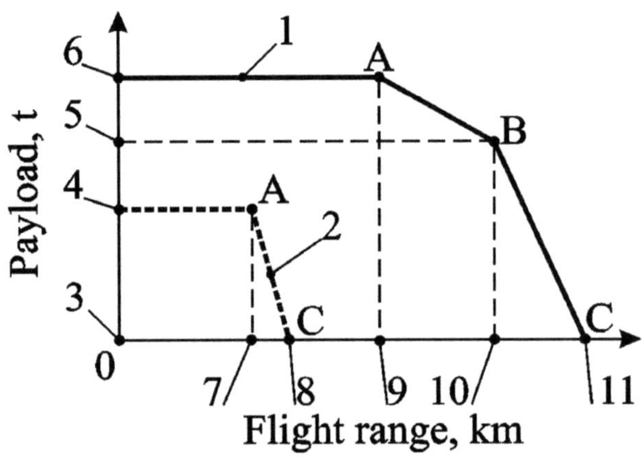

Fig. 3.11 A typical payload graph for an electrically powered compared to a conventional powered flying vehicle [10]: 1—Graph for a flying vehicle with a conventional fuel-powered propulsion; 2—Graph for a flying vehicle with an electrical propulsion; 3—No payload; 4—Maximum payload for a vehicle with an electrical propulsion; 5—Reduced payload for a vehicle with a conventional fuel-powered propulsion; 6—Maximum payload for a vehicle with a conventional fuel-powered propulsion; 7—Maximum range with full battery charge and payload for a vehicle with an electrical propulsion; 8—Maximum range with full battery charge for a vehicle with an electrical propulsion; 9—Maximum range with full fuel and payload for a vehicle with a conventional fuel-powered propulsion; 10—Maximum range with full fuel and reduced payload for a vehicle with a conventional fuel-powered propulsion; 11—Maximum range with full fuel for a vehicle with a conventional fuel-powered propulsion

a full fuel capacity; however, it is devoid of any payload. This case is emblematic of a case involving the relocation of a flying vehicle; it does not pertain to the generation of revenue.

The flight range characteristics of a flying vehicle equipped with an electric propulsion system exhibit significant variability. Assuming a level playing field, the maximum payload capacity of such flying vehicle is notably lower than that of conventional vehicle powered by traditional fuel. This phenomenon can be attributed to the high specific gravity of batteries, as illustrated in Fig. 3.12. Furthermore, the potential flight range is comparatively limited, a consequence of the reduced energy expenditure characteristic of contemporary electric batteries. It is evident that decreasing the payload results in a reduction in energy consumption. However, such adjustment only marginally enhances the flight range. This is attributable to the fact that the total payload share is already negligible, thereby constraining the potential for significant range augmentation.

3.3.4 Engine Weight Efficiency

The overall weight efficiency of a UAV is contingent upon various factors, including the weight efficiency of its power plant [1, 3]. Consequently, the weight of the engine has a substantial impact on the overall mass of the UAV. A critical evaluation of the weight efficiency of diverse engine types is imperative to ascertain the most

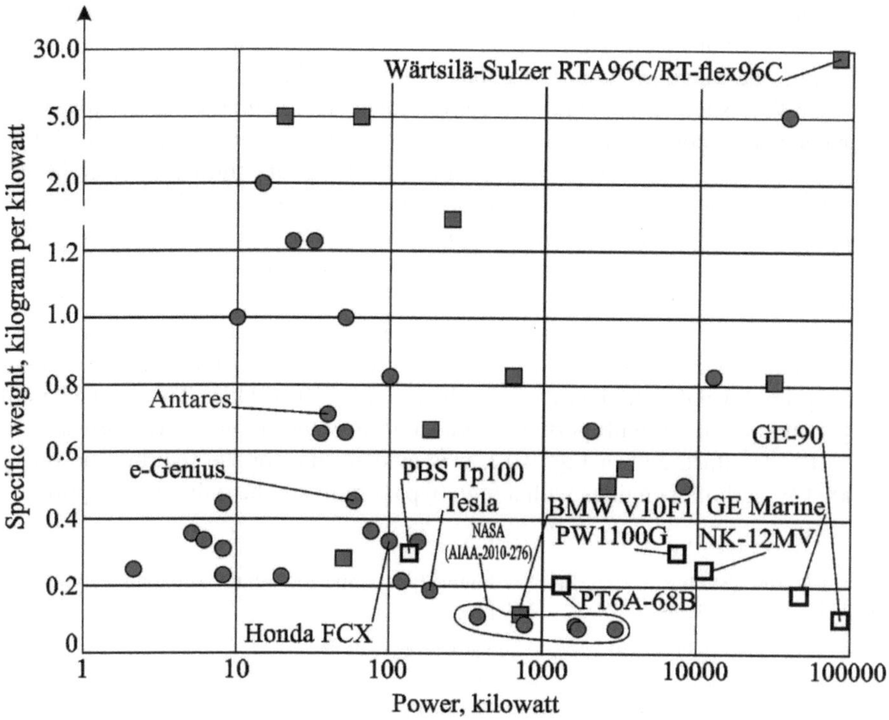

Fig. 3.12 The specific weight of different engine types [1, 4, 5, 7, 8, 10]: ⚬—electric; ▪—piston; ▫—gas-turbine

efficacious solution. The specific gravity of the engine is a useful metric for this purpose.

Piston engines possess a weight advantage over turboprop engines. However, when taking into account the combination of thrust efficiency and engine specific weight, it is determined that, for engines requiring more than 500 horsepower, turboprop engines are preferable to piston engines [7]. Piston engines with a capacity exceeding 500 horsepower demonstrate suboptimal weight efficiency. Consequently, it is recommended that a UAV with a required engine power of more than 500 horsepower use a turboprop engine. For example, the Textron Lycoming IO-720-A [12] eight-cylinder four-stroke piston engine with air cooling and a dry weight of 271 kilograms develops a maximum power of 400 horsepower. In contrast, the Allison 250-B17 [13] turboprop engine generates 420 horsepower while maintaining a weight of 88.4 kg.

A turbojet engine possesses a weight advantage over a turbofan engine. For example, the SNECMA Atar 9 K 50 turbojet engine [14], equipped with an afterburner combustion chamber, has a dry weight of 1582 kilograms and generates a thrust of 70.6 kN. The Pratt & Whitney F100-220P turbofan engine [15], with a dry weight of 1526 kg, has a thrust of 74.3 kN without afterburner and 120.1 kN with afterburner.

A comparison of an electric propulsion system with a reciprocating engine reveals that, in general, the former is lighter than the latter. However, a comparison of the weight of an electric motor plus its battery pack with that of a similar piston engine plus the weight of fuel reveals that the total weight of the electric power plant is greater. For example, the mass of a conventional electric motor with a capacity of 2 horsepower is approximately 300 g. However, to operate for a duration of 15 min, it necessitates a battery weighing approximately 400 g. In contrast, the weight of a 2 horsepower single-cylinder piston is approximately 400 g. The fuel requirement to operate this engine for 15 min is approximately 250 g. Therefore, in regard to micro drones, it is imperative to utilize a piston engine in order to guarantee the drone's minimum mass.

As illustrated in Table 3.1, the mean specific weight values of contemporary aviation engines are presented. It is important to note that electric motors demonstrate optimal weight efficiency within the range of 10,000 kW. The particulars of electric motors, along with pertinent indicators for other motors, are illustrated in Fig. 3.12. As illustrated in Fig. 3.12, it is evident that for the required power of 10,000 kW, turbofan engines with a high bypass ratio are the only viable option.

Table 3.1 The specific weight for various engine types [1, 4, 5, 7, 8, 10]

Engine type	Specific weight, kilogram per kilowatt
Piston	0.4
Turboprop	0.24
Turbofan	0.12
Turbojet	0.18
Electric	0.13

3.4 Examples of Power Plant Options

3.4.1 Examples of the Use of Power Plants for UAVs

Now, let's look at some examples of calculations to help us select the parameters of the power plant.

Example 3.1 Initial data: The flight range is 1000 km and the take-off weight of the UAV is 1000 kg. There are no special requirements for take-off or landing conditions.

Based on the initial data and the above-described criteria, it is reasonable to select a power plant with propellers.

For such UAVs, it is possible to apply the traditional statistical method of selecting engines for UAVs at the initial stage. This can be done according to the required starting thrust-to-weight ratio. For such power plants and light aircraft, a starting thrust-to-weight ratio of 0.15 kW/kg is typical. In this case, the total required starting thrust-to-weight ratio is:

$$P_{\text{to}} = 0.15\,\text{kW} / \text{kg} \times 1000\,\text{kg} = 150\,\text{kW}.$$

Specifically, for a UAV with the given initial data, it is essential to choose an engine with a power output of 150 kW. The following engines are suitable for this capacity:

1. The Rotax 915 i S piston engine [16] with 141 kW of power and a dry weight of 142 kg.
2. The Walter M601E turboprop engine [17] with 150 kW of power and a dry weight of 200 kg.
3. The MGM Compro High-Power BLDC/PMSM Series electric motor [18] with 150 kW of continuous power and a mass of 60 kg.

Now, let's estimate the required fuel and battery mass for these engines. To simplify the calculations, we will not consider the navigation fuel reserve or the additional battery capacity.

The specific fuel consumption for a reciprocating engine is 0.29 kg/kWh. Aircraft with this type of engine typically fly at about 220 km/h. Thus, the flight time will be:

$$t = L / V = 1000\,\text{km} / 220\,\text{km} / \text{h} = 4.6\,\text{h},$$

where L is the flight range, kilometers; V is the flying velocity, kilometers per hour.

Once the flight time has been determined, it will be easy to calculate the necessary fuel mass for the UAV:

$$m_{\text{fuel}} = P \times t \times q_{\text{fuel}} = 141\,\text{kW} \times 4.6\,\text{h} \times 0.29\,\text{kg} / (\text{kW} \cdot \text{h}) = 188.1\,\text{kg},$$

where q_{fuel} is the specific fuel consumption, kilogram per kilowatt per hour. For the engine Rotax 915 iS it equals 0.29 kg/(kW·h).

A similar calculation can be performed for the Walter M601E turboprop engine. Its specific fuel consumption is 0.35 kg/kWh. Aircraft with this type of engine typically travel at speeds of about 400 km/h. Thus, the flight time will be:

$$t = 1000\,\text{km} / 400\,\text{km} / \text{h} = 2.5\,\text{h}.$$

After determining the flight time, we will determine the required mass of fuel for the UAV:

$$m_{\text{fuel}} = 150\,\text{kW} \times 2.5\,\text{h} \times 0.35\,\text{kg} / (\text{kW} \cdot \text{h}) = 131.3\,\text{kg}.$$

The required voltage for the electric motor is 800 V, and this engine was used on experimental aircraft that could reach speeds of about 450 km/h. Therefore, the flight time will be:

$$t = 1000\,\text{km} / 450\,\text{km} / \text{h} = 2.2\,\text{h}.$$

After determining the flight time, we will determine the required battery capacity for the UAV:

$$Q_{\text{bat}} = \frac{P \times t}{k \times U} = 150\,000\,\text{W} \times 2.2\,\text{h} / (0.7 \times 800\,\text{V}) = 590\,\text{A} \cdot \text{h},$$

where k is the battery utilization factor, which is typically around 0.7, and U is the voltage in the circuit, which in this case is 800 V.

The equation for determining the mass of the battery for such a capacity is:

$$m = \frac{E}{e} = \frac{U \times I}{e} = (800\,\text{V} \times 590\,\text{A} \cdot \text{h}) / 250\,\text{V} \cdot \text{h} / \text{kg} = 1890\,\text{kg},$$

where E is the stored energy, volt ampere hour; e is the energy density, volt hour per kilogram, for the Li-Ion battery it is about 250 V·h/kg.

For a battery with this capacity at a voltage of 800 V, excluding the weight of the housings and cooling system, the total battery weight will be approximately 1890 kg. This contradicts the original data, which set the take-off weight at 1000 kg.

As the calculations above show, modern electric motors have a lower mass. However, electric motors require power sources that are significantly heavier than the fuel needed for combustion engines.

Similar results can be obtained for VTOL drones with comparable baseline data.

Example 3.2 Initial data: the flight range is 10 km, the take-off weight of the UAV is 1 kg, and the take-off and landing must be vertical.

Based on the initial data and the above criteria, it is reasonable to conclude that a multicopter design is the best choice. Using the same method as in the previous example, we will calculate the required power plant.

Such drones typically fly at a speed of about 10 m/s (36 km/h). The flight time will be approximately 17 min (0.28 h).

The power required to keep the multicopter in the air is determined by the equation:

$$P_{\text{hover}} = \frac{mg\sqrt{\dfrac{mg}{2\rho A}}}{\eta},$$

where m is the mass of the drone, which is taken to be 1 kg. The acceleration of gravity, g, is then calculated for low altitudes, with a value of 9.81 m per second squared being the standard assumption. The density of the air, ρ, is also taken into consideration, with a constant of 1.225 kg per cubic meter being utilized to simplify the calculation. The efficiency of the engine and propeller, η, is determined using the data from Sect. 3.3.3, 0.7. Finally, the disk area of all rotors is A, measured in square meters.

As demonstrated by the equation, the primary issue pertains to the selection of propellers parameters, specifically their area and efficiency. For a 1-kg multicopter, the typical hovering power is approximately 50–100 watts for all engines.

To ascertain the required power when the drone is moving at cruising speed, the following equation is employed:

$$P_{\text{cruise}} = \frac{0.5\rho V^3 C_{\text{D}} A_{\text{front}}}{\eta},$$

where the drag coefficient, denoted by C_{D}, is a measure of a drone's aerodynamic properties. For micro drones, this coefficient typically ranges from 0.3 to 0.5. The frontal area of the drone, denoted by A_{front}, is a function of the drone's dimensions and structure. For the purposes of this analysis, we will assume A_{front} to be 0.1 m^2.

Utilizing the aforementioned data, it is possible to attain the requisite power within the range of 30–60 watts.

The next step involves calculating the power consumption of the power plant.

$$E_{\text{total}} = P_{\text{cruise}} \times t = 60\,\text{W} \times 0.278\,\text{h} = 16.67\,\text{W}\cdot\text{h}.$$

Subsequent to this, it is imperative to ascertain the required battery capacity. Assuming a voltage of $U = 11.1$ V (3S LiPo), the following equation is derived:

$$Q_{\text{bat}} = E_{\text{total}} / U = 16.67\,\text{W}\cdot\text{h} / 11.1\,\text{V} = 5.4\,\text{A}\cdot\text{h}.$$

Given the capacity and assuming the battery is of the LiPo type, the mass of the battery can be estimated.

$$m = \frac{E}{e} = 16.67\,\mathrm{W\cdot h} / 200\,\mathrm{V\cdot h/kg} = 0.08\,\mathrm{kg},$$

where $e = 200$ V·h/kg for the LiPo type battery.

The results of this example indicate that the total required engine power in hover mode is between 50 and 100 watts, while the required engine power in cruising mode is between 30 and 60 watts. It is also assumed that the peak power, which would be reached in the event of acceleration due to wind, is up to 150 watts. In the event that the quadcopter configuration is accepted, the recommended engine power is as follows:

$$4\,\text{engines} \times 40\,\mathrm{W} = 160\,\mathrm{W}\,(\text{with reserve}).$$

The T-Motor F60 Pro II (1750KV) [19] is an exemplary representation of the aforementioned category of engines. The engine possesses a power output ranging from 50 to 150 watts, a thrust range of 300 to 500 g, and a voltage configuration of 3S–4S LiPo. Its weight is recorded at 31.5 g.

The subsequent stage will be of a multicopter with equivalent initial data specifications, but with a power plant design that utilizes reciprocating engines. The requisite power and flight duration remain equivalent.

The determination of the fuel rate can be achieved through the utilization of the following equation:

$$c_t = \frac{P_{\text{cruise}}}{\eta Q} = 60\,\mathrm{W} / \left(0.25 \times 12{,}000\,\mathrm{W\cdot h/kg}\right) = 0.02\,\mathrm{g/s} = 1.2\,\mathrm{g/min},$$

where η is the efficiency of the engine and propeller, and for reciprocating engines it is 0.25; Q is the specific energy of petrol, which is equal to 12,000 watt hour per kilogram.

Therefore, the required mass of fuel for the flight:

$$m_{\mathit{fuel}} = c_t \times t = 1.2\,\mathrm{g/min} \times 17\,\mathrm{min} = 20.4\,\mathrm{g}.$$

For the purpose of this case, a reciprocating multicopter will be utilized, with a single engine of the OS Engines FS-26 Surpass model [20] being employed. The device possesses a power range of 60 watts to 100 watts, a thrust of 300 g, and a dry weight of approximately 260 g.

As demonstrated by the calculations above, the mass of a power plant that utilizes electric motors is marginally lower than that of a piston engine power plant (206 g compared to 281 g). However, the calculations did not take into account the mass of additional systems, such as the mass of the cooling system for electrical components, the mass of the mounting systems, the mass of the fuel components for the piston engine circuit, the mass of the transmission, and so forth.

A close examination of the examples provided reveals that the type of power plant is contingent on numerous criteria. It is incumbent upon the drone developer to first conduct a statistical analysis of existing drones. Accordingly, based on the aforementioned analytical findings and taking into account the criteria previously described, the most suitable type of power plant should be selected for the projected UAV.

As examples, we will explain some types of power plants and types of UAVs with such power plants.

1. Electric Motor (Battery or Hybrid)

   An example of a UAV using this type of power plant is the DJI Matrice 300 RTK multicopter [21]. The motor used in this UAV is a brushless electric motor (BLDC). The energy source for this motor is lithium polymer (LiPo) batteries.

Advantages of this power plant include:

- Low noise levels
- Easy construction and maintenance
- High reliability
- Environmental friendliness

However, there are some disadvantages to this type of system, including:

- Limited flight time, usually up to 1 h
- Dependence on ambient temperature

This type of power plant is suitable for short missions such as aerial photography, monitoring, and small cargo delivery.

2. Internal combustion engine

   UAV Example: Bayraktar TB2 [22].
   Engine Type: Piston Opposed (Rotax 912 or Similar).
   Fuel: AI-95 Petrol.

Advantages:

- Long flight duration (Up to 24+ Hours)
- Relatively low cost of operation
- Good fuel efficiency

Disadvantages:

- Vibration and noise
- Requires regular maintenance
- Sensitivity to fuel quality

This type of power plant is suitable for Medium-Altitude Long-Flying UAVs for reconnaissance, surveillance and strike missions.

3. Turboprop Engine

Example: MQ-9 Reaper (Predator B) [6, 23].
Engine type: Honeywell TPE331 or Rotax 914 in smaller UAVs.
Fuel: Aviation kerosene (Jet-A1).

Advantages:

- High power output with moderate fuel efficiency
- Good altitudes (up to 15 km)
- Long operational lifespan

Disadvantages:

- Expensive
- Maintenance can be difficult
- Higher weight compared to piston engines

This type of power plant is suitable for long-range, high-altitude reconnaissance, strike, and patrol missions.

4. Turbojet Bypass Engine
Example: RQ-4 Global Hawk [6, 24].
Engine Type: Rolls-Royce AE 3007H with high BPR.
Fuel: Aviation kerosene (Jet-A1).

Advantages:

- Very high speed (Up to 900+ km/h)
- Long range (Over 20,000 km)
- Ability to operate at high altitudes (18+ km)

Disadvantages:

- High cost
- High fuel consumption
- Complex operation

This type of power plant is suitable for strategic reconnaissance, High-Altitude Long-Term Missions.

3.5 Conclusions

In this chapter, the primary specifications of various power plant types and the fundamental criteria for their selection were examined. A thorough analysis was conducted to identify the various limitations associated with distinct categories of power plants. A detailed review of the combustion and electric power plants was conducted, accompanied by elucidations concerning the features of the UAV. Two examples were provided for the power plant calculations.

References

1. Tiniakov D., Li L., Su Y. Airworthiness design of civil aircraft systems and structures. Datian Aviation Beijing Beixue Publishing House. 2018.
2. Tiniakov D., Li L. Airworthiness engineering basis. Datian Aviation Beijing Beixue Publishing House. 2020.
3. Marshall D, Barnhart R, Shappee E, Most M. Introduction to unmanned aircraft systems. New York, NY: CRC Press; 2016.
4. Karthik A, Chiniwar D, Das M, et al. Electric propulsion for fixed wing aircrafts—a review on classifications, designs, and challenges. J Eng Sci. 2021;16:129–45. https://doi.org/10.30919/es8d573.
5. Da Silva F, Fernandes J, Da Costa Branco P. Barriers and challenges going from conventional to cryogenic superconducting propulsion for hybrid and all-electric aircrafts. MDPI Energies. 2021;14:6861. https://doi.org/10.3390/en14216861.
6. Li L., Kretov A., Tiniakov D. Aeronautics— – Characteristics and Emerging Technologies. Chapter: UAVs and Hydrogen Technologies are the Future of Civil Aviation [Internet]. IntechOpen; 2025. https://doi.org/10.5772/intechopen.1008352.
7. Becker L, editor. A NASA guide to engines. Cleveland: NASA; 2007.
8. Wyatt D, Tooley M. Aircraft electrical and electronic systems. London: Routledge; 2018.
9. Kadyk T, Winnefeld C, Hanke-Rauschenbach R, Krewer U. Analysis and design of fuel cell systems for aviation. MDPI Energies. 2018;11:375. https://doi.org/10.3390/en11020375.
10. Leishman G. Introduction to aerospace flight vehicles. Beach, FL: Embry-Riddle Aeronautical University; 2025. [Internet]. https://eaglepubs.erau.edu/introductiontoaerospaceflightvehicles/. Accessed 15 Sept 2025
11. Torenbeek E. Advanced aircraft design: conceptual design, analysis, and optimization of subsonic civil airplanes. Chichester: Wiley; 2013.
12. Lycoming. IO-720 series operator's manual. Williamsport: Lycoming Engines; 2006. [Internet]. https://www.lycoming.com/sites/default/files/attachments/IO-720%2520Operator%2520Manual%252060297-19.pdf. Accessed 15 Sept 2025
13. Rolls–Royce. Rolls-Royce Corporation M250-B17. Rolls–Royce. 2020. [Internet]. https://customers.rolls-royce.com/cs/idcplg?IdcService=PSA_GET_FILE&RevisionSelectionMethod=LatestReleased&Rendition=Web&noSaveAs=1&dDocName=RRC026283. Accessed 15 Sept 2025
14. Gunston B. Turbojet: history and development 1930–1960, vol. 2. Crowood Press; 2007.
15. Prisacariu V, Ciubotaru A. Aspect regarding of the Pratt & Whitney F100 jet engine performances. In: Conference Scientific Research and Education in the Air Force; 2017. https://doi.org/10.19062/2247-3173.2017.19.1.20.
16. BRP-Rotax. Installation Manual for Rotax® Engine Type 915 i A Series. BRP-Rotax. 2019. [Internet]. https://www.rotax-owner.com/pdf/IM_915iASeries_ED0_R1.pdf. Accessed 15 Sept 2025
17. Novák M, Šlofar J. Practical training for WALTER M601 engine. MAD Mag Aviat Dev. 2013;1(3):23–7. https://doi.org/10.14311/MAD.2013.03.05.
18. MGM. MGM BLDCin 200kW. [Internet]. https://mgm-compro.com/electric-motor/mgm-bldcin-200kw/. Accessed 15 Sept 2025
19. Wojcik S. Stallion user manual. Flightory. 2025. [Internet]. https://flightory.com/wp-content/uploads/2025/02/STALLION-MANUALV2.pdf. Accessed 15 Sept 2025
20. Lee C. OS FS-26 surpass. Model engine tests. [Internet]. https://sceptreflight.com/Model%20Engine%20Tests/OS%20FS-26%20Surpass.html. Accessed 15 Sept 2025
21. DJI. Matrice 300 RTK. User manual. DJI; 2022. [Internet]. https://www.dji.com/cn/downloads/products/matrice-300#tuning_params. Accessed 15 Sept 2025

22. Peria-Peigne L. TB2 Bayraktar: big strategy for a little drone. French Institute of International Relations; 2023. [Internet]. https://www.ifri.org/en/memos/tb2-bayraktar-big-strategy-little-drone. Accessed 15 Sept 2025
23. Zountouridou E, Kiokes G, Kiokes G, et al. A guide to unmanned aerial vehicles performance analysis—the MQ-9 unmanned air vehicle case study. J Eng. 2023;(6):e12270. https://doi.org/10.1049/tje2.12270.
24. Petrescu F., Petrescu R. Northrop. Norderstedt: Books on Demand GmbH. 2012.

Chapter 4
Some Aspects About Aerodynamic Performance

4.1 Introduction

In the design of UAVs, well-known aerodynamic principles are generally employed; however, it is imperative to consider the features inherent in unmanned vehicles. The fundamental distinction between UAVs and manned aircraft lies in the absence of a human pilot aboard the former. Concurrently, the absence of a pilot generally enables the miniaturization of flying vehicles. Miniaturization is a viable solution for numerous tasks where the mass and dimensions of the device's payload are less than the mass of a typical human being. Examples of such applications include piloting and surveillance.

The transition to smaller dimensions and lower weight renders it feasible to reduce the cost of completing the task in comparison with using traditional manned aircraft. When comparing different classes of flying vehicles, it is often convenient to use the similarity parameter Reynolds number [1]

$$\mathrm{Re} = \frac{\rho V b}{\nu}, \tag{4.1}$$

where ρ is the density of air at flight altitude; V is the characteristic flow velocity (for drones with a wing, it is equal to the flight speed, for drones with propellers, the flow velocity around the propeller blade); b is the characteristic size (aor drones with wings, this is the mean aerodynamic chord (MAC) of the wing, for drones with propellers, this is the MAC of the propeller blade); ν is the kinematic viscosity of air.

The dynamics of small flying vehicles are governed by low Reynolds numbers. Concurrently, the low values of Reynolds numbers, which are less than 0.5×10^6, are significantly less than the typical Reynolds numbers for large scale aircraft, which are approximately 20×10^6. This differentiation is attributed to a pronounced increase in the coefficient of aerodynamic drag with a decrease in the Reynolds

L. Li, D. Tiniakov, *Unmanned Aircraft Design*,
https://doi.org/10.1007/978-981-95-7167-3_4

number, a phenomenon referred to as the drag crisis. In contrast to the transonic drag crisis, the drag coefficient increases with decreasing speed. This phenomenon, however, cannot cause phenomena similar to the "sound barrier." Nevertheless, it has the potential to significantly reduce the lift-to-drag ratio of the flying vehicle.

At the dawn of the twentieth century, experimental studies revealed that flow rates and the corresponding Reynolds numbers were not particularly elevated, aligning with the characteristic values observed in contemporary large- and small-sized UAVs. However, this research trajectory did not undergo substantial development, as aircraft speeds and dimensions underwent rapid advancements. New challenges and tasks emerged, particularly in the context of high speeds.

The issues concerning low speeds and Reynolds numbers resurfaced during the 1980s. The inquiries concerning low Reynolds numbers have been addressed by both aircraft modelers and researchers in the domain of wind energy. The subsequent advancements in this domain of aerodynamics are contingent upon the proliferation of diminutive drones, a prospect that has become feasible due to the advent of microelectronics. The relationship between the domain of aviation model sports and the field of unmanned aircraft technology has been a persistent feature throughout the evolution of the former. A considerable number of aircraft modelers worldwide are engaged in the research, design, and production of UAVs. In the production of UAVs, technologies developed on aircraft models are frequently employed [2].

A substantial number of research studies have recently been dedicated to examining the aerodynamics of small-sized UAVs, thereby validating the significance of this subject. In the United States, the development of the aerodynamics of vehicles with low Reynolds numbers has received significant attention due to the increased focus on wind energy, the design of prototypes of solar-powered vehicles, vehicles that utilize human muscular strength (Daedalus prototype), and small-sized drones [3].

The impact of the Reynolds number on the maximum lift-to-drag ratio of wings with different aspect ratios and propeller blades has been a subject of extensive research for a considerable period. There exist methodologies for recalculating the aerodynamic characteristics obtained in wind tunnels with relatively low values of the Reynolds number for full-scale ones.

The relationship between the geometric parameters of the lift-creating surfaces (wings or propeller blades) and their aerodynamic characteristics is altered, to a certain extent, by low Reynolds numbers.

When the dependence of the profile drag on the Reynolds number is taken into account, the lift-to-drag and power parameter are no longer subject to infinite increase with increasing wing or blade aspect ratio. As illustrated in Fig. 4.1 [4], the drag coefficient of a "smooth" airfoil exhibits characteristic behavior in response to variations in the Reynolds number. In the range of Reynolds numbers of the order of 10^5, a sharp change in the drag coefficient of the airfoil becomes evident. It is evident that alterations in airfoil characteristics exert a substantial influence on the aerodynamic performance of wings or blades, particularly with regard to minimizing drag in a specific operational mode. It is imperative that such effects be considered at all stages of the design process for the wing or propeller of small UAVs.

In the case of drones equipped with wings, the primary parameters of the wing, such as its target function, payload weight, and the required range of flight speeds,

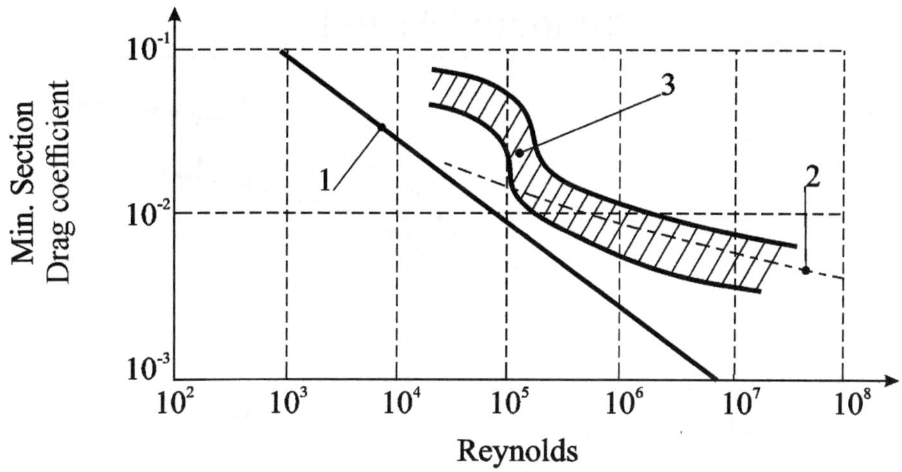

Fig. 4.1 The characteristic behavior of the drag coefficient when the Reynolds number is altered: 1—Laminar flat plate; 2—Turbulent flat plate; 3—"Smooth" airfoils

are considered during the design process. The necessity to address particular challenges is manifest in the geometric and, consequently, the aerodynamic characteristics of the flying vehicle. The reduction of an UAV's dimensions can result in a deviation from conventional designs and necessitate the implementation of alternative solutions to maintain aerodynamic efficiency. A potential necessity arises from the necessity to reduce the wing aspect ratio excessively. Such a reduction would result in a decrease in lift-to-drag ratio and, consequently, flight range. Consequently, at constant flight speeds and assuming equivalent drone dimensions, the take-off weight may undergo an increase.

The necessity of employing dome-mounted controlled thermal imaging and video cameras, in conjunction with other payloads housed in containers that are challenging to streamline sufficiently to minimize the drag of the UAV, necessitates an augmentation of the lift coefficient to ensure the attainment of acceptable values for maximum lift-to-drag ratio and flight duration. In such instances, the most advantageous lift coefficient for ensuring the maximum duration of flight is often found near the maximum lift coefficient, C_{Lmax}. Consequently, the utilization of wing profiles characterized by a high lift coefficient becomes imperative.

However, as the Reynolds number decreases, the maximum lift coefficient undergoes a concomitant change. Conversely, a decline in Reynolds number often corresponds to a decrease in maximum lift coefficient. As illustrated in Fig. 4.2, the maximum coefficient of lift from the Reynolds number exhibits a characteristic behavior [4].

These effects are significantly related to the behavior of the laminar-turbulent transition, local and global flow separation. Consequently, a comprehensive evaluation of the airfoils, wings, propeller blades, and layouts necessitates an analytical approach that can account for the influence of the laminar-turbulent transition, ascertain its position, and model local separation bubbles.

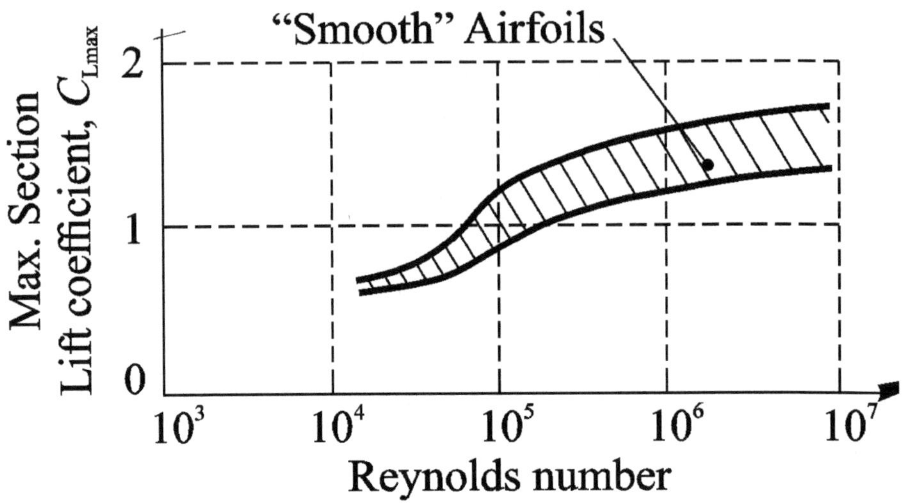

Fig. 4.2 The characteristic behavior of the lift coefficient when the Reynolds number is altered

In the context of designing the wing shape of a small-sized UAV, it is imperative to consider not only the well-known solution for optimal circulation [5], but also the impact of the drag crisis. The lift-to-drag ratio of a wing in a certain mode can be enhanced by reducing aerodynamic drag. The utilization of an elliptical circulation distribution is a strategic approach employed to minimize the induced component of the drag. An elliptical circulation distribution can be obtained through the use of an elliptical wing chord distribution or a specially selected wing twist distribution with an arbitrary chord distribution.

Propeller-type UAVs (helicopter or multicopter schemes) are distinguished by the integration of rotors, which can occupy intricate spatial positions and, consequently, generate a substantial induced effect on the lift-creating surfaces (for VTOL with a wing), as well as on each other (multicopter scheme), particularly at elevated translational flight speeds. This impact must be considered when developing UAVs in general and control systems for them in particular.

In this chapter, an exploration of select aspects of aerodynamics for drones of various configurations will be undertaken.

4.2 Basic Concept of Aerodynamic

4.2.1 Aerodynamic Forces Acting on the Unmanned Aerial Vehicle

As the book discusses various types of UAVs that generate lift based on different aerodynamic principles, Fig. 4.3 presents two potential options. As illustrated in Fig. 4.3a [1], Option 1 is a diagram of the action of forces on a drone with a fixed wing. Option 2 (Fig. 4.3b [6]) is a diagram of the action of forces on the propeller, which creates vertical thrust for copter types of drones.

A thorough analysis of the first option will be conducted to facilitate a comprehensive understanding of its implications. Due to the forward movement, an air flow movement is created around the wing of the drone. This flow typically moves around the wing with a certain angle of attack. The flow's velocity guarantees the formation of a composite set of aerodynamic forces acting upon both the wing and the complete UAV body.

Two systems of axes are utilized to ascertain aerodynamic forces.

One such component is the drone body axis system, which is associated with the UAV body. The second is a system of axes relative to the flow, related to the velocity of the incoming flow.

The normal and axial forces depicted in Fig. 4.3a are derived from the drone's body axis system and remain constant in direction relative to the flow direction.

The lift and drag depicted in Fig. 4.3a are determined by the direction of the incoming flow and remain constant in the direction relative to the drone's orientation.

The figure also illustrates the gravity vector, which is the primary factor determining the lift in steady flight. In order to achieve level flight, it is imperative that equilibrium between gravity and lift is maintained.

The conversion of forces exerted within the body's axial system into those acting upon the flow axis system can be achieved through the utilization of vector conversion techniques, specifically the angle of attack.

The moment of pitch relative to the center of gravity, as illustrated in Fig. 4.3a, demonstrates a positive value when measuring this moment (i.e., the nose of the drone moves up).

The second option will be considered in greater detail below. The propeller is permitted to move translationally in the direction of its axis with a velocity V_0. The element of the blade located at a radius r with a chord c and a thickness dr is considered to be a wing with an area of $b \times dr$ (see Fig. 4.3b). When the propeller rotates with angular velocity ω, the propeller-blade velocity of the element is $U_0 = r \times \omega$. The principle of motion reversal is employed, whereby it is assumed that a flow with axial velocity V_0 and propeller-blade velocity U_0 comes into contact with the stationary element of the blade. Consequently, the resulting velocity with

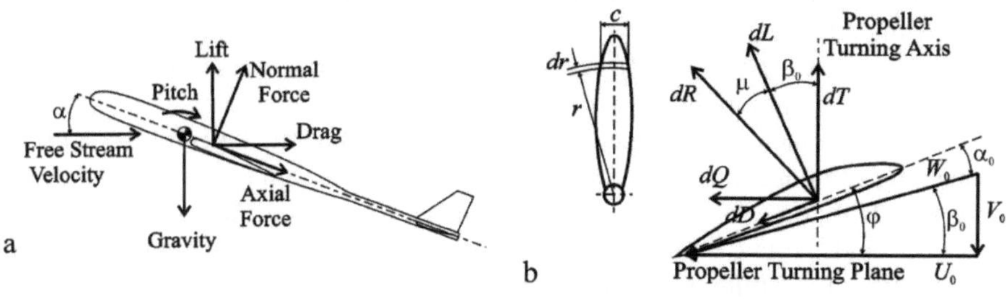

Fig. 4.3 Two options for lift generation: (**a**) force diagram for a fixed-wing drone; (**b**) force diagram for a propeller blade

which the blade element flows around will be equal to the vector sum $W_0 = V_0 + U_0$. This phenomenon is referred to as the relative velocity.

The vector W_0, with the airfoil chord of the blade section, forms the angle of attack of the blade section, designated as α_0. When combined with the propeller turning plane, it forms the angle of inflow, denoted as β_0. The aggregate of the angles α_0 and β_0 is equivalent to the angle of the blade section, designated φ, given by $\varphi = \alpha_0 + \beta_0$.

The blade element's circulation generates distributed loads on its surface, leading to the resulting dR, which is defined as the elementary total aerodynamic force.

The projections of the total aerodynamic force dR on the turning axis and on the turning plane are, respectively, the elementary thrust dT and the elementary force of resistance to turning dQ. Conversely, the force dR can be expressed in terms of projections on the direction of velocity W_0 and perpendicular to it. The respective values obtained are dD, which is elementary drag, and dL, which is elementary lift. The vector of total aerodynamic force, denoted dR, is deflected from the vector of lift, represented by dL, by an angle μ.

As illustrated in Fig. 4.3a and b, and as outlined in the description of the forces acting on the wing and the elementary section of the propeller blade, a common principle exists between these two options for generating aerodynamic forces, despite the disparate movements of the aforementioned elements (wings and propeller blades).

4.2.2 Drag Forces Acting on the Unmanned Aerial Vehicle

The aerodynamic drag force, defined as the braking force acting on aт UAV, is contingent upon the velocity of the incoming flow, exhibiting directionality. The concept of drag can be categorized into several distinct classifications, as outlined below [1].

The phenomenon of profile drag is attributed to the presence of undesirable pressure gradients, which are contingent upon the configuration of the drone's body.

The phenomenon of friction drag originates from the viscous friction forces that manifest within the boundary layer in proximity to the drone's surface.

Induced drag is defined as the increase in drag associated with the formation of wing (blade) lift and vortices, which are the result of pressure differences at the tips of the wings (blades).

Wave drag is a phenomenon resulting from shock waves and other compressibility effects at high flight speeds.

For UAVs, which are equipped with propellers to generate thrust or lift, an additional drag component must be considered. This phenomenon is known as rotational drag [6]. This force is defined as the resistance to rotation that occurs when the propeller blades rotate through the air. It is directed parallel to the direction of rotation, thereby preventing the propeller from rotating.

4.2.3 Turning Flight Conditions

A UAV engaged in a turning maneuver follows a methodical sequence of steps to execute this maneuver. The initiation of a turn is achieved by deflecting the control surfaces (or by modulating the operating mode of the propellers) that generate the moment of the drone's turn. This particular moment of rotation corresponds to the angle of inclination (φ) of the drone relative to a designated axis. Upon attaining the desired maneuver angle, the control surfaces (propellers) revert to the neutral position. At this stage, the drone maintains neutral stability in relation to the maneuver and, therefore, remains in the set maneuver angle.

As an illustration, Fig. 4.4 [1] depicts the drone roll maneuver. At a given roll angle, the lift of the UAV can be divided into two components. The vertical component is responsible for supporting the drone's weight, while the horizontal component provides the centripetal force necessary for executing the roll maneuver. In the case of a UAV equipped with propellers, the collective thrust exerted by all the propellers can be decomposed into distinct components. To ensure the continuity of horizontal flight, the drone must satisfy next condition

$$L \times \cos j = W. \tag{4.2}$$

In this turn mode, the drone utilizes the elevators to maintain the necessary angle of attack (and consequently lift) to ensure level flight. Concurrently, the rudders (propellers) will be utilized to ensure steering coordination in the event of a slip or skid.

4.2.4 Nondimensional Coefficients

In order to draw comparisons between the aerodynamic forces (described in Sect. 4.2.1) of different UAVs and conditions of flight, it is necessary to employ dimensionless parameters. This dimensionless determination is carried out using the following coefficients [1]

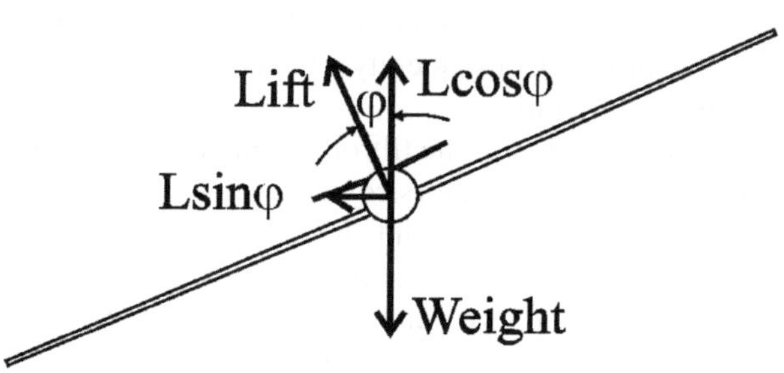

Fig. 4.4 Dependences of several aerodynamic performance on the wing sweep

$$C_L = \frac{L}{0.5 \times \rho \times V^2 \times A_{ref}}, \tag{4.3}$$

$$C_D = \frac{D}{0.5 \times \rho \times V^2 \times A_{ref}}, \tag{4.4}$$

$$C_M = \frac{M}{0.5 \times \rho \times V^2 \times A_{ref} \times c_{ref}}, \tag{4.5}$$

where, C_L and C_D are lift and drag coefficients; C_M is aerodynamic control moment coefficient; L and D are lift and drag; M is control moment; V and ρ are the free stream velocity and its density; A_{ref} and c_{ref} are the reference area and lever respectively.

The aerodynamic perfection of an object can be evaluated by its lift-to-drag ratio (C_L/C_D). An elevated lift-to-drag ratio has been demonstrated to enhance a drone's fuel efficiency and augment its climbing performance.

The utilization of these parameters facilitates the comparison and evaluation of diverse drone designs.

4.3 Discussion About UAV's Aerodynamic Efficiency

The aerodynamic characteristics of a flying vehicle are contingent upon its configuration.

For instance, the UAV with a fixed-wing configuration exhibits the highest lift-to-drag ratio among all other configurations [7]. However, it is important to note that this category of UAV is not capable of taking off or landing vertically. Conversely, UAV of the propeller type (helicopter or multicopter configurations) possess the capacity for vertical takeoff and landing; however, the lift-to-drag ratio of these configurations is exceedingly poor. A position intermediate between the two extremes is occupied by wing-mounted UAVs with lifting propellers or tilt-rotor drone. Conversely, the lift-to-drag ratio of the former will be inferior to that of a purely fixed-wing configuration due to the propellers or their fairings protruding into the airflow. Consequently, the aerodynamic drag increases, thereby decreasing the lift-to-drag ratio.

A consideration of the tiltrotor vehicle reveals the existence of additional issues. Firstly, the primary factors contributing to this phenomenon are the intricacy of the design and the concomitant increase in weight. The increased complexity of the design is associated with additional moving parts, such as propellers or propulsion in whole. This, in turn, leads to a decrease in the reliability of the power plant and the vehicle as a whole. Secondly, there are also difficulties with optimizing the propellers for such an UAV. The generation of lift (takeoff/landing modes) and thrust necessitates the manipulation of various propeller parameters. The compromises made in this case result in a decrease in efficiency for specific flight modes.

Therefore, it is logical to consider the aerodynamic characteristics of each possible UAV configuration separately.

As is evident from the aforementioned relations, the designer's objective is to minimize the area A_{wet} (wet area) and optimize the aerodynamic coefficients and the area A_w (wing area).

From the standpoint of minimizing the area A_{wet}, the flying wing and Blended Wing Body (BWB) configurations are optimal [8]. These configurations are characterized by issues with controllability and balancing in the vertical plane. However, when considering the dimensions of the UAV and the absence of humans on board, these issues are less consequential in comparison to those encountered in conventional aircraft. In the conventional configuration, the surface area streamlined by the flow is larger due to the streamlined surfaces of the aircraft components entering the flow, such as the fuselage, tail, nacelles, etc.

The optimization of aerodynamic coefficients constitutes a distinct scientific problem with numerous potential avenues for exploration. For instance, the wing profile and shape in the plan can be optimized, and various types of drag can be reduced.

Given the sheer scope of aerodynamic research, which precludes its inclusion in a single book, and the focus of the proposed book on UAV-related subjects, the present study will address particular aerodynamic dependencies that contribute to the configuration of the UAV.

4.3.1 Influence of the Wing Parameters

As previously stated, for flying vehicles operating on the basis of aerodynamic flight principles, the wing constitutes the primary component responsible for generating lift. Consequently, the aerodynamic efficiency of a fixed-wing drone is contingent upon the geometric parameters of the unit. A subsequent examination of these parameters and their impact on the aerodynamics of the drone will follow.

The geometric parameters of the wing are listed below: span, area, sweep, wing incidence, airfoil parameters, twist (geometric and aerodynamic), aspect ratio, taper ratio, and wing-tip parameters.

4.3.1.1 Wing Span

The wingspan is defined as the distance between the extreme tips of the wings.

For fixed-wing flying vehicles, an increase in the span results in improved aerodynamic characteristics [8]. However, to comprehensively evaluate the impact of the span, it is imperative to consider the elongation of the wing, a subject that will be explored in the following section.

As the wingspan increases, the effect of the fuselage on the wing sections close to the wing tips decreases, thereby improving their effectiveness.

In the case of a finite-span wing, the pressure difference in the sections near the wing tip decreases due to the influence of air flow from the high-pressure zone (under the wing) to the low-pressure zone (above the wing). The lifting force is most significant in the root section, with a decline towards the wing's tips. Consequently, the wing area in the tip of the wing decreases, leading to a reduction in the losses experienced during the generation of lift.

The efficacy of the control surfaces (ailerons) located at the wing tips is enhanced as the shoulder for aerodynamic force on these surfaces is augmented. Consequently, given a certain degree of maneuverability, the area of the aileron can be reduced.

However, an increase in wingspan is associated with a significant increase in wing loads. This results in an increased demand for strength and rigidity. These issues will be discussed in the following sections.

4.3.1.2 Wing Area

The wing area is defined as the area of the projection of the wing onto a horizontal plane.

The magnitude of lift is directly proportional to the surface area of the wing. This result is derived from the wing lift Eq. (4.3).

It is evident that increasing the wing area will result in a concomitant increase in drag [1, 7]. This phenomenon can be attributed to the fact that drag is composed of two distinct components: drag proportional to the cross-sectional area of the wing perpendicular to the flow, and drag resulting from the friction of the flow against the surface of the wing. In the case of a wide wing, the boundary layer—defined as the region of the flow adjacent to the wing's surface that is hindered due to friction forces—becomes more substantial. This increased thickness of the boundary layer, in turn, leads to an augmentation in drag.

The flow stall, a term used in the field of aeronautics, occurs when the critical angle of attack is reached. This occurs when the airflow experiences a sudden change in direction, leading to a cessation of smooth flow around the wing's airfoil (leading edge). At this point, the airflow breaks off from the upper surface of the wing. An augmentation in the wing area has the potential to induce flow stall at an earlier stage, thereby diminishing lift due to a reduction in the pressure differential between the upper and lower wing surfaces.

4.3.1.3 Wing Sweep

The sweep angle of a wing is defined as the angle between the wing's leading edge (or, in some cases, the quarter-chord line) and a perpendicular line to the drone's longitudinal axis (fuselage centerline).

The reduction of wave drag is achieved through the implementation of a higher sweep wing angle [1, 7, 9, 10]. This configuration allows for the generation of subsonic flow around the leading edge, leading to a decrease in wave drag. Consequently, this reduction in wave drag results in a decrease in the overall drag experienced by the wing.

The impact on stability and controllability is evident at high transonic flight speeds. The spatial flow surrounding the swept wing, characterized by the drainage of the boundary layer along the span to the tip sections, and the flow disruption at the wing's tips, known as terminal stall, result in an augmentation of drag and a diminution of aerodynamic perfection (lift-to-drag ratio).

Figure 4.5 presents several graphs of the aerodynamic dependences on wing sweep angle [1, 7].

4.3.1.4 Wing Aspect Ratio

One of the key parameters determining the aerodynamic efficiency of a flying vehicle is wing aspect ratio. Its influence is complex and often contradictory [8].

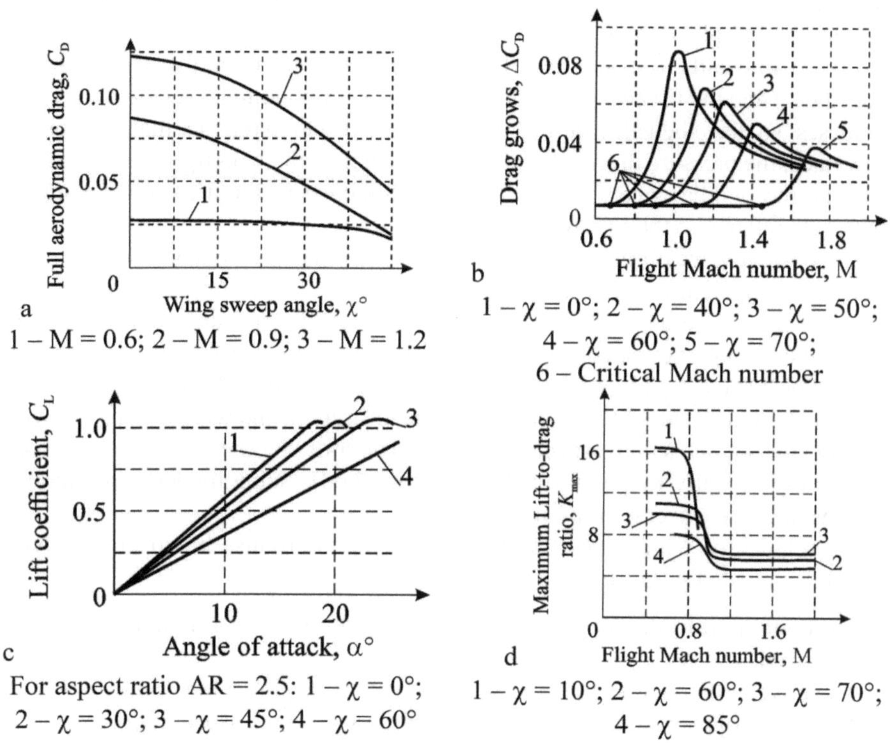

Fig. 4.5 Dependence of several aerodynamic characteristics on wing sweep angle: (**a**) total drag vs. wing sweep angle; (**b**) drag rise factor vs. Mach number; (**c**) lift coefficient vs. angle of attack; (**d**) maximum lift-to-drag ratio vs. Mach number

Wing aspect ratio (AR) is defined as the square of the wingspan divided by its area

$$AR = L^2 / A_w, \tag{4.6}$$

where L is the wingspan and A_w is the wing area.

A wing with a high aspect ratio is long and narrow, like a glider or high-altitude drone, while a wing with a low aspect ratio is short and wide, like a supersonic jet.

The aspect ratio has the greatest effect on the induced drag of the wing. Induced drag is the drag caused by lift generation. At the wing tips, air flows from the high-pressure area under the wing to the low-pressure area above the wing. This flow creates tip vortices. These vortices create a bevel in the airflow, "twisting" it downward. As a result, the total aerodynamic force is deflected backward, creating a component known as induced drag.

The higher the wing's aspect ratio, the lower the induced drag (Fig. 4.6a). In a long wing, it is more difficult for the vortexes at the tips to reach the middle of the wing. The downwash flow they create is smaller. Consequently, the aerodynamic force is deflected back less, and the induced component of drag is reduced.

A high wing aspect ratio is a significant advantage for flight modes requiring high lift (takeoff, landing, and maneuvering) and for vehicles where efficiency is critical (civil airplanes and gliders).

Figure 4.6b and c illustrate the effect of wing aspect ratio on the wing lift-curve slope (C_L^α) and the lift coefficient dependences correspondingly. As the aspect ratio increases, so does the value of C_L^α . This increase is due to the fact that, at subsonic flow rates, the pressure drop at the wingtips is zero, so the wingtip zones create less lift than the middle zones. As the aspect ratio increases, the area of the wingtips decreases relative to the total wing area.

Concurrently, we observe a reduction in the critical angle of attack and the maximum lift coefficient as the wing aspect ratio increases (Fig. 4.6c).

Since induced drag accounts for a large portion of total drag during cruising flight, increasing the aspect ratio improves aerodynamic efficiency (lift-to-drag ratio). For instance, gliders have an aspect ratio of 30–40 to fly the maximum distance with minimal height loss. However, due to strength and weight issues, it is not possible to use such a high aspect ratio for civil aircraft. For these aircraft, the aspect ratio conventionally ranges from 7 to 10 (Fig. 4.6d). Some modern civil aircraft designs have an aspect ratio of up to 11.5.

The choice of wing aspect ratio is always a compromise based on the flying vehicle's intended purpose:

- For civil aircraft, the focus is on efficiency and range, so a high aspect ratio is preferred.
- Maneuverability and supersonic speeds (fighters) require a low aspect ratio.
- Versatility (general-purpose aircraft) requires an average aspect ratio.

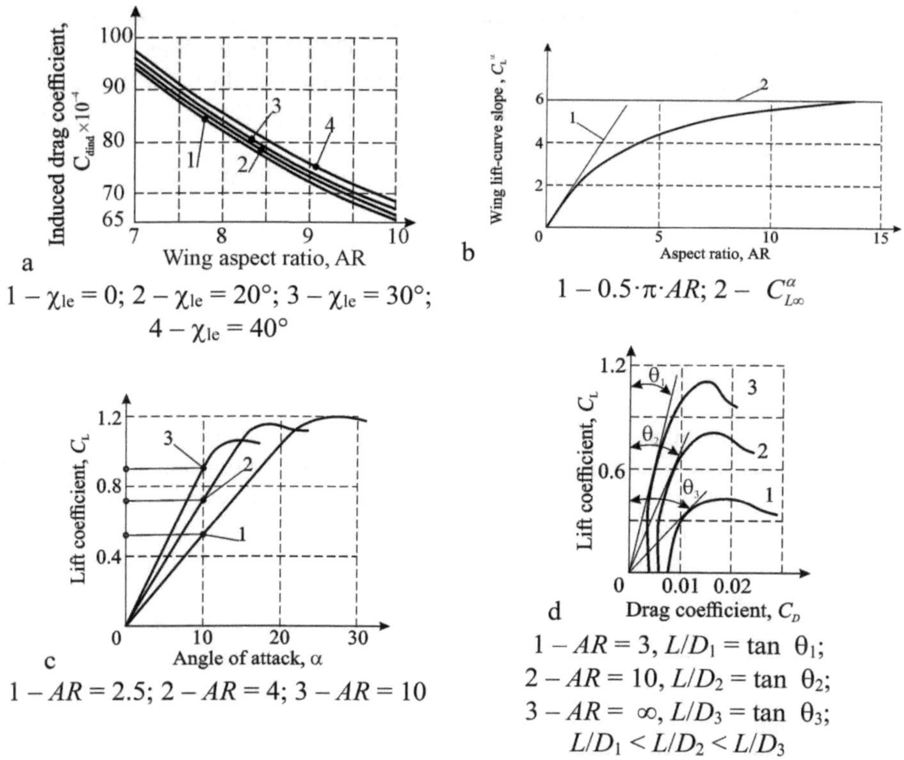

Fig. 4.6 Dependence of aerodynamic characteristics on wing aspect ratio [1, 7, 10, 11]: (**a**) induced drag vs. aspect ratio; (**b**) lift-curve slope vs. aspect ratio; (**c**) lift coefficient vs. angle of attack; (**d**) drag polar

4.3.1.5 Wing Taper Ratio

The wing taper ratio (λ) is defined as the ratio of the length of the wing tip chord to the length of the root chord. Consequently, the potential values of the taper ratio span from negative infinity to one.

An intriguing observation is that in the countries that were previously part of the USSR, a specific convention exists regarding the determination of the taper ratio of the wing. This convention stipulates that the ratio of the length of the root chord to the length of the tip chord is inversely proportional. Consequently, the potential narrowing values for this case range from 1 to infinity. It is imperative that the reader be aware that the values of the taper ratio can be determined in a variety of ways, as indicated in the numerous references. However, this is not a significant problem, since both values can be recalculated by reverse division. In this book, the traditional method of defining the taper ratio will be employed.

The effect of the wing taper ratio on drag is primarily related to its impact on induced drag [1, 7, 9]. The tapered wing ($\lambda < 1$) distributes lift more evenly over the span. This distribution of lifting force is optimal because it minimizes induced drag. This is because it provides the lowest possible pressure drop between the upper and lower surfaces at the tips of the wing, thereby reducing vortex generation. Figure 4.7a shows the relationship between the growth factor of the induced drag and the taper ratio. It seems that the rational value of the taper ratio should be in the range of 0.2–0.4. A high taper ratio leads to increased induced drag. This is because the stall effect increases with the taper ratio.

The tapered wing has a higher lift-to-drag in cruising flight modes due to a significant reduction in induced drag (Fig. 4.7b).

4.3.1.6 Wing Incidence

The wing's arrangement in a vertical plane relative to the fuselage can be classified into three categories: low-wing, medium-wing, and high-wing [2].

For most flying vehicles, the choice of wing layout is primarily related to layout considerations. The need for free evaluable space inside the fuselage precludes the medium-wing scheme because it is impossible to pass the wing's center section through the fuselage. In the case of using a wing without a center section, with the wing's consoles attached to the reinforced frame, is impractical due to weight concerns.

In contrast to the mid-scheme, the high- and low-wing schemes do not interfere with the creation of a single cargo compartment. When selecting between these options, a high-wing configuration is typically preferred. From an aerodynamic perspective, high-wing designs offer distinct advantages over low- and medium-wing configurations. The high-wing configuration enables a circulation distribution that approaches an elliptical shape, maintaining the same wing shape in the plan, without incurring a loss in fuselage area. This characteristic is in contrast to the low- and medium-wing designs, which experience a loss in fuselage area. Concurrently, the high-wing's inherent interference drag, though higher than that of a medium-wing but lower than a low-wing, enables the attainment of a high lift-to-drag ratio for aircraft constructed in accordance with this configuration. It has been demonstrated that, when the wing is positioned in the low position, the drag is greater than when

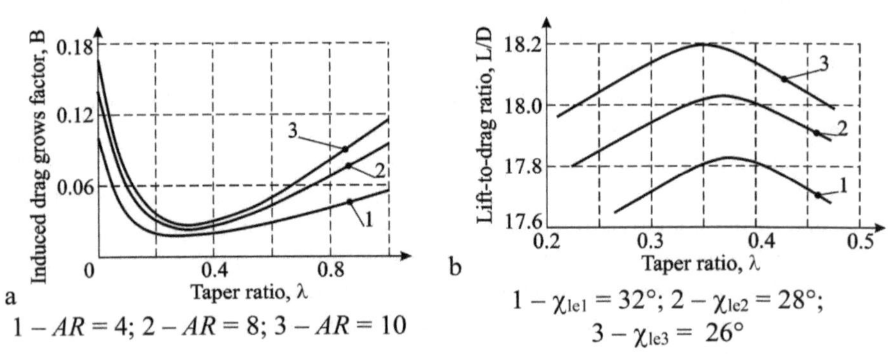

Fig. 4.7 Dependence of aerodynamic characteristics on wing taper ratio [1, 9]: (**a**) induced drag grows factor vs. taper ratio; (**b**) lift-to-drag ratio vs. taper ratio

the wing is positioned in the mid or high position, particularly at speeds less than 0.7. The following polar plots, which depict the wing layouts on the fuselage, reveal a salient feature.

As illustrated by the polar graphs (C_L vs. C_D) for the three wing configurations on the fuselage, the C_{D0} (for $C_L = 0$) for the low-wing configuration is higher than that of the mid-wing and high-wing configurations (see Fig. 4.8 [1, 7]).

4.3.1.7 Wing Airfoil Parameters

The aerodynamic airfoil of a wing is characterized by two primary parameters: camber and thickness (see Fig. 4.9a) [2]. The determination of aerodynamic dependencies is typically accomplished through the utilization of specific parameters: camber (*CR*) and thickness (*TR*) ratios.

Camber is defined as the degree of deviation of the midline of an airfoil from a straight chord. As the camber ratio of the airfoil increases, the lift coefficient experiences a substantial rise. The positive camber engenders a non-zero lift force even at zero angle of attack ($\alpha = 0°$). Consequently, the value of $C_{L\max}$ (maximum lift force coefficient) increases (Fig. 4.9b) [1, 7]. It has been demonstrated that this leads to an increase in C_{L0}, which is defined as the value of the lift force at $\alpha = 0°$.

Concurrently, the drag coefficient is undergoing an increase. The camber of the surface disrupts the "ideal" streamlined shape, primarily increasing the profile drag (pressure and friction drag). This phenomenon is particularly evident at low angles of attack, where induced drag is minimal. Consequently, the minimum value of $C_{D\min}$ increases.

Consequently, this results in a contradictory effect on the lift-to-drag ratio (*L*/*D*). During takeoff and landing, the *L*/*D* increases due to a substantial gain in lift and a comparatively modest rise in drag (Fig. 4.9c [1, 7]). In cruising modes (low angles of attack, where C_L is low), the *L*/*D* may decrease, as the "extra" profile drag from

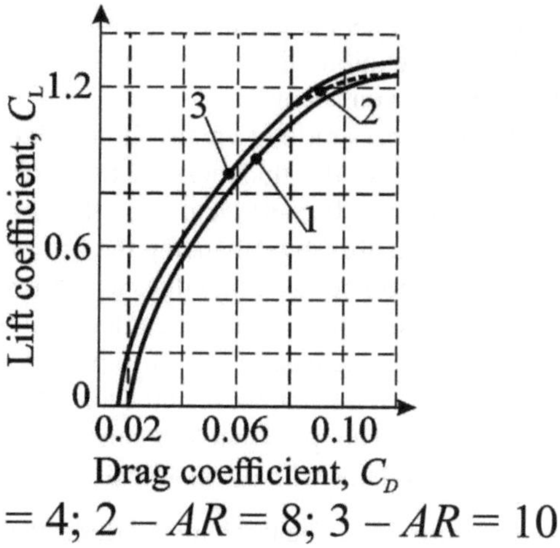

Fig. 4.8 Dependences of Lift and Drag coefficients on the wing's arrangement in a vertical plane relative to the fuselage

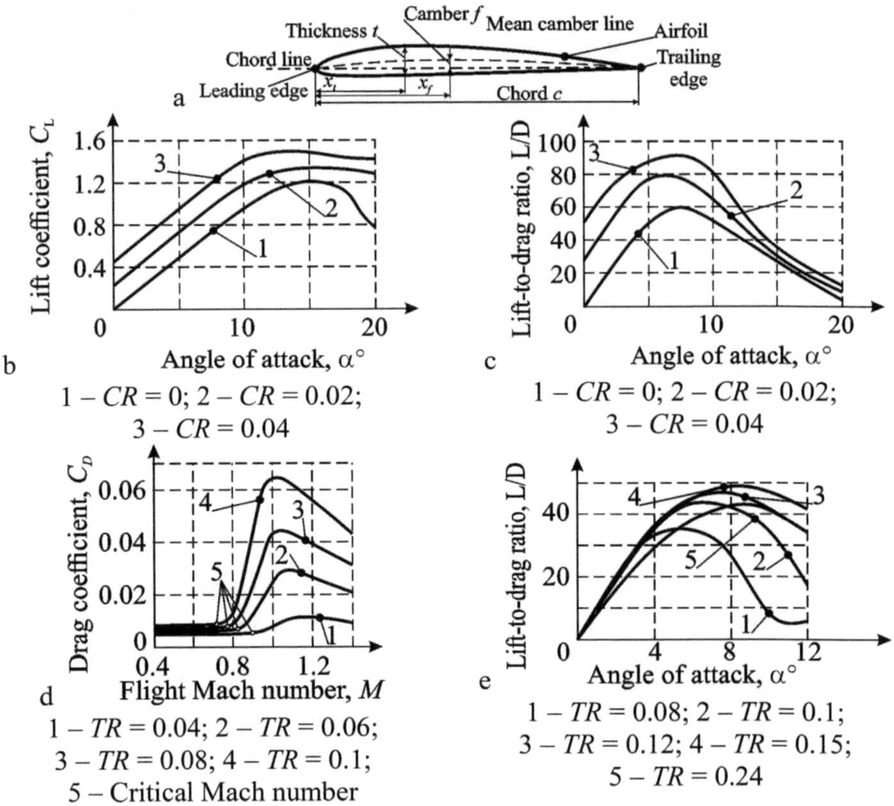

Fig. 4.9 Dependence of aerodynamic characteristics on airfoil parameters: (**a**) airfoil geometry; (**b**) lift coefficient vs. angle of attack; (**c**) lift-to-drag ratio vs. angle of attack for different camber ratios; (**d**) drag coefficient vs. Mach number; (**e**) lift-to-drag ratio vs. angle of attack for different thickness ratios

camber becomes the dominant factor. The maximum lift-to-drag ratio, denoted as L/D_{max}, can exhibit slight variations depending on the particular characteristics of the airfoil in consideration.

The thickness ratio of an airfoil is defined as the ratio of its maximum thickness to its chord. As the thickness of the airfoil is increased, there is a concomitant increase in the lift coefficient. The thicker airfoil exhibits a higher camber of the upper surface, thereby enhancing the lift force. Additionally, thick airfoils generally possess a smoother leading edge, which results in the displacement of flow stall to high angles of attack. It has been demonstrated that the C_{Lmax} and the critical angle of attack (α_{crit}) increase. The airfoil provides high angles of attack before the stall.

However, as the thickness of the airfoil increases, the coefficient of drag increases significantly. This phenomenon constitutes the primary effect. The thickness ratio directly correlates with the profile drag:

- Pressure drag. The thick airfoil deflects the flow more strongly, creating a higher pressure difference between the leading and trailing edges.
- Friction drag. The surface area exposed to the action of the flow increases.

At transonic and supersonic speeds, the increase can lead to severe consequences. Consequently, the wave drag experiences a marked increase (Fig. 4.9d [1, 7]). High thin profiles ($\bar{t}$ = 3–5%) are utilized for high speeds. Consequently, the $C_{D\min}$ increases significantly.

For the L/D, the effect of increasing the thickness of the airfoil is contradictory, but still more negative (Fig. 4.9e [1, 7]). In cruising modes characterized by low angles of attack, there is a substantial increase in drag, which consequently leads to a decrease in maximum L/D_{max}. However, in take-off and landing modes (close to α_{crit}), a thick airfoil may offer a distinct advantage, given its high C_L and the delay of flow stall.

4.3.1.8 Wing Twist

The following is a concise exposition of the twist design technique.

Geometric twist is defined as the physical twisting of a wing in which the angle of attack (α) varies in span (Fig. 4.10a).

Aerodynamic twist is defined as a variation in the wing airfoil along the span, independent of alterations in installation angle, which emulates the effect of twisting (Fig. 4.10b [1]).

The two types of twisting are designed to control the distribution of lift over the wingspan. This control improves the aerodynamic efficiency and stability of the aircraft.

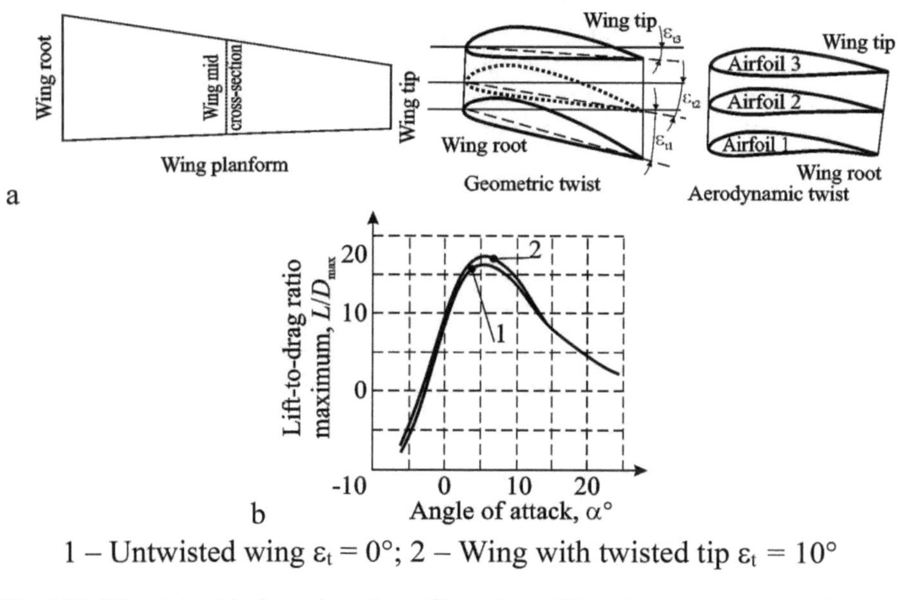

Fig. 4.10 Wing twist: (**a**) wing twist options; (**b**) maximum lift-to-drag ratio vs. twist angle

The presence of the twist results in a reduction of the induced drag [10]. The induced drag is directly proportional to the uneven distribution of the lift force (or, more precisely, the magnitude of the tip vortices). The primary objective of utilizing twisting is to minimize this drag. Achieving an optimal twisting distribution enables the minimization of C_{Di} for a given lift coefficient, C_L.

The effect is achieved in a variety of ways. In the case of geometric twist, the required local angle of attack is determined by the angle of airfoil installation in a specific wing section. Typically, a negative angle of the wing airfoil is applied at the wing's tip. In the case of aerodynamic twist, this is achieved by varying the airfoils. It has been demonstrated that an airfoil with higher camber (e.g., at the root) generates higher lift at the same angle of attack than an airfoil with lower camber (e.g., at the tip). Consequently, the lift distribution is also aligned with the elliptical, thereby reducing the induced drag.

The geometric twist, particularly in its negative form, signifies that the wingtips function with negative or zero lift during cruising. This results in a negative lift force at the tips, which is not rational.

The aerodynamic twist facilitates the utilization of airfoils that maintain optimal efficiency at the cruising angle of attack ($\alpha \approx 0$). This reduction in drag contributes to an enhancement in overall aerodynamic efficiency, particularly in the context of main cruising flight modes.

4.3.1.9 Wing Tips

The primary objective of any tip (Fig. 4.11a [11]) is to mitigate the induced drag, which exerts a dominant influence on cruising flight modes (at low angles of attack). This objective is realized through the control of the tip vortices.

To illustrate, in the absence of wingtips on a wing, air moves from a high-pressure area (below the wing) to a low-pressure area (above the wing). This phenomenon gives rise to the formation of robust tip vortices. These vortices, or rotating flows of

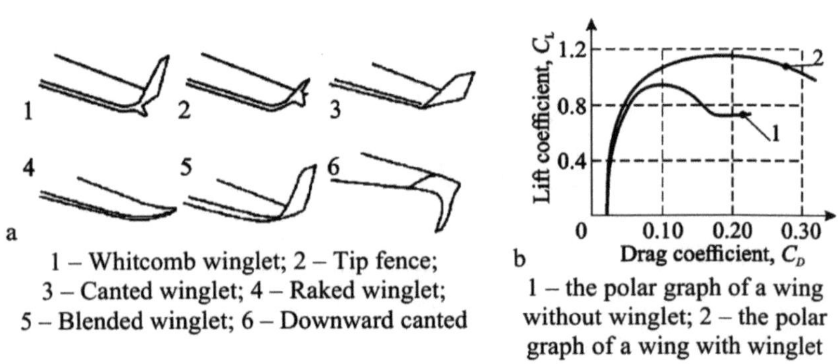

Fig. 4.11 Wing tips: (**a**) wingtip configurations; (**b**) drag polars with and without winglets

air, interact with the flow behind the wing, causing a tilt in the flow and inducing an additional drag force, known as induced drag. In cruising mode, the induced drag coefficient can reach up to 30% of the total drag [12].

In the case of a wing with tips mounted at an angle, an aerodynamic force is created. The aforementioned model can be decomposed into two components:

- The transverse component is directed inwards. This component functions to directly counteract the flow of air, thereby reducing the intensity of the tip vortices.
- The longitudinal component is directed forward. This component functions in opposition to the incoming flow, thereby generating a thrust.

Consequently, the induced drag is reduced by 5–15% (depending on the type of tip, flight mode, and initial wing specifics). This phenomenon directly results in an enhancement of the lift-to-drag ratio (Fig. 4.11b [1]).

This solution has also contributed to an effective increase in wing aspect ratio. This phenomenon also contributes to a reduction in induced drag. Furthermore, the turned-back tips augment the effective sweep of the wing, thereby delaying the onset of a wave crisis at high speeds.

At present, a multitude of types of wingtips are in existence. Figure 4.11a presents a depiction of some of the most frequently utilized drone options. The expediency and configuration of the tips are contingent upon the objective of the UAV.

4.3.1.10 Examples of Drones with Different Purposes and Wing Parameters

As previously stated, the wing's parameters of a drone are contingent upon its intended function. The following examples are provided to illustrate the range of existing drones.

1. The RQ-7 Shadow UAV (Fig. 4.12a) [13] is characterized by a rectangular wing with a high aspect ratio. The sweep angle is set at 0, and the taper ratio is 1. These wing parameters are particularly well-suited for UAVs operating at low flight speeds over short ranges.
2. The MQ-9 Reaper UAV (Fig. 4.12b [3]) is characterized by a straight tapered wing with a low sweep angle, a high aspect ratio, and a moderate taper ratio. The selected wing parameters are in accordance with the objective of this UAV, which is to achieve long-range, medium subsonic flight speed.
3. The MQ-20 Avenger UAV (see Fig. 4.12c [14]) is characterized by a tapered wing with a high sweep angle, a high aspect ratio, and a moderate taper ratio. The selected parameters, similar to those of the previous UAV, are well suited for long-range flight. The higher sweep angle is corroborated by the higher flight speed, which is approximately twice as high.
4. The WZ-8 UAV (Fig. 4.12c [15]) is characterized by a delta-shaped wing, exhibiting a high sweep angle, a low aspect ratio, and a low taper ratio. The following parameters are characteristic of a supersonic flying vehicle.

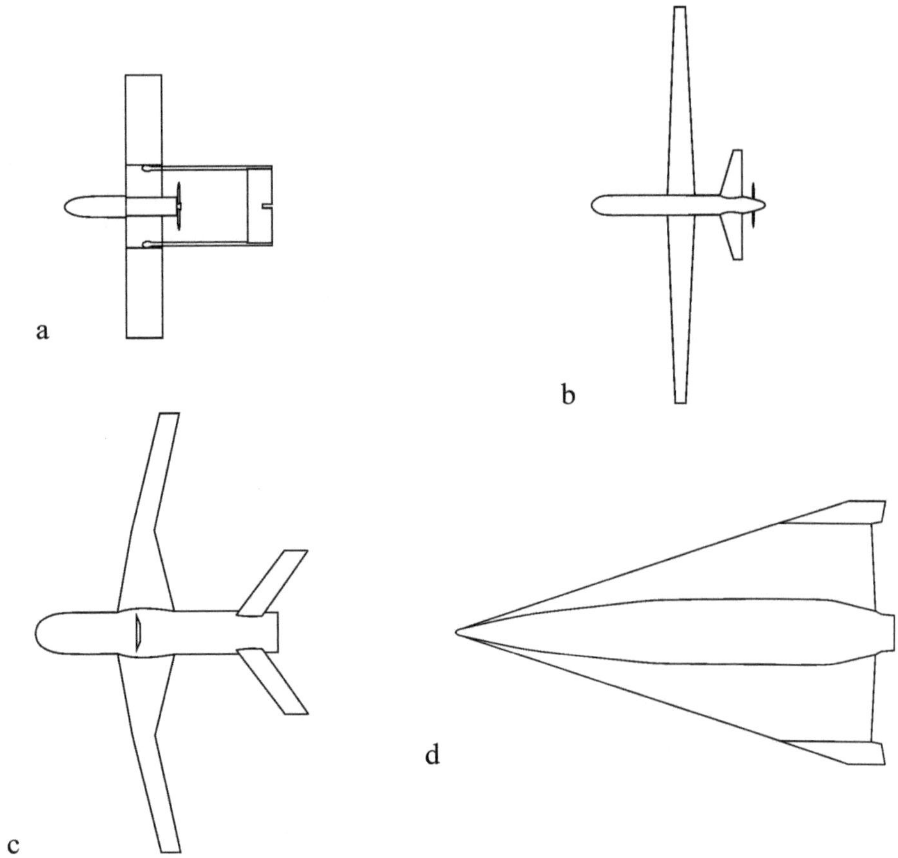

Fig. 4.12 Wing planform views for some UAVs: (**a**) RQ-7; (**b**) MQ-9; (**c**) MQ-20; (**d**) WZ-8

4.3.2 Influence of the Propeller Parameters

Another equally important component in the design of the UAV is the propeller. This structural component is present in the majority of drones, with the exception of those equipped with turbojet or rocket engines.

In the context of UAVs, propellers play a pivotal role in generating thrust for horizontal flight, particularly in aircraft equipped with piston and turboprop engines. In the context of helicopter or multicopter UAV configurations, propellers play a pivotal role in generating lift [7, 13]. In the context of hybrid UAV configurations (VTOL), propellers play a pivotal role in generating both thrust and lift.

Therefore, it can be concluded that the aerodynamic efficiency of the propellers, in conjunction with the efficiency of the wing, is paramount for the majority of UAV types. The ability to optimize propellers is paramount to ensuring the efficient operation of drones.

To illustrate this point, we will employ an analogy with the wing to elucidate the main parameters of the propellers that impact their aerodynamic efficiency. The following parameters will be enumerated: blade length, blade area, blade aspect ratio, blade taper ratio, blade sweep, blade twist, blade tips, and blade airfoil parameters. As the reader will observe, the vast majority of the parameters align with the geometric parameters of the wing. Furthermore, an analysis of the blade pitch control can be conducted.

A more thorough explanation of these parameters can be facilitated by analogy with the parameters of the wing.

It is important to acknowledge that there are discrepancies in methodologies when designing propellers for the purpose of generating lift and thrust.

The main rotor of a helicopter-type UAV generates an upward lift force, thereby lifting and holding the UAV in the air [7, 16]. The flying vehicle's operation occurs in a vertical plane, thereby providing both lift and control. In such instances, the propeller functions in markedly challenging conditions, encompassing hovering, maneuvering, and flight at varied velocities. The aerodynamic loads experienced by the blades exhibit variability due to the variation in angles of attack and flow rates at distinct turning stages.

The propeller of a fixed-wing type of UAV generates thrust in a direction that is forward-oriented, thereby propelling the UAV horizontally [7, 13]. In this scenario, the propeller functions in a more stable mode, generating thrust at a relatively constant translational velocity of the UAV. The operational conditions of the propeller exhibit reduced variability in comparison to those of the main rotor.

These and other features will be considered during the analysis of the parameters of the blades.

4.3.2.1 Blade Span

Typically, the blades of the main rotor are longer in length. In this instance, there are length restrictions; however, these restrictions are less stringent than those imposed on the blades utilized for thrust generation. The utilization of longer blades has been demonstrated to result in reduced induced drag, akin to the aerodynamic performance of a wing [7, 17].

For a propeller designed to generate thrust, a mandatory limitation exists on the length of the blade [7, 17]. This phenomenon can be attributed to the proximity of the ground's surface. In the least severe instance, when a rapidly rotating blade collides with the ground, only the blade will deform. However, the repercussions can be considerably more severe. Consequently, an increase in rotational speed is required to generate the necessary thrust force. Blades of shorter length have higher drag, which results in greater energy losses.

4.3.2.2 Blade Area

The required area of the blades is determined based on the required thrust or lift force, the diameter of the propeller, and the permissible load on the swept area [7, 16, 17]. This process is iterative in nature, whereby the area of the blades is matched with other parameters (RPM, blade shape, engine power) to achieve optimal efficiency.

The propeller's efficacy in generating force is directly proportional to its capacity to capture and bite a significant volume of air. The mass air flow rate is directly proportional to the area of the disk that the propeller sweeps (swept area) and the flow rate.

It is at this point that the primary parameter is revealed—the load on the swept area, also known as disk load:

A decrease in the disk load is indicative of a larger propeller diameter at a given force level. This results in the propeller's operation with substantial volumes of air, thereby imparting a marginal acceleration. This is a highly effective mode.

A high disk load results in a small propeller diameter, which enables the propeller to accelerate small volumes of air to high speeds. This mode exhibits reduced efficiency and increased power requirements. However, it permits the reduction in the size of the propeller.

The area of the blades is directly related to the swept area. There is a direct correlation between the size and width of the blades and the disk area that results in coverage. The higher the blades number and/or the width of the blades (the chord), the larger the disk area that is covered will be.

The area of the thrust-generating blades is selected so that, with the selected diameter and available engine power, the calculated thrust is provided with maximum efficiency, without exceeding the strength and speed limits at the tips of the blades.

Hovering mode is critical for a helicopter drone (or multicopter), and the area of the blades directly determines its ability to take off.

The area of the blades is not the primary value in this calculation, which is determined using the provided equation. This outcome is the result of a complex optimization process in which engineers select the diameter, number of blades, and their shape based on the required force and constraints to find the perfect balance between efficiency, weight, strength, and dimensions.

4.3.2.3 Blade Aspect Ratio

Analogous to a wing, a higher aspect ratio (longer and narrower blades) diminishes the induced drag caused by tip vortices, thereby enhancing the propeller's efficiency in hover mode and at lower speeds [7, 16, 17]. However, it is imperative to consider the limitations imposed by load and stiffness. In contrast to the wing, the blade exhibits a lower degree of rigidity, a deficiency that is particularly problematic.

For the main rotor, blade aspect ratio is typically within the range of 8–15, a compromise between aerodynamic efficiency and rigidity. The employment of a higher aspect ratio is a design choice that is particularly salient in the context of flying vehicle that are characterized by a reduced weight.

The typical aspect ratio for thrust-producing propellers is in the range of 6–10. As previously mentioned, limitations are imposed on the length of the blade, and blades that exceed these restrictions have been shown to increase the frontal drag of the aircraft. In this instance, the propellers are engineered to operate at optimal efficiency within a designated speed range. In the case of low-speed drones, the aspect ratio may be increased to enhance thrust at lower RPM. In the case of high-speed drones, such as turboprops, the aspect ratio may be reduced to ensure the blades can withstand high loads and tip speeds that approach the speed of sound.

The blade aspect ratio for the lift generating propellers is a critical component in the design of these systems. How is this aspect ratio selected in practice? This approach is informed by a consensus and the intended function of the drone.

Drones for carrying heavy loads and hovering. Efficiency in hover mode is critical for such drones. Consequently, they utilize blades that exhibit high aspect ratio, characterized by an extended length and relatively narrow width. The primary objective is to achieve maximum power (lift) while minimizing power consumption.

High-speed and maneuverable drones. In this instance, the strength and rigidity of the blades are particularly salient, as they are engineered to withstand elevated aerodynamic loads at high velocities and during maneuvers. In many cases, a lower aspect ratio is selected. The primary objective of this case study is to assess the strength, stability, and operational capacity of the aircraft at elevated flight speeds.

Multi-purpose and lightweight drones. An optimal compromise is being sought for such drones. The average aspect ratio is selected. These drones are often equipped with 2–4 blades of moderate width, a configuration that offers a satisfactory balance between efficiency, strength, weight, and cost. The primary objective is to prioritize versatility, cost-effectiveness, and the simplicity of the design.

The process of blade aspect ratio is not selected in isolation; rather, it is part of the integrated design of the entire propeller(s) (number of blades, shape, airfoil, and twist). This is done in accordance with specific technical requirements for the drone, where its main purpose is a key factor.

The methodology employed in determining the optimal blade aspect ratio for engine propellers is of particular interest. The underlying principle remains consistent, as does the compromise and the purpose of the drone.

Fixed pitch propellers (simple, lightweight drones). Concurrently, they aspire to achieve high aspect ratio, thereby maximizing efficiency in cruising mode.

Heavy turboprop drones. In such an instance, it becomes imperative to effectively absorb the substantial power output of the engines. In order to avert the propeller's excessive diameter growth, multi-blade propellers with moderate or low aspect ratio are employed. The blades, being shorter and wider, facilitate the "collection" of a greater disk area, thereby ensuring that the diameter remains within acceptable limits. In this context, the emphasis is placed on the strength of the blade rather than on its theoretical efficiency.

Modern sweptback blade propeller. The configuration of these blades is a deliberate strategy to circumvent the innate mechanisms of the natural environment. The curved tip, resembling a "saber," serves to mitigate the shockwave at the extremities of the blades at high speeds (transonic) and to diminish the strength of the tip vortices. This enables the utilization of higher circumferential speeds, thereby facilitating enhanced efficiency in cruising flight modes.

The selection of the optimal propeller is contingent upon the engine's power, the cruising speed, and the diameter restrictions. It has been demonstrated that as the power and speed of a drone increase, the number of blades also increases, and the individual aspect ratio of each blade decreases.

4.3.2.4 Blade Taper Ratio

The taper ratio of the blade, which involves a reduction in its chord from the root to the tip, constitutes a form of aerodynamic twist [7, 16, 17]. It is imperative to address the primary issue associated with the propeller, namely the uneven induced velocity (see Fig. 4.13a [6]). The inducted velocity of an air propeller is defined as the velocity of the air flow that is caused by the interaction of the propeller's rotation with the surrounding air. The propeller accelerates the air, imparting additional velocity to the surrounding medium. The phenomenon under scrutiny exhibits a propensity to emit a forceful expulsion, accompanied by a distinct twisting of the flow.

The movement of the blade gives rise to two components of the flow velocity:

- The propeller-blade velocity U. It increases in proportion to the radius.

$$U = r \times \omega. \tag{4.7}$$

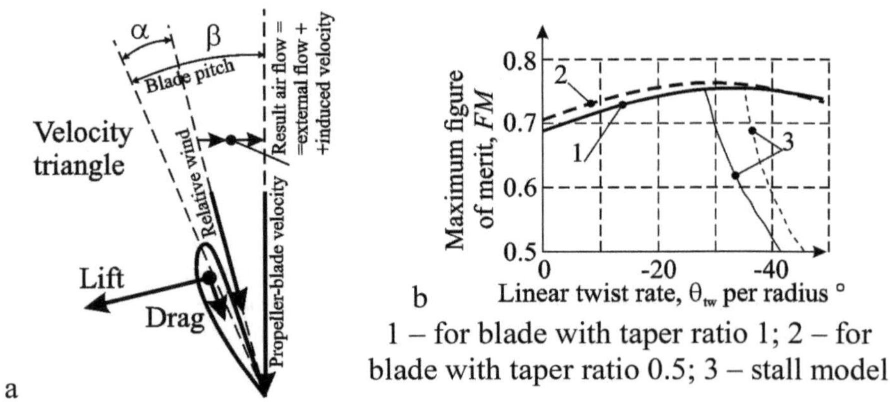

Fig. 4.13 Taper ratio application to blades: (**a**) forces and angles on a propeller blade in airflow; (**b**) maximum figure of merit vs. linear twist rate

- The pitch velocity *V*. The velocity of the external flow is defined as the airspeed for thrust or the axial vertical velocity for lift.

The effective angle of attack of the blade section is determined by the angle difference between the vector of the resulting flow velocity *W* and the chord of the airfoil. As the propeller-blade velocity increases towards the extremity of the blade, the resulting velocity *W* also increases in a robust manner towards the extremity. In the absence of constriction, that is, the implementation of a constant chord along the entire length of the blade, then:

- The propeller-blade velocity at the root (short radius) is minimal, resulting in the airfoil operating at low Reynolds numbers and failing to generate sufficient force.
- At the maximum radius of the tip, the circumferential velocity is elevated, resulting in a substantial increase in the force exerted by the airfoil. This, in turn, can lead to a number of adverse outcomes, including the overload of the blade structure, the disruption of flow, and a precipitous decline in efficiency.

The employment of the blade, in conjunction with a specific taper ratio, effectively addresses this issue by enhancing the load distribution along the span.

It is imperative to consider the utilization of a propeller for the purpose of generating lift, particularly in the context of helicopter or multicopter configurations. Concurrently, the primary objective is to generate maximum lift with minimal power consumption and to ensure flow stability over the entire range of modes.

Consider the effect of blade taper ratio:

- Reduction of wave drag. At the tips of the blades, the relative flow velocity approaches the velocity of sound, even at moderate speeds of a drone. The narrow tip of the propeller creates less wave drag, thereby delaying the onset of a wave crisis to higher speeds of rotation of the propeller or flying vehicle flight. This is a critical factor in maintaining efficiency during high-speed flights.
- Noise reduction. Narrow tips generate vortexes with reduced intensity, thereby mitigating vortex noise, which constitutes one of the primary sources of rotor noise.

Therefore, the application of blade taper ratio is a pivotal technique for enhancing aerodynamic efficiency across the full spectrum of flight modes. This is achieved by balancing the load and mitigating induced and wave losses, thereby optimizing the aerodynamic performance of the lift-generating propeller.

The subsequent discussion will address scenarios where a propeller generates thrust for a flying vehicle. In this case, the primary objective is to generate maximum thrust while minimizing power consumption (i.e., achieving high efficiency) in the design flight mode (typically cruising).

It is imperative to ascertain how the taper ratio of the blade impacts the aerodynamic characteristics of this particular propeller type (Fig. 4.13b):

- Increase the efficiency of the propeller. As with the rotor for the lift generating, the primary objective is to minimize induced losses. At the tips of the blades, air flows from the high-pressure area behind the blade to the low-pressure area in

front of it, forming terminal vortices. These vortices are based on kinetic energy, which represents induced losses. The utilization of a narrow tip engenders a less intense vortex, thereby reducing these losses and enhancing efficiency.
- Optimization of operation at the design speed. The propeller has been developed to operate within a particular flight mode. The taper ratio allows for the blade to be "loaded" with aerodynamic force in a more rational way. The sections at the root operate at low velocities, and the wide chord provides the necessary level of thrust. The sections at the tips operate at high speeds, and the narrow chord allows them to generate thrust efficiently, without stalling, and without creating excessive drag.
- Reduction of rotational resistance. The frontal drag of the blade is composed of two components: form drag and induced drag. The narrow tip has a smaller surface area, which directly reduces the form drag of the tip itself. Given its operation at the highest speeds, this results in a substantial enhancement.
- Reduction of compressibility. In the context of high-speed drones, the taper ratio of the tip is of paramount importance in delaying the growth of wave drag when transonic speeds are attained at the blades' tips.

Consequently, for a thrust-generating propeller, the taper ratio of the blade is the primary method of maximizing its efficiency and minimizing losses in tip vortices. This, in turn, directly affects thrust and fuel consumption. This configuration enables the propeller to operate with high efficiency at the design cruising speed.

In light of the aforementioned evidence, it can be concluded that the taper ratio of the blade constitutes a primary aerodynamic technique for optimizing the load distribution over the span.

In both cases, the taper ratio constitutes more than a mere "lightening" of the blade; it is the primary engineering technique for creating an efficient and operable propeller system.

4.3.2.5 Blade sweep

The taper ratio of the blade, which involves a reduction in its chord from the root to the tip along its span.

The sweep is introduced to delay a shock stall, defined as a sharp increase in drag when the critical Mach number (M = 0.7–0.8) is reached in the blade sections [6, 7, 16, 17]. As the linear velocity of the blade increases from root to tip, the local flow velocity (the external velocity of the incoming flow plus the propeller-blade velocity) reaches supersonic values at the blade tips. This results in the following conditions:

1. A marked increase in wave resistance is observed.
2. The displacement of the flow separation point to the trailing edge is presented.
3. The occurrence of strong shock waves has a significant impact on the efficiency of the propeller, resulting in a reduction of its effectiveness. This phenomenon can also lead to the onset of flutter.

The implementation of a sweep, defined as a deviation of the line 25–50% of the chords back, has been identified as a solution to these problems due to two key effects:

- A reduction in the effective Mach number. The aerodynamic force is contingent upon the normal component of the flow velocity relative to the leading edge line ($V_n = V \times \cos(\chi)$), wherein χ denotes the sweep angle. This theoretical framework enables the postponement of the shock stall to a higher value of the overall velocity.
- The redistribution of air pressure along the blade span. The sweep contributes to a more uniform distribution of pressure over the blade span, thereby "smear-out" the compaction surge and reducing its intensity.

The sweep is generally applied to the tip sections of the blade, where there is a risk of a shock stall. This approach enables the deferral of airflow stall, thereby augmenting the critical angle of attack of the blade section. Furthermore, the aerodynamically swept tip exhibits a similar behavior when its span is increased, while the induced drag experiences a slight decrease.

4.3.2.6 Blade Twist

As with the taper ratio and sweep described above, the primary function of twisting is to ensure an even distribution of the load over the blade span and to achieve the rational angle of attack for each section in the design flight mode [6, 16, 17]. This configuration is designed to optimize efficiency and thrust/lift ratios. However, it should be noted that the twist is used more often due to its higher efficiency. The principle of twisting is described in detail in the wing twisting section. In this section, the focus will be on the impact of twist angles on the aerodynamic characteristics of the propeller.

In the context of propellers that generate lift, the application of twisting has been demonstrated to yield the following effects:

- Optimization of the lift along of the propeller disc. For the hover mode and at low speeds without blade twisting, the lift would be generated primarily at the tips of the blades (where the propeller-blade velocity is high), and at the root it would be virtually absent. This approach is suboptimal due to its potential to generate a substantial bending moment on the blades, which could result in their deformation. The twist facilitates the "additional loading" of the root part, thereby substantially augmenting the lift coefficient of the entire main rotor.
- Reduction of induced losses. The uniform distribution of load along the blade serves to minimize the induced drag.
- Redistribution of loads between the blades. It is imperative to note that the blade's movement in both the forward (advancing side) and backward (retreating side) directions during flight with translational velocity is subject to markedly distinct conditions. The combination of twist and flapping motion serves to

equalize the lift along the propeller disc, thereby preventing stall on the blades moving in a backward direction and reducing the wave drag on the blades moving in a forward direction. This is a critical factor in achieving maximum flight velocity.

In the case of thrust-generating propellers, the twist has been shown to provide the following benefits:

- Maximizing efficiency for the design mode. The primary aim is to optimize the angle of attack for each section of the blade, from the root to the tip, at the optimal cruising velocity. This configuration ensures that the propeller generates maximum thrust with minimal power consumption, thereby operating at the peak of its aerodynamic efficiency.
- Reduction of losses by the tip vortices. The uniform distribution of the load along the span has been shown to reduce the intensity of the tip vortices and, consequently, the induced drag.
- Preventing stall effect. For take-off mode, characterized by low flight speed and high RPM, the twist functions to prevent the stall at the external, fast-moving sections of the blade.
- Compromise for different flight modes. A fixed-pitch propeller invariably represents a compromise. The twisting approach has been meticulously engineered to operate within a single mode. During the takeoff mode, the outer sections of the wing will maintain a near-optimal angle of attack, while the root sections will exhibit a reduced angle. At the cruising speed, the root sections of the blade achieve the designed mode, while the tip sections may exhibit an insufficient angle of attack. Variable pitch propellers enable the control of the entire blade, thereby adjusting the drone to different flight modes.

The effect of the twist angle on the aerodynamic characteristics of the blade is demonstrated in Figs. 4.13b and 4.14 [6]. As illustrated in Fig. 4.14, the local induced inflow ratio of the blade is contingent on its linear twisting along the blade. It is evident that as the value of this ratio decreases, the induced losses also decrease. As demonstrated in the accompanying graphs, for each section of the blade, the appropriate angle of twist can be selected to ensure the minimum value of this ratio.

The blade's geometry is characterized by a pivotal twist, an integral element that enables the propeller to function effectively. This approach enables the utilization of the entire operational area of the blade to generate force, thereby markedly enhancing the aerodynamic perfection and efficiency of the propeller as a whole.

4.3.2.7 Blade Tips

The configuration of the blade tip constitutes a pivotal aspect of the air propeller design, exerting a direct influence on the efficiency, acoustic output, and operational stability of the flying vehicles.

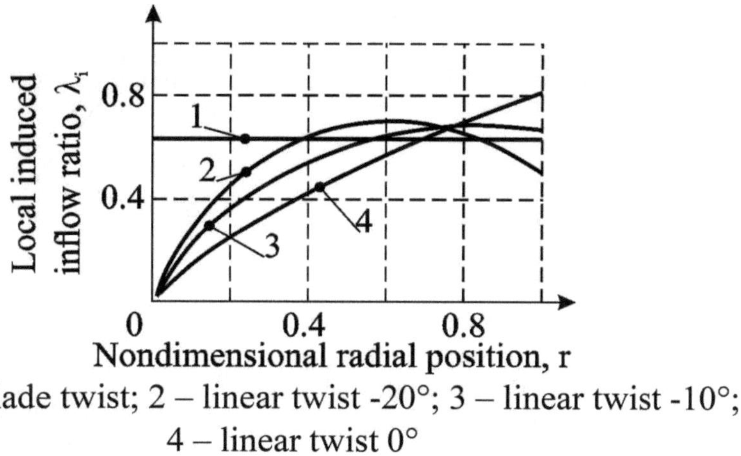

Fig. 4.14 Spanwise distribution of inflow ratio in a rotor blade foe different linear twist rate

The primary objective of the tip is to mitigate the induced losses incurred due to the airflow from the high-pressure area to the low-pressure area of the blade, traversing through the tip of the blade.

Let us look at some of the most commonly used tips (Fig. 4.15a [18]).

Straight/Sheared Tip This type of tip is characterized by its classic, readily manufacturable configuration, wherein the blade chord tips abruptly at the extremity of the blade.

This configuration of the tip generates strong, concentrated tip vortices. This phenomenon results in significant induced losses and a comparatively modest aerodynamic perfection of the blade. The efficiency with which thrust/lift is generated is the lowest among the range of possible options.

Additionally, it generates substantial noise, particularly at elevated rotational velocities, due to the pronounced interaction of vortices with the external airflow.

Historically, such tips have been employed on numerous propellers of older flying vehicles due to their ease of production. This technology has become increasingly prevalent in low-speed and economical drones.

Tapered Tip The blade's taper is characterized by a linear decrease in the chord, with the blade's tip being the point of minimum chord. The pivotal parameter is the sweep angle along the leading or trailing edge.

Tapered Tip with Low Sweep The sweep angle is minimal, typically measuring less than 15°. The shape of the blade is similar to that of a straight tip, but with a smooth taper.

In comparison with a straight tip, a smooth decrease in the chord results in a more gradual change in circulation along the span in the tip section. This

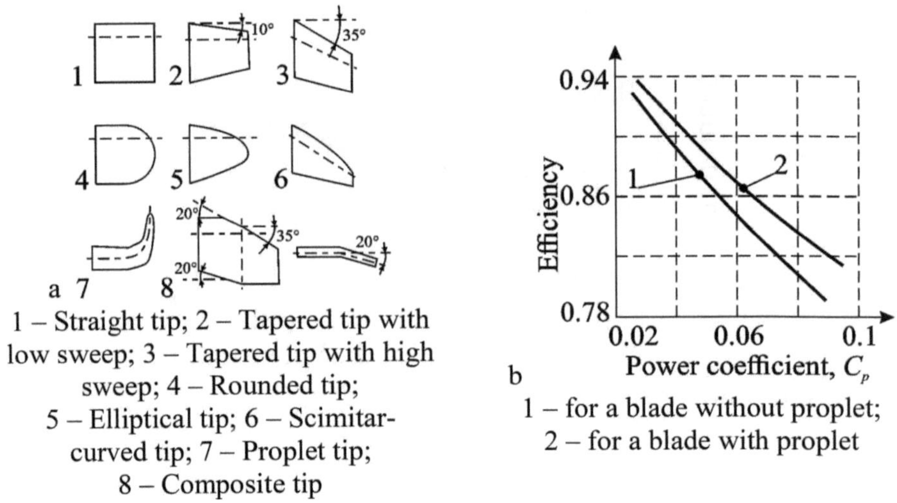

Fig. 4.15 Blade tip configurations [6, 18]: (**a**) blade tip designs; (**b**) efficiency vs. power coefficient

phenomenon results in a reduction in the intensity of the tip vortex, given that the airflow manifests with a diminished "jump" along the chord. This approach has been shown to reduce induced drag and enhance the aerodynamic performance of the blade.

This type of the tip generates less noise than a straight tip due to a less intense vortex, yet more noise than specialized shapes, such as scimitar-curved tip or winglet-shaped designs.

The low sweep has been shown to have a negligible effect on the shock stall. Upon attaining critical Mach numbers, a precipitous surge in wave drag is anticipated at the extremity of the blade.

This component is a prime example of a "workhorse" for helicopter drone rotors. The optimal compromise between ease of production, durability, and aerodynamic perfection at cruising and low flight speeds has been achieved.

Tapered Tip with a High Sweep The sweep angle is of consequence (exceeding 20–25°). The shape of the tip becomes more pronounced.

The phenomenon of weakening the vortex persists and intensifies. However, a purely tapered shape with a high sweep introduces a novel effect: a shift in the aerodynamic center to the trailing edge. This results in a high torque for twisting of the blade, which can lead to aeroelastic vibrations (flutter).

The high sweep results in a substantial increase in the critical Mach number at the tip of the blade. The resultant effect is analogous to that of a swept wing.

In addition to mitigating noise generated by vortices, a high sweep frequency modifies the acoustic spectrum, thereby shifting it to a less perceptible region for the human ear and reducing pulse noise.

The potential applications of this design solution extend to the domain of transport and high-speed drones, where the issue of shock stall control is of paramount importance. The utilization of this technology has also been observed in contemporary helicopter drones, where achieving high flight speeds is imperative.

Rounded Tip In this instance, the blade is characterized by a smooth, rounded tip. The formation of a vortex is distributed over a larger area, resulting in a slight reduction in intensity compared to a straight edge. This results in a marginal reduction in induced drag.

This option generates a lower decibel level than a straight tip, yet it produces more noise than specialized forms.

This configuration is a prevalent choice, particularly in the context of helicopter-style drone rotors and numerous propellers, where reliability and simplicity are paramount.

Elliptical Tip The shape of the blade's tip is congruent with half of the ellipse. While this configuration is aerodynamically optimal for the wing, it is not necessarily the most advantageous solution for the propeller.

The result of this design solution is a distribution of lift along the span, at which point the induced drag is minimal. This configuration is theoretically optimal for minimizing losses due to the formation of vortices.

The manufacturing process is complex and costly, which contributes to the overall expense of the product. The principle underlying this tip has served as a foundational basis for numerous contemporary blade designs.

Scimitar-Curved Tip The tip segment of the blade is canted rearward relative to the direction of rotation.

This solution is regarded as the primary method for enhancing efficiency at high speeds. The aerodynamic configuration of the scimitar-curved shape results in an increase in the critical Mach number, thereby delaying the onset of shock stall and significantly mitigating the occurrence of a sharp increase in drag at high circumferential velocities. It also contributes to the "stretching" and weakening of the tip vortex.

The effectiveness of such tips in reducing noise is particularly pronounced at high flight speeds.

The manufacturing process is complex and costly, which contributes to the overall expense of the product. This type of blades is utilized for the high-speed fixed-wing transport drones and contemporary helicopter drones.

Proplet Tip A small proplet is mounted at the tip of the blade, often at an angle to the plane of rotation.

This configuration is the most effective design solution for suppressing tip vortices. The proplet exerts a force that is oriented in opposition to the twisting moment of the vortex, thereby effectively "destroying" it. This results in a substantial reduction in induced drag and an enhancement in the propeller's efficiency (see Fig. 4.15b).

This results in a substantial increase in the complexity and cost of the blade. Its use has seen a marked increase in modern propellers and the aviation industry, where it is prioritized for its contributions to fuel economy.

Composite Tip This option comprises a combination of several approaches, including a scimitar-curved tip with a small proplet.

This configuration enables the integration of multiple advantages, including the scimitar-curved shape, which delays the shock stall, and the proplet, which counteracts the vortices. This design is the most advanced and effective.

This option is applicable for modern and future developments in the field of propellers and main rotors.

A combined approach is frequently observed. For instance, the main rotor blade is characterized by a rectangular root, a tapered middle section with a low sweep, and a tapered tip with a high sweep or scimitar-curved shape. This configuration enables the integration of the benefits derived from diverse shapes, thereby demonstrating an effective utilization of design principles in the context of rotor blade optimization.

Consequently, the evolution of blade tips has followed a trajectory of increasing complexity, aimed at maximizing the suppression of harmful vortex effects and adapting to increasingly high flight speeds.

4.3.2.8 Blade Airfoil Parameters

A propeller blade is not merely a plate with a constant cross-section. The design of the blade is predicated on compromise, with consideration given to the evolving parameters of the scope. As previously mentioned, twist along the span is a prevalent technique employed in the fabrication of blades (see Fig. 4.16 [6]). The twist, akin to that of a wing, can be geometric and aerodynamic. A thorough examination of the airfoil parameters was conducted in the book section about the wing. In this book section, we will delve into the discussion of which airfoils are deemed suitable for different blade sections along the span.

Root Section (Close to the Hub) The following conditions are characteristic of this section: low circumferential velocity, yet the necessity to generate lift/thrust.

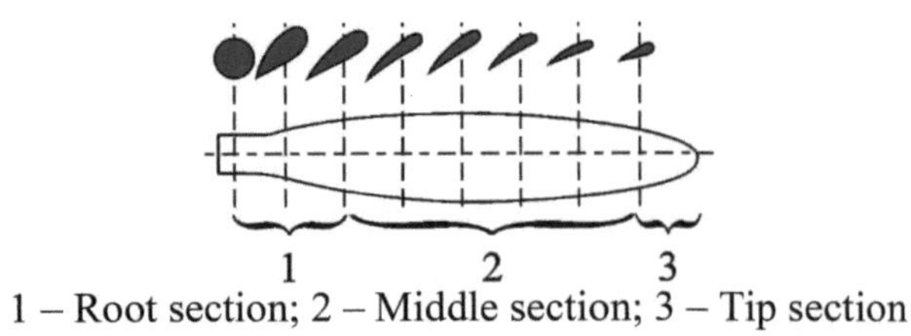

Fig. 4.16 Spanwise airfoils application for a blade of an air propeller

The next airfoil is more suitable: high thickness and chamber. The high $C_{L\max}$ serves to counterbalance the low velocity. The necessity of high thickness is two-fold: it is essential for the purpose of fastening and ensuring durability. Airfoils such as NACA 44xx and 23xx are frequently employed in such applications.

Middle Section The following conditions are characteristic of this section: optimal operational conditions and the main part of the lift/thrust generation.

The following criteria are considered acceptable for the airfoil characteristics: moderate thickness and chamber with maximum aerodynamic perfection (lift-to-drag ratio). Modern laminar airfoils are frequently employed in this context.

Tip Section The operational parameters of the section include a maximum circumferential velocity of 0.8–0.9 M. The probability of a shock stall is high.

The following characteristics are indicative of an acceptable airfoil: a thin airfoil, a low chamber, or even symmetry. The primary objective is to postpone the manifestation of shock stall (supersonic flow zones), that is, to mitigate wave drag. The use of specialized high-speed airfoils, such as the NACA 16-series or contemporary supercritical airfoils, is a key element for this section.

The selection of an aerodynamic airfoil for a propeller blade is not the choice of a single airfoil; rather, it is the creation of a complex structure, where each section of the blade is optimized for specific aerodynamic and strength conditions. Concurrently, the overarching objective entails attaining optimal integrated aerodynamic efficiency across all operational conditions (takeoff, cruising, landing) while ensuring the safety and stability of operations.

4.3.3 Influence of the Fuselage Parameters

As delineated in Chap. 3, UAVs encompass components that extend beyond the mere generation of lift. The aerodynamic properties of drones are profoundly influenced by the fuselage (Fig. 4.17) [1, 2, 7]. As with other aircraft, the fuselage of a drone primarily creates drag in most aerodynamic schemes. The sole exception to this rule pertains to the flying wing scheme or the blended wing body scheme that incorporates the wing and fuselage components. However, for such schemes, the primary design parameters have been previously delineated in the section about the wing.

In this section, the fuselage will be considered as a separate component. Such scheme is currently the most widely used for both drones and conventional aircraft, including airplanes and helicopters.

In this configuration, the fuselage generates a negligible lifting force, which can be disregarded, while exhibiting substantial drag. Consequently, the objective of optimizing the geometric parameters of the fuselage is to minimize its aerodynamic drag. In this instance, it is advisable to draw upon the accumulated knowledge from the background of conventional aviation. To facilitate a comprehensive analysis, it

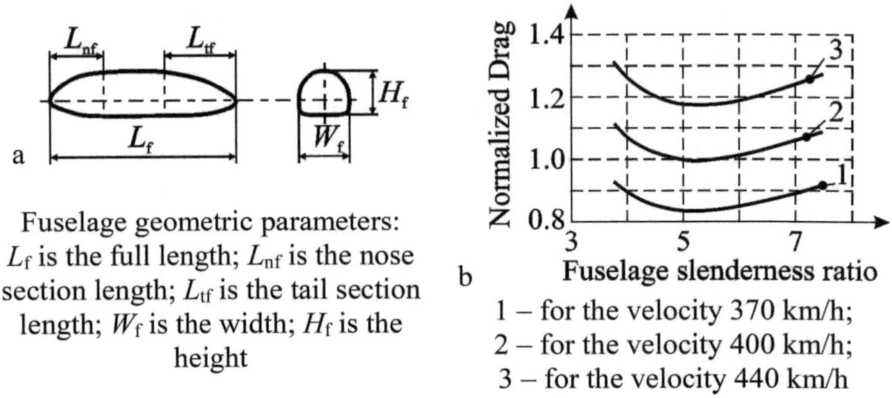

Fig. 4.17 Fuselage parameters [1, 2]: (**a**) fuselage geometry; (**b**) normalized drag vs. slenderness ratio

is imperative to consider the geometric parameters of the fuselage in a manner that differentiates between fixed-wing drones and copter drones.

4.3.3.1 Fuselage for a Fixed-Wing Drone

The geometry of the fuselage of a fixed-wing UAV represents a critical compromise between aerodynamics, payload layout, and stability.

The geometric parameters of the fuselage can be categorized into several groups, as follows:

1. The parameters of the fuselage as a whole are hereby defined.
2. The parameters of the nose and tail parts of the fuselage.
3. Layout parameters.

A thorough examination of each of the aforementioned points is warranted.

The parameters of the fuselage as a whole include length, diameter, length-to-diameter ratio, and cross-sectional shape (see Fig. 4.17a) [2].

The length of the fuselage is defined as the distance from the nose to the tail of the drone. Assuming that all other variables remain constant, an increase in length is associated with a reduction in drag. The elongated fuselage features more gentle angles and a smooth flow, which reduces pressure drag (the main component of fuselage drag). Furthermore, it has been demonstrated to enhance directional stability. The elongated fuselage functions as a stabilizing surface situated behind the center of mass, thereby enhancing stability along the trajectory. Conversely, the incorporation of such a fuselage results in an increase in both mass and moment of inertia, which can have a detrimental effect on the drone's maneuverability.

It is imperative to note that the diameter or width and height are the primary determinants that specify the maximum dimensions of the cross-section. Increasing the diameter results in a substantial increase in drag. The area of the midsection (the largest cross-section) is directly proportional to the pressure resistance ($D \sim S_{mid}$) [1, 7]. This is the most significant factor. Furthermore, longitudinal stability may be compromised due to an increased displacement of the center of pressure of the fuselage.

The combined ratio between length and diameter is the slenderness ratio of the fuselage. Fuselage slenderness is defined as the ratio of length to a characteristic transverse dimension (e.g., diameter or average width) ($\lambda_f = L_f/D_f$) [2].

High slenderness ($\lambda_f > 8$–10) is a typical characteristic of high-speed and high-altitude UAVs. The drone's design, characterized by its smooth contours, has been shown to significantly reduce pressure drag. UAVs with a low slenderness fuselage ($\lambda_f < 5$–6) exhibit high drag, yet a substantial internal volume with a reduced length (see Fig. 4.17b).

The aerodynamic drag of the fuselage is significantly influenced by its cross-sectional shape.

The round or elliptical shape of the fuselage provides optimal aerodynamics when the angle of attack and the sliding velocity are varied. Additionally, it exhibits minimal drag at zero angle of attack.

A rectangular shape with rounded corners has been shown to have suboptimal aerodynamics. However, it possesses a more streamlined layout and manufacturing process.

The configuration of the flat shape provides augmented lift (supporting the fuselage), yet it may concomitantly compromise directional stability.

The dimensions of the nose and tail sections of the fuselage are instrumental in determining the tapering of the fuselage towards the nose and tail.

The length of the nose of the fuselage is equivalent to the distance from the nose section to the midsection. A longer and more pointed nose has been demonstrated to reduce drag during cruising. The reduction in the length and width of the nose simplifies the placement of antennas and cameras; however, it concomitantly increases drag.

The length of the tail is defined as the distance from the midsection to the tail tip of the fuselage. The long tail section plays a critical role in ensuring directional stability. This configuration enables the tail unit to function efficiently by shifting the fin and stabilizer away from the center of mass. A short tail has been demonstrated to result in diminished stability.

The narrowing of the nose section at a sharp angle results in a disruption of the flow and an increase in drag.

A sharp angle of narrowing of the tail section has been shown to cause disruption of the flow and the formation of vortices, which increases drag and causes shaking.

4.3.3.2 The Fuselage for a Copter Scheme Drone

The aerodynamics and geometry of the fuselage for helicopter and multicopter drones differ significantly from those for fixed-wing aircraft [7, 16, 17]. The primary objectives of this field are the optimization of compactness, the mitigation of mutual influence among flows, and the assurance of stability in hovering and transient modes.

These schemes are characterized by a set of common features. It is evident that these are lower speeds in comparison to those of fixed-wing drones. The predominant mode of operation for these UAVs is either hovering or flying at low speeds. The issue of drag aerodynamics is of secondary importance when compared to the problem of vortex flow control.

Another common feature is its compactness. The fuselage's design objective is to optimize compactness around the power plant and payload. However, the design frequently incorporates a frame or components that serve to safeguard the propellers and equipment in the event of contact with obstacles.

Despite the similarities in the underlying principle of creating lift between helicopter and multicopter drones, it is crucial to consider the specific number of propellers when assessing the performance of these vehicles. A multicopter configuration with a high number of propellers exhibits distinctive aerodynamic characteristics.

The helicopter scheme features a fuselage design that evokes the conventional fuselage configuration found in fixed-wing drones. However, it is important to note that there are also specific characteristics that are unique to this drone type. For instance, the capacity to maneuver in diverse directions while maintaining the integrity of the fuselage's orientation. Therefore, it is imperative that the fuselage exhibit as close to aerodynamic characteristics as possible in different flight directions. This consideration assumes particular significance in the context of coaxial propellers.

The length of the fuselage is an important design parameter, particularly in the case of a single-rotor aircraft. A fuselage that is both longer and narrower exhibits reduced drag during forward flight. The primary objective, however, is to position the tail boom with the tail rotor at a sufficient distance from the center of mass to generate a yaw torque (counteracting the torque of the main rotor). A beam of inadequate length will result in the tail rotor's ineffectiveness.

The cross-section of the fuselage is more conducive to drones with elliptical or circular shapes. An ellipsoid or sphere is precisely the shape that enables the equalization of drag for different flight directions.

The configuration of the tail boom, including its shape and cross-section, exerts a degree of influence. A beam with a profiled section (e.g., akin to a wing) has the capacity to generate lift by unloading the main rotor in forward flight, thereby markedly enhancing efficiency. The circle section of the beam is suboptimal in terms of aerodynamics; however, it is more straightforward to manufacture. During the flow around the beam, vortices can be generated (the Karman's vortex wake). In the event that the frequency of generated vortexes corresponds with the natural oscillation frequency of the beam, a powerful vibration will ensue.

The position of the tail rotor is contingent upon the requisite level of efficiency. The tail rotor must operate in a "clean" flow, and not in a slipstream from the fuselage or main rotor. The arrangement of the fin is often the standard practice, while other installations, such as the fenestron, are employed in cases where the geometry of the tail boom is to be altered. The duct serves two primary functions: it protects the propeller and slightly increases its efficiency during hovering.

It is important to note that the correct shape of the upper fuselage surface behind the main gearbox/main rotor hub helps reduce drag. A zone of severe turbulence and rarefaction is formed behind the main rotor hub. A smooth, raised chamber on the fuselage (behind the rotor) facilitates unobstructed flow in this area, thereby reducing drag and preventing stall.

The multicopter scheme is characterized by its unique specifications for the fuselage. In this particular instance, the fuselage often functions as the frame itself, to which the arms with motors are attached. It is possible that a compartment for equipment and payload is situated at the center of the frame. Additionally, the payload is frequently situated within specialized suspensions beneath the central drone body.

The central body should ideally possess a symmetrical shape. In contrast to the helicopter scheme, which possesses a single main rotor, the tail boom is not a single component in this case. This suggests that when moving in different directions, it is possible to equalize the aerodynamic drag. Ellipsoidal or spherical shapes are particularly well-suited for this approach.

The configuration and diameter of the frame in this scheme are of paramount importance. They are crucial not only to reduce the drag of the fuselage, but also to reduce the mutual influence of the propellers. The dimensions are typically determined by the diameter of the circumscribed circle that covers all the propellers (Disk Area). The larger diameter facilitates the use of larger propellers, thereby increasing static thrust and hovering efficiency. This results in a reduction of the rotational speed required to generate the same level of thrust, thereby leading to a decrease in noise levels. However, this configuration results in an increase in size, mass, and moment of inertia, thereby reducing maneuverability. The reduction in diameter enhances the rigidity and maneuverability of the frame, yet it necessitates the utilization of smaller and more rapid propellers, thereby diminishing their efficiency and augmenting noise levels.

The configuration of the arms, including their shape and orientation, is determined by the cross-section, length, and direction of the motors to which they are affixed.

An arm with an aerodynamic shape, or "airfoil section," as it is otherwise known, experiences significantly less drag during forward movement than a circle or rectangular cross-section. This is a critical factor in increasing flight duration and maximum velocity. The turning arms in front results in a reduction of drag during forward movement. The length of the arms is directly proportional to the overall diameter of the frame.

The compactness and clean topology of the upper surface are critical factors in determining the distance at which the fuselage elements (frames) are situated beyond the range of the rotors. Any object positioned above or below the plane of rotation of the propeller will be drawn into its propeller stream (downwash/upwash). This phenomenon is known to result in a decline in efficiency. The propeller is operating in a state of stalling, characterized by a decline in thrust and an escalating rate of power consumption. For instance, the placement of a battery above a propeller can result in a reduction of thrust by up to 10–20%. The "ideal frame" is defined as a through fuselage, wherein the stream is able to pass freely without encountering any obstructions.

The presence of protective fairings is indicative of the enclosure of electronics, batteries, and other onboard equipment. A properly designed fairing has been shown to reduce drag during design flight conditions. The presence of an "incorrect fairing" can exacerbate the already intricate flow patterns, leading to the formation of turbulent zones and the potential destabilization of the drone.

The protective ring on the propeller represents a clear compromise. This configuration has been demonstrated to result in a substantial increase in drag, thereby impeding the airflow through the propeller. However, it has also been shown to offer a protective barrier against collisions with surrounding objects.

A salient attribute is the solidity ratio. This is the ratio of the area of all structural elements (arms, body) to the disk area of the propellers. As the ratio increases, the area under the drone is increased and the loss of interaction with the propeller stream is elevated. Consequently, the overall aerodynamic efficiency decreases.

In the case of most UAVs, the layout is prioritized, resulting in the frequent use of short fuselages with a relatively large diameter. Fairings and the maximum possible smoothing of all edges are employed to reduce drag.

In all schemes, aerodynamic drag at high speeds becomes a limiting factor, which forces engineers to use streamlined shapes, even on compact vehicles.

Fixed-wing drones possess a distinct advantage in terms of flight speed. Consequently, the reduction of drag becomes a paramount concern for these drones, in contrast to drones designed with copter schemes. It is important to note that copter schemes are inherently incapable of achieving high flight speeds. Furthermore, the presence of a large number of propellers can generate additional vortex drag. Consequently, the requirements for fuselage aerodynamics are of secondary importance for these schemes.

4.3.4 Influence of the Landing Gear

In contemporary civilian aircraft, the landing gear is invariably designed to be retractable [2]. Consequently, the impact of the landing gear on the aircraft's aerodynamics is considered negligible. Conversely, a non-retractable landing gear configuration is a prevalent design choice for numerous UAVs. In this section, the impact of non-retractable landing gear and the circumstances necessitating retractability will be analyzed. A thorough analysis is imperative to assess the impact of the landing gear on the aerodynamics of the UAV, and the circumstances under which the retractable landing gear is deemed essential.

The landing gear, a component arranged on the exterior of a flying vehicle, is streamlined by the air flow. It has a multifaceted detrimental impact on the aerodynamic characteristics of an UAV [1, 7, 17]. A substantial increase in profile drag is observed. The landing gear, comprising struts, wheels, and braces, has been observed to generate vortices and turbulence, particularly when the elements possess intricate geometry. Simultaneously, interference drag levels rise. At the juncture where the landing gear is attached to the fuselage or wing. Additional interference drag is observed due to the interaction of aerodynamic flows surrounding the various sub-components of the landing gear.

Consequently, in order to ensure sustained cruising speed and altitude, the power plant must operate with higher power, which directly results in a decrease in range and duration of flight.

The landing gear has been shown to increase the drag coefficient, while the lift coefficient remains unaffected. Consequently, the aerodynamic perfection of the UAV is noticeably diminished. Simultaneously, the drone's ascension is accompanied by a decrease in its maneuverability, a widening of its turning radius, and a diminution in its rate of climb.

Additionally, there is a violation of aerodynamic balancing and stability principles. The landing gear is situated beneath the center of gravity, functioning as a "fin" that extends beneath the fuselage. This configuration has the potential to induce an additional roll or yaw moment, particularly during flight in the presence of a crosswind. Consequently, the autopilot must exert more effort to counteract these disturbances, which, in turn, results in elevated energy expenditure.

The non-retractable landing gear generates high-frequency and low-frequency noise from the incoming flow, thereby unmasking the UAV on an acoustic basis.

The elements protruding from the drone's body serve to increase the effective Radar Cross-section Area (RCS), thereby rendering the UAV more visible on radars, i.e., its degree of invisibility is reduced. This is a salient consideration in the context of military drones.

Notwithstanding the aforementioned disadvantages associated with non-retractable landing gear, this configuration remains the prevailing standard for the majority of drones.

The following discussion will examine the primary criteria that influence the determination of the applicability of retractable and non-retractable landing gear. The following text is intended to provide a comprehensive overview of the subject matter.

Ensuring the range and duration of flight is a critical component of this process. If design calculations demonstrate that a decrease in drag resulting from the retraction of landing gear leads to an increase in range or flight time by 15–20% [7, 17] or more, then the complexity of the design is deemed justified. For high-speed and high-altitude UAVs, this gain can be decisive.

In this analysis, the primary criteria influencing the determination of the suitability of retractable and non-retractable landing gear will be considered.

The subsequent criterion pertains to the velocity of the flight. The magnitude of the drag force is directly proportional to the square of the velocity. At low speeds (up to 200 km/h on average [1, 6, 17]), the effect of additional landing gear drag can be tolerable. However, at speeds ranging from 250 to 300 kilometers per hour, the losses become substantial [1, 6, 17]. For UAVs whose cruising speed exceeds these thresholds, the implementation of a retractable landing gear is nearly obligatorily mandated. It should be noted that this critical velocity level varies according to the

UAV's type. The determination of this level of velocity of feasibility is contingent on the dimensions and mass of the drone in question. For each specific project, it is necessary to conduct independent research.

It is evident that stealth requirements play a pivotal role in the operational capabilities of a specific category of drones. For military reconnaissance and attack UAVs, such as the MQ-9 Reaper, reducing radar and acoustic visibility is a key tactical requirement. The retractable landing gear has been demonstrated to have a substantial impact on RCS, thereby offering a significant reduction in this critical parameter.

An additional criterion that may be considered is the presence of margin energy capacity and weight. In the event that the UAV design has been engineered to include a sufficient margin of safety and engine power to "pull" the additional weight and complexity of the retractable system without a significant reduction in the payload capacity, the option of a retractable landing gear can be considered.

The majority of civilian drones are constructed according to the multicopter scheme. In this particular instance, the velocity of the drone does not exceed 100 kilometers per hour, except for certain specialized drones. In the case of an electric power plant, limitations on battery capacity are also in effect. The issue of excess energy capacity is not a matter of concern. Therefore, it can be concluded that the utilization of retractable landing gear for this particular drone model is not contingent upon any specific prerequisites.

In the case of fixed-wing drones, conducting detailed studies based on the required flight speeds and the capabilities of the power plant is imperative.

The incorporation of retractable landing gear in UAVs is advantageous, particularly for drones that prioritize aerodynamic efficiency, range, speed, and flight duration, as these factors often supersede the disadvantages associated with increased design complexity, weight gain, and cost. For the majority of small and medium-sized UAVs operating at moderate speeds, a non-retractable landing gear is typically the preferred option due to its simplicity, minimal weight, and reliability.

4.4 Conclusions

This chapter is devoted to an exposition of the base concepts and principles of drone aerodynamics. A thorough examination of the designs of drones and their components, along with their effect on the aerodynamic characteristics of the UAV, was conducted. Examples of graphical dependencies that confirm the described patterns are provided.

References

1. Liu P. Aerodynamics. Beijing: Science Press/Springer; 2022. https://doi.org/10.1007/978-981-19-4586-1.
2. Tiniakov D, Li L. Airworthiness engineering basis. Datian Aviation Beijing Beixue Publishing House; 2020.
3. Li L, Kretov A, Tiniakov D. Aeronautics—characteristics and emerging technologies: UAVs and hydrogen technologies are the future of civil aviation. London: IntechOpen Limited; 2024. https://doi.org/10.5772/intechopen.1008352.
4. Mueller T. Aerodynamic measurements at low Reynolds numbers for fixed wing Micro-air vehicles. Development and operation of UAVs for military and civil applications. Belgium: VKI; 1999.
5. Prandtl L. Applications of modern hydrodynamics to aeronautics: report national advisory committee tor aeronautics. 1923:116.
6. Obukhovsky A. The aerodynamics of the propeller. Novosibirsk: Novosibirsk State Technical University Press; 2016.
7. Marques P, Da Ronch A. Advanced UAV aerodynamics, flight stability and control. Hoboken, NJ: Willey; 2017.
8. Kretov A, Tiniakov D. Evaluation of the mass and aerodynamic efficiency of a high aspect ratio wing for prospective passenger aircraft. MDPI Aerospace Issue. 2022;9:497. https://doi.org/10.3390/aerospace9090497.
9. Tiniakov D, Utenkova V. Analysis on tapered wing shapes of partial criteria of their effectiveness. Aerospace technic and technology. Kharkov Natl Aerosp Univ. 2012;96(9):54–60.
10. Tiniakov D. Determination of the geometrical twist angles of trapezoidal wing local chord by the base of the factor growth induced drag. Open information and computer integrated technologies, vol. 59. Kharkov National Aerospace University; 2013. p. 14–20.
11. Bycko N, Gamanukha T, Grebenikov A, Turev V. The effect of wing aspect ratio on the aerodynamic characteristics of a subsonic transport aircraft in cruise flight mode. Open information and computer integrated technologies, vol. 59. Kharkov National Aerospace University; 2006. p. 14–20.
12. Kretov A, Tiniakov D. Evaluation of wing structures at the conceptual stage of transport category aircraft projects. Aviation. 2022;26(4):235–43. https://doi.org/10.3846/aviation.2022.18041.
13. Marshal D, Barnhart R, Shappee E, Most M. Introduction to unmanned aircraft systems. NW: CRS Press; 2015.
14. General Atomics Aeronautical Systems. Predator C Avenger. Powey: General Atomics Aeronautical Systems Inc. 2025. [Internet]. http://www.ga-asi.com/products/aircraft/pdf/Predator_C.pdf. Accessed 15 Sept 2025
15. Mathur A. Drones and counter-drone systems. New Deli: KW Publishers Pvt. Ltd.; 2023.
16. Moiseev V. Unmanned helicopters. Current state and development prospects. Kazan: Shkola; 2019.
17. Bulat P, Dudnikov S, Kuznetsov P. Fundamentals of aerodynamics of unmanned aircraft. Moscow: Sputnik+; 2021.
18. Brocklehurst A, Barakos G. A review of helicopter rotor blade tip shapes. Prog Aerosp Sci. 2012;56:35–74. https://doi.org/10.1016/j.paerosci.2012.06.003.

Chapter 5
Structural Features

5.1 Introduction

In order to execute their designated missions, contemporary UAVs are required to possess elevated mechanical reliability. In order to accomplish this objective, it is necessary to select a load-bearing configuration, structural elements, and material that will ensure the stresses in these elements do not exceed their destructive limits under all possible loads. Furthermore, the drone must be designed in such a way that self-exciting vibrations do not occur, which would inevitably lead to the destruction of the structure. The analysis of strength involves the application of stress and strain calculation methods to classical force elements, including beams, rods, frames, plates, and shells.

In this chapter, an analysis is conducted of the loads acting on the flying vehicles in order to ascertain their designed values and to determine the special methods for the calculation of stresses and deformations in real structures. It is important to note that real structures are usually complex combinations of many bearing elements.

Depending on the specific purpose of the UAV, there is a distinction in not only the design, but also the nature of the operating loads. Consequently, fixed-wing drones are predominantly influenced by transverse loads, while hybrid unmanned aerial vehicles are subject to the effects of both longitudinal and transverse loads. For drones equipped with propellers, vibration loads and dynamic loads on the propellers will be characteristic. It has been demonstrated that discrepancies in operating conditions give rise to a number of additional loading characteristics.

The methods employed to calculate the strength of manned and unmanned aircraft bear a significant degree of similarity. However, the discrepancy in design, the particulars of flight conditions, and ground-based operating modes contribute to the selection of tasks and the methods of their realization.

The methods employed to calculate the strength of complex structures such as the wing and body of an UAV, and blades, especially when operating beyond the

L. Li, D. Tiniakov, *Unmanned Aircraft Design*,
https://doi.org/10.1007/978-981-95-7167-3_5

limit of elasticity, are mathematically cumbersome or insufficiently developed from the point of view of analytical dependencies [1]. The attainment of relatively accurate strength analysis results is only possible through the implementation of computer simulation. The implementation of this objective necessitates the development of a UAV design that has been thoroughly and adequately developed. In order to facilitate comprehension of the interdependencies presented in this chapter, many of the original methods have been substituted with alternative calculation approaches that are less intricate but more visually intuitive. These alternative methods are based on a calculation scheme of relatively elementary nature, which reflects the primary characteristics of the stress state of the structure.

In the context of evaluating the structural integrity of complex systems, it is not always feasible to obtain numerical values that accurately mirror their operational behavior. Consequently, the necessity arises to verify calculations through a combination of simulation and experimental methods. This relationship between calculations and experiments is, in essence, a characteristic dependence of modern engineering tasks.

The calculation examples are hypothetical and serve to illustrate the application of the theory.

5.2 Basic Concept of Structural Analysis

5.2.1 Relation with Other Engineering Stages

Conducting a strength analysis is unfeasible in the absence of knowledge regarding the loads. Furthermore, the intricacies of their analysis are on par with the complexity inherent in strength analysis. The interconnection between strength analysis and other engineering stages is facilitated by external loads. This relationship, which cuts across various scientific disciplines, is illustrated in Fig. 5.1 [2].

Following the selection of the UAV's scheme and preliminary design, the configuration, overall dimensions, and weights of its main parts and compartments have

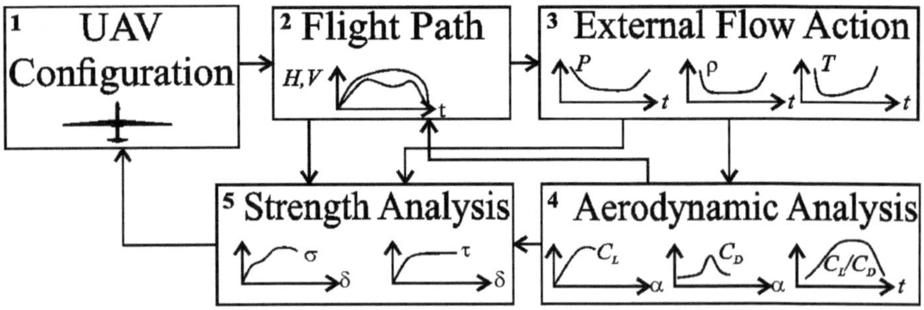

Fig. 5.1 Interconnection between strength analysis and other engineering stages

been determined. In the figure, this design stage is designated with the number 1. Subsequently, the trajectory of movement should be calculated (stage 2). Then, the external flow effects should be analyzed (stage 3). After that, the aerodynamic characteristics of the drone should be determined (stage 4). Finally, the structure should be analyzed for strength and stability (stage 5).

5.2.2 Design and Verification Analysis for Strength

In the process of developing any engineering device, with a particular emphasis on aircraft, two primary strength computational tasks must be addressed: design and verification.

The objective of the design strength analysis is to ascertain the basic dimensions of the structure in accordance with its specified overall dimensions and external loads. Consequently, the design strength analysis yielded the estimated dimensions. It is imperative to acknowledge that prior to the execution of the design analysis, the specific configuration and the overall dimensions of the UAV are only known in a preliminary state.

Conversely, in the verification strength analysis, a fully finished structure of the UAV with a specific appearance is employed, and the dimensions of all its structural elements are already known. The objective of the verification analysis is to evaluate the extent to which the completed structure exhibits desirable levels of weight and strength. At this stage, the factors of safety margin and stability are determined. The verification analysis is based on the most modern and accurate methods used in strength analysis science (for issue, the engineering simulation).

At the stage of the design analysis for strength, the load-bearing layout scheme of the structure is known, but the dimensions of its component parts are unknown or require clarification during many iterations. Therefore, the dimensions of these components are calculated in the following order:

1. The operating conditions are analyzed, and the methods of load determination are selected.
2. The design cases are selected, and the loads acting on the structure in these design cases are determined.
3. The structural components are analyzed to determine the type of loading, and the structural materials for these components are selected or specified.
4. The analyzing of external loads is achieved through the application of safety factors.
5. A systematic approach is formulated, delineating an idealized representation of the design, through the elimination of superfluous factors and the compartmentalization of a multifaceted undertaking into more manageable components. The brief framework underlying the analysis can be delineated as a sequence of steps, as illustrated below:

(a) The initial step in the process is the division of the initial tasks into smaller, more manageable components.
(b) The next step is the simplification of the loading. This can be achieved by neglecting its separate components, reducing various loads to equivalent ones, simplifying the type of load distribution, or replacing a distributed loading with a concentrated one.
(c) The third step is the simplification of the geometric shape of the structure. This is achieved by reducing the structure to one of the following geometric shapes: a rod, a beam, a plate, a shell, or a combination thereof.
(d) The fourth is the idealization of material properties, with isotropy serving as a salient example.

6. According to the specified loadings, the dimensions of the structural components are determined. In complex cases, additional iterations are incorporated into the strength analysis. These iterations occur during the initial setup of the structure's dimensions and configuration. Subsequently, the stress field is determined, and its comparison with the permissible stresses is conducted. The application of strength theories enables the conversion of the stress field into a one-dimensional stress state.
7. The resulting dimensions of the structural components are approximated to the nearest recommended standard dimensions.

   Conversely, in the verification strength analysis, the structure is comprehensively designed and all its features are meticulously delineated, down to the minutest details. All its dimensions are systematically determined and selected in accordance with established requirements and standards. The primary stages of the verification analysis differ from the design analysis by incorporating an additional step.
8. The determination of the real factors of safety margin and stability constitutes a critical component of the verification analysis. In circumstances in which the factors in issues are not readily apparent, these factors are determined through the conduct of tests.
9. A thorough review of the verification and design analyses reveals the paramount importance of reliably determining the external loads acting on the structure.
10. In order to develop a more nuanced understanding of the matter, further consideration must be given to the issue of rationing external loads using the safety factor and the selection of design cases.

5.2.3 Margins of Safety and Safety Factors

In the domain of strength analysis, two primary methods are conventionally employed to assess the bearing capacity of a structure: the method of permissible stresses and the method of breaking loads. Prior to delving into the intricacies of these methodologies, it is imperative to establish a common understanding of the terminology employed.

Unlimited (or unsafe) stresses are those stresses that result in structural fracture or plastic deformation.

Permissible stresses refer to the maximum stresses that are deemed acceptable within a structure, ensuring its reliable and safe operation.

Design stresses are stresses that occur in a structure under the action of loads applied to it.

The method of permissible stresses is widely employed in the field of general mechanical engineering, wherein the minimum weight requirement is of secondary importance when compared to the necessity for safe and durable operation of the structure.

The permissible stresses are a fraction of the elastic limit, σ_{pl}, or yield stress, $\sigma_{0.2}$, which are considered to be the unlimited (unsafe) stresses, i.e.,

$$\sigma_{\text{per}} = \sigma_{0.2} / MS, \tag{5.1}$$

where *MS* is the margin of safety for various structures the range of *MS* falls between 1.5 and 5. The following reasons influence the selection of this factor:

1. The degree of accuracy in the determination of operating loads and the design methods employed.
2. The degree of uniformity of the structural materials utilized, their sensitivity to mechanical treatment, and the extent of dispersion of physical and mechanical properties must be considered.
3. The significance of the specific components.

Consequently, when employing the method of permissible stresses, the structure invariably functions within the limits of elastic deformations, and the design stresses are less than permissible, i.e.

$$\sigma_{\text{d}} \le \sigma_{\text{per}}. \tag{5.2}$$

Given the lack of plastic deformations within the structure, it can be concluded that the safety factors for loads and stresses will remain consistent with the aforementioned method. As illustrated in Fig. 5.2, the stress-strain diagram of the steel sample, this coincidence is discernible on the linear section of the curve, extending up to the limit of proportionality, denoted as σ_{lp}.

Beyond the limit of proportionality, a redistribution of stresses occurs, resulting in a violation of the proportionality between stresses and load. Consequently, within the domain of structural plastic deformation, the stress safety margin does not permit the assessment of safety margin by its loads. In the design of any flying vehicles, the mass factor plays a crucial role. Therefore, the structure is designed so that the strength properties of the structural material are fully utilized. Consequently, the permissible stresses in aeronautical engineering are defined as the proportionality limit or yield strength (if residual deformations are permissible), rather than being

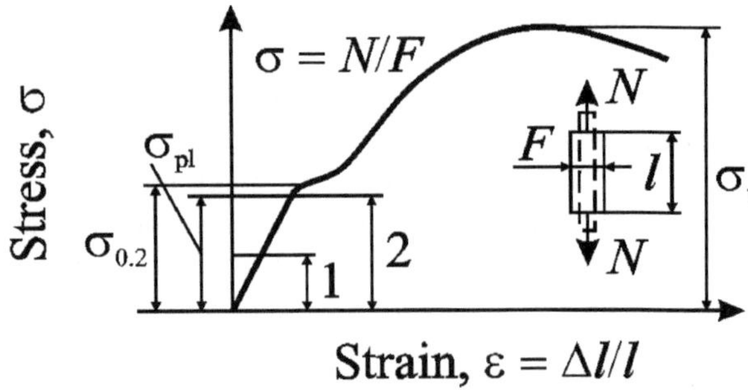

Fig. 5.2 Stress-strain curve for a steel sample: 1—σ_{per} for the method of permissible stresses; 2—σ_{per} for the method of breaking loads

considered as a part of them, as is typically the case in general mechanical engineering.

In this case, the strength margin encompasses the area in which the stress and load factors of the safety margin will differ. Consequently, the design is analyzed based on the breaking loads that most accurately characterize its performance.

The factor of safety margin in the breaking loads method is equal to:

$$MS = \frac{N_{br}}{N_{per}}, \tag{5.3}$$

where N_{per} is the permissible load, i.e., ensuring the occurrence of safe stresses (in this case, the proportionality limit).

Should the concept of design load N_d be introduced, the load acting upon the structure should be understood as such:

$$N_d \leq N_{per} = \frac{N_{br}}{MS}. \tag{5.4}$$

However, it is not feasible to ascertain a breaking load of N_{br} through computational means, except the most elementary scenarios. Consequently, the method for determining breaking loads is as follows.

The factor of safety margin *MS* is hypothesized to be equivalent to one, and the requisite margin is incorporated into the design load, which is subsequently multiplied by the safety factor *f*. To differentiate between the design load that has been estimated in this manner and the load that exerts influence on the structure, the latter is designated as the operational load N_o, i.e., now

$$N_d = f \times N_o. \tag{5.5}$$

This load is employed for the analysis of structural strength through the method of breaking loads.

In the general case, the factor of safety margin can be determined through experimental means subsequent to the determination of N_d. So,

$$MS = \frac{N_{\mathrm{br}}}{N_{\mathrm{d}}} \approx 1, \tag{5.6}$$

however, its value must be no less than 0.98. The role of it is to assess the degree of perfection of the designed structure.

Theoretically, ascertaining the value of the safety factor is quite challenging, as it is impossible to identify all the factors influencing its value. In essence, the safety factor's primary function is to serve as a compensatory tool:

1. There are discrepancies between the deterministic analysis and the actual random representations of external loads, internal forces, and the structural load-bearing capacity.
2. There are some differences in the design and loading conditions of the UAV compared to the actual loadings.

This necessitates the collection of statistical data on safety factors, which will be utilized in the future for the design of new UAVs.

In the following section, an analysis of the typical design cases utilized in the design of various UAV components will be presented.

5.3 Loads Categories

The estimation of loads remains a critical area of analysis, as any error or faulty assumption may result in the UAV exceeding its designed weight capacity or structural failure when confronted with actual loads during flight.

In the context of UAV loads, those associated with high-g maneuvering are particularly salient. It is imperative to acknowledge that maneuvering loads constitute merely a fraction of the total loads that must be accommodated by the UAV structure.

In the ensuing sections, an analysis will be conducted to elucidate the transformation of external loads into various types of internal loads within diverse drone components. In the majority of cases, these internal loadings can be categorized as known types.

However, external loads for different types of UAVs are found to vary significantly in nature. For fixed-wing UAVs, the predominant source of loading is expected to be aerodynamic loads. However, for propeller-driven UAVs, aerodynamic loads will only be relevant for their propellers. In the case of the body of such drones, aerodynamic loads are practically negligible. However, the loads from their propellers, including vibrations, have a significant impact on their strength.

It is imperative to acknowledge that two distinct categories of external loads exist. These loads can be categorized as either static or dynamic. A static load is defined as a constant or slowly changing load. In this case, the deformations and stresses in the bearing element are also constant or change gradually, in proportion to the load, without frequencies processes. Dynamic loads are defined as variable loads that are subject to constant changes in magnitude and direction, occurring at specific frequencies.

Static loads determine the peak strength of a structure, i.e., the maximum load that the structure can withstand before failure. Dynamic loads, on the other hand, determine a structure's fatigue life, defined as the number of cycles the structure can withstand before failure. Finally, resistance to fluctuations is determined by dynamic loads.

Let's consider the design load options for different types of UAVs.

5.3.1 Fixed Wing UAVs

Table 5.1 presents a comprehensive overview of the major load categories encountered by fixed wing UAV [3].

For each structural component of the UAV, one of the loads enumerated in Table 5.1 will predominate. As illustrated in Fig. 5.3 [3], typical critical loads for a drone are shown. It is imperative to acknowledge that the lifting surfaces invariably play a pivotal role in high-g maneuver conditions.

The static loads on a fixed wing UAV encompass aerodynamic, weight, and power loads, which undergo negligible change, thereby enabling the structure to be analyzed for strength and rigidity at a designated "frozen" moment in time (e.g., at the point of maximum overload during the most challenging maneuver).

Dynamic loads are defined as loads that occur during sudden maneuvers, such as a sharp deviation in the control surfaces of an UAV. They can also include sudden gusts of wind (Fig. 5.3), which may be caused by turbulence. Additionally, dynamic

Table 5.1 Fixed wing UAV design loads

Category of load	Type of load	Category of load	Type of load	Category of load	Type of load	Category of load	Type of load
Air loads	Maneuver	Landing	Vertical load factor	Power plant	Thrust	Other	Bird strike
	Gust		Spin-up		Torque		Actuation
	Control deflection		Spring-back		Gyroscopic		Crash
	Component interaction		Crabbed		Vibration		
	Buffet		One strut		Duct pressure		
Inertia loads	Acceleration		Arrested				
	Rotation		Braking				
	Dynamic	Take-off	Catapult				
	Vibration		Aborted				
	Flutter						

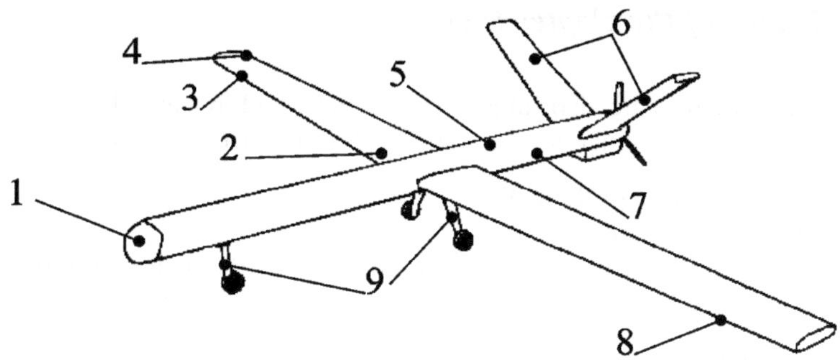

Fig. 5.3 Fixed wing UAV typical critical loads: 1—gust; 2—positive maneuver and static gust; 3—positive dynamic gust; 4—aileron roll; 5—negative maneuver; 6—yaw and pitch maneuvers, and lateral and vertical gust; 7—lateral maneuver; 8—negative gust; 9—landing

loads can arise during landing due to impact on the landing surface (Fig. 5.3), such as through landing gear struts on the fuselage or wing frames. Vibrations from the power plant, flutter, and buffeting are also considered dynamic loads and pose significant hazards to aircraft safety.

5.3.2 Propellers

When assessing the strength of the propeller blade, both static and dynamic loads must be considered. This necessity arises from the fact that the blades constitute the most stressed structural elements. Consequently, it is imperative that their strength guarantees correct operation within the established limits.

The following static loads are some of the most salient [4]:

- The centrifugal forces of the blades, the magnitude and distribution of which along the blade span are contingent upon the aerodynamic characteristics of the airfoil, the set angle, the shape of the planform, and other factors.
- The air load per blade is determined by taking into account the operational load and safety factor.
- The weight of the blade structure.
- The following list of dynamic loads comprises sources of dynamic loads [4]:
- The dynamic interaction between the flow's direction and velocity as it impacts the blades during forward flight is a crucial aspect of this process. This phenomenon results in the oscillatory movement of the blades, which oscillate in both the vertical plane and the plane of turning.
- The variable Coriolis forces resulting from the blade's oscillatory movement act in the plane of turning.
- The effect of wind on non-rotating blades, as observed in a parking lot, is the result of aerodynamic forces that generate bending stresses.

5.3.3 *Frame of the Copter UAV*

Propeller-driven UAVs are also subject to both static and dynamic loads.

The following list enumerates the static loads that must be considered [5]:

- The structural weight of the UAV.
- The weight of the payload.
- The loads from the lifting propellers in hovering mode or in long-term translational flight must be taken into account.

Dynamic loadings encompass a range of forces, including [5]:

- The inertia forces during maneuvers. These forces are particularly salient in drones, which exhibit high maneuverability, thereby rendering maneuverable loads a significant factor in determining strength.
- The engine torque, defined as reactive torque from the engines, exerts a torque that tends to turn the frame in the opposite direction of the propellers' rotation.
- Vibration loads are also a consideration, stemming from the imbalance of the propellers and imperfections of the motors.
- Shock loads, which occur during landing or collision, are another critical factor.
- Thermal impact from heating the components of the power plant is a particular concern for small drones.

5.3.4 *Air Loads*

5.3.4.1 Maneuver Loads

The greatest air loads on an UAV generally result from the generation of lift during high-g maneuvers [3]. It is noteworthy that the structural dimensions of the fuselage are predominantly determined by the lift generated by the wing (or propeller) rather than by the air pressure exerted directly on the fuselage itself.

The flying vehicle load factor n, is defined as the ratio of the UAV's maneuvering acceleration to the standard acceleration due to gravity ($g = 9.81\ m/s^2$). At lower speeds, the maximum load capacity of an UAV is constrained by the maximum lift that can be achieved.

At higher velocities, the maximum load factor is constrained to a certain limit, which is determined based on the anticipated utilization of the drone. The Wright Brothers designed their Flyer to a 5-g limit load. This remains a reasonable limit load factor for many types of aircraft. For aircraft that are operated by a crew of human operators, the primary limitation is that of human health. The aforementioned problem is not a concern in the case of UAVs. However, an additional constraint exists, namely the onboard equipment's sensitivity to overloads. A notable aspect of contemporary electronic equipment pertains to its heightened sensitivity to elevated loads. It is important to note that the required negative load factors are typically much less in magnitude than the positive values.

There is no single mandatory overload requirement for drones. The regulations are of an advisory and procedural nature. The current state of standards for UAVs will be discussed in a separate chapter.

However, based on the extant global experience in drone engineering, several recommendations can be put forth:

1. Recreational and commercial drones (EASA category C1/C2, up to several kilograms) [6, 7]:

   – Maximum load factor: +3.0 to 4.0 g / −1.0 to −1.5 g.
   – Operational load factor: +2.0 to 2.5 g / −0.5 to −1.0 g.

These drones are not designed for abrupt maneuvers.

2. Racing Drones (FPV Racing Drones) [8]

   – Maximum load factor: can reach +10 to 15 g or more.

These drones are subjected to the most extreme loads due to abrupt maneuvers and frequent collisions. The design of racing drones prioritizes the reduction of weight and the enhancement of maintainability, often at the expense of durability.

3. Industrial and commercial UAVs (category C3/C4, up to 25 kg or more) [6, 7]

   – Maximum load factor: +4.0 to 6.0 g / −2.0 to −3.0 g.

   The utilization of UAVs for the purposes of photography, cartography, and monitoring has become a prevalent practice in recent times. These systems necessitate higher reliability indicators.

4. Military and tactical UAVs (MALE, HALE) [9]

   – Ultimate load factor: for Normal Category: +3.8 g / −1.52 g; for more maneuverable: +6.0 g / −3.0 g and above.

   The requirements for these drones are designed to mirror those of manned military aircraft as closely as possible. These UAVs are assigned a maneuverability category (for example, the Normal Category or the Utility Category), as is the case with aircraft.

5. UAVs for cargo delivery [10]

   The loads exhibited a strong dependence on the weight and location of the load (i.e., drone balance). It has been demonstrated that the critical factor in this context is not flight overloads, but rather landing shock loads.

The V-n diagram is a graphical representation of the relationship between the limit load factor of a fixed-wing drone and its airspeed. The V-n diagram of Fig. 5.4 is characteristic of a general fixed-wing drone [3]. It is noteworthy that the maximum lift load factor reaches 1.0 at the point of level-flight stall speed, as would be anticipated. The aircraft is susceptible to stalling at higher speeds when attempting to exceed the available load factor, such as during a steep turn.

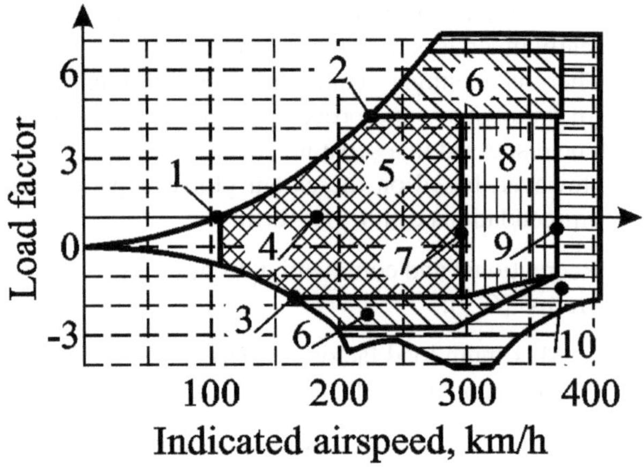

Fig. 5.4 V-n diagram (maneuver): 1—Normal stall speed; 2—Maneuvering speed; 3—Maximum airspeed that allows full maneuverability in a negative lift case; 4—Level flight at 1 g; 5—Normal operating range; 6—Structural damage; 7—Maximum structural cruise speed; 8—Caution range; 9—Never exceed speed; 10—Structural failure

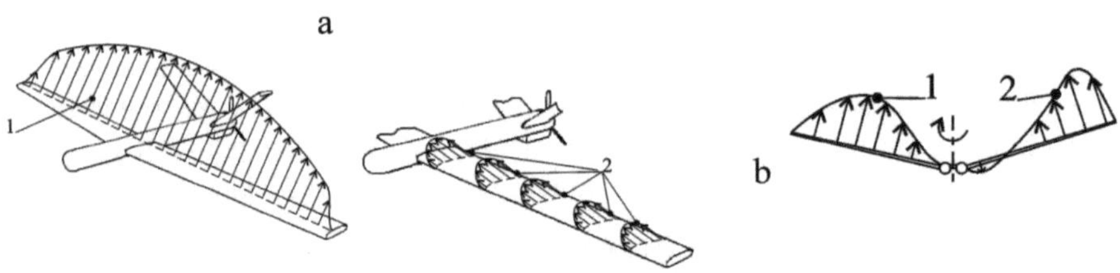

Fig. 5.5 Lift distribution: (**a**) along the wing span and chord; (**b**) along the propeller blade radius

5.3.4.2 Gust Loads

In certain instances, the loads experienced by the UAV when confronted with a strong gust may exceed the maneuver loads. For drones, encountering conditions such as gusts near thunderstorms or high-altitude "clear air turbulence" can result in load factors that range from negative 1.5 to positive 3.5 g or more [6, 7],. When a fixed-wing UAV encounters a gust, the resultant effect is an alteration in the angle of attack.

5.3.4.3 Air Loads on Lifting Surfaces

An initial approximation may be obtained by calculating a simple summation of wing and tail moments about the fixed-wing UAV's center of gravity (c.g.). This calculation should be performed without consideration of effects such as downwash and thrust axis.

Subsequent to ascertaining the total lift on the wing and tail, the spanwise and chordwise load distributions can be determined (see Fig. 5.5a) [3].

Same approximation can be applied and for blades of propellers (see Fig. 5.5b) [11].

The utilization of wind-tunnel and aerodynamic panel program data is contingent upon their availability. In the context of the initial design, classical approximation methods have been shown to yield reasonably good results.

In accordance with classical lift-generating theory, the spanwise lift (or load) distribution along the wing or blade is proportional to the circulation at each span station. The calculation of the spanwise lift distribution is achieved through the implementation of a vortex lifting-line calculation.

5.3.4.4 Air-Loads Due to Control Deflection

The operation of the control surfaces generates air-loads through multiple approaches. The most significant impact is observed in the effect of the elevator on angle of attack and, consequently, load factor. The rudder's influence on yaw angle can impose substantial loads. The deflection of control surfaces generates additional loads directly upon the wing or tail structure.

The term "maneuver speed" or "pullup speed V_p" refers to the maximum velocity at which a ground control station pilot can fully deflect the controls of an UAV without causing damage to either the frame or the controls themselves. For the majority of drones, the maneuver speed is less than the maximum level cruise speed, V_L. Maneuver speed is either established in the design requirements or determined using an empirical relationship.

5.3.5 Inertial Loads

Inertial loads, on the other hand, are defined as the resistance of mass to acceleration. It mathematically expressed as $F = m \times a$, where F is force, m is mass, and a is acceleration. The various accelerations resulting from maneuvers and gusts, as previously outlined, establish the stresses experienced by the aerodynamic surfaces.

The total forces exerted on an object within the UAV are equivalent to the object's mass multiplied by the UAV's load factor. This results in the generation of additional stresses throughout the UAV, which must be determined. It is evident that the weight of the wing (blade) structure will induce torsional loads on the wing (blade), in addition to the aerodynamic torsional loads.

Furthermore, inertial loads resulting from turning maneuvers must be taken into account. For instance, the equipment at the wing tips of a drone rolling at a high rate will experience an outward centrifugal force. The resultant force is equivalent to an outward load factor that is calculated by multiplying the distance from the drone's center of gravity by the square of the turning radius, subsequently dividing this value by the gravity acceleration.

A tangential acceleration force is produced throughout the UAV by a turning acceleration, such as that caused by a gust, an abrupt elevator deflection, or by landing gear strut impact. This force is equivalent to the distance from the UAV's center of gravity multiplied by the angular acceleration, divided by the gravity acceleration.

The loads produced by vibration and flutter are, in essence, acceleration forces of a distinct nature. Adequate design should mitigate flutter and reduce vibrations to an imperceptible level.

5.3.6 Powerplant Loads

It is evident that the engine (motor) mounts must possess the capacity to withstand the thrust of the engine, in addition to its drag when the engine is idling or undergoing gust. Additionally, the mounts are required to provide vertical support for the combined weight of the engine and the design load factor. The engine mounts are typically engineered to support a lateral load that is one-third of the vertical design load. The mounts must be designed to withstand the gyroscopic loads caused by the turning gear (and propeller) at the maximum pitch and yaw rates [12].

For a propeller-driven UAVs equipped with a reciprocating engine, the engine mounts must be capable of withstanding the torque of the engine, multiplied by a safety factor based on the number of cylinders [12]. This phenomenon is particularly evident in engines with a limited number of cylinders, where the failure of a single cylinder can result in a significant increase in engine jerkiness.

For an engine with two cylinders, the safety factor is 4.0; for engines with three cylinders, the safety factor is 3.0; and for engines with four cylinders, the safety factor is 2.0. It has been determined that an engine with five or more cylinders necessitates a safety factor of 1.33. These safety factors are then multiplied by the maximum torque in normal operation to obtain the design torque for the engine mounts.

In the context of jet engines, the air loads within the inlet duct must be taken into consideration, as they frequently constrain a portion of the flight envelope [12]. A pressure surge, known as a "hammershock," is particularly severe.

For an engine mounts system of any type, it is especially important to take into account the inertial loads from maneuvering and landing, since these loads are directly associated with the presence of moving parts in the power plants [12].

Vibration loads from the power plant are also critical for any UAV. The vibrations are caused by imbalances in the propellers, cylinders, bearings, shafts, and other components. The proper balancing of the propellers' blades, in conjunction with the utilization of damping materials such as rubber or silicone anti-vibration pads positioned between the motor and the frame, has been demonstrated to result in a substantial reduction of hazardous vibrations [12].

5.3.7 Landing Loads

During the landing process, the UAV's frame undergoes substantial dynamic loads. In the event that the UAV is equipped with traditional landing gear, a portion of these loads can be mitigated by the landing gear dampers. These components can be utilized as full-featured shock absorbers for heavy drones [12]. In addition to the aforementioned approach, the drone can be equipped with a specialized landing gear strut that possesses the capability of elastically deforming during the landing process. This design component serves to mitigate the impact of the landing, thereby reducing the load exerted on the landing gear and the drone's frame.

In the absence of traditional landing gear—such as in the case of parachute deployment—additional dynamic loads are also present [13]. Initially, the fittings at which the parachute connects to the drone's frame will experience dynamic loads resulting from the parachute's abrupt aerodynamic braking force. Furthermore, upon contact with the landing surface, the drone's frame will experience shock loads.

As demonstrated above, dynamic loads during landing are invariably present, irrespective of the landing method. It is incumbent upon the designer to devise methodologies for the mitigation of these loads, while also incorporating their consideration during the assessment of the drone frame's structural integrity.

To illustrate this point, consider the example of a traditional wheeled landing gear.

To comprehensively assess the range of potential landing gear loads, a series of landing scenarios must be systematically analyzed. The approach encompasses a level landing, a tail-down landing, a one-wheel landing, and a crabbed landing. For the purpose of certification, the aircraft may undergo drop tests, wherein an actual aircraft is dropped from a certain altitude. The requisite drop distance is generally determined to be 3.6 times the square root of the wing loading [12].

At the point of tire-to-ground contact, rotation is absent. During the brief fraction of a second it takes for them to spin up, a large rearward force is exerted by friction with the runway. This spin-up force can account for up to half of the total vertical force exerted during landing.

The load on the retractable landing gear during retraction is typically determined by the sum of the air loads and the assumption that the aircraft is undergoing a two-g turn [12]. While other landing gear loads, such as those associated with taxiing and turning, are typically of lesser importance, they must nevertheless be given full consideration during the detailed design of the landing gear and its supporting structure.

5.4 Loadings for UAVs Parts Analysis

The structural elements belonging to the airframe primarily fulfill load-bearing functions. The present moment is suitable for deliberation on the loads exerted on structures and components.

In this section, a review of typical loads and load cases for an UAV will be conducted. These loads can be classified into two categories: concentrated and distributed. Each of these categories can further be subdivided into static or dynamic loads, as previously defined.

5.4.1 Externally Loaded Airframe

The assessment of flight performance necessitates the delineation of the principal forces that act on an UAV in horizontal steady flight. These forces include lift in equilibrium with weight and thrust in equilibrium with drag. For a comprehensive illustration of these forces, refer to Figs. 4.3 and 5.5 [3]. The aforementioned equilibriums are subject to variation when subjected to various maneuvers; however, they are typically predicated on the major forces.

The airframe is typically subjected to a combination of loads, or load cases, which may be attributed to various factors such as drone maneuvers, gusts, landing, and so forth. The loads exert forces on the structure, either in a concentrated or distributed manner. For instance, the thrust is considered as a concentrated load, while the wing lift or blade lift is represented as a distributed force acting on the wing or blade (Fig. 5.5). In reality, the engines apply the thrust to the fuselage via a limited number of fittings (concentrated loads), while the lift is a force resulting from the multiplication of pressure and the area it acts on (distributed force). The consequences of concentrated or distributed loading on a structure are illustrated by the example of a beam in bending, as shown in Fig. 5.6 [3].

It is imperative to consider the beam illustrated in Fig. 5.6, which has been rigidly clamped at one side and loaded at the other side by either a concentrated load at the tip of the beam or a distributed load decreasing towards the beam's tip. In the event that both load cases are equivalent in terms of the total force *P* applied in the vertical direction, the deflection of the beam for the concentrated and distributed case can be calculated using the following dependencies [3]:

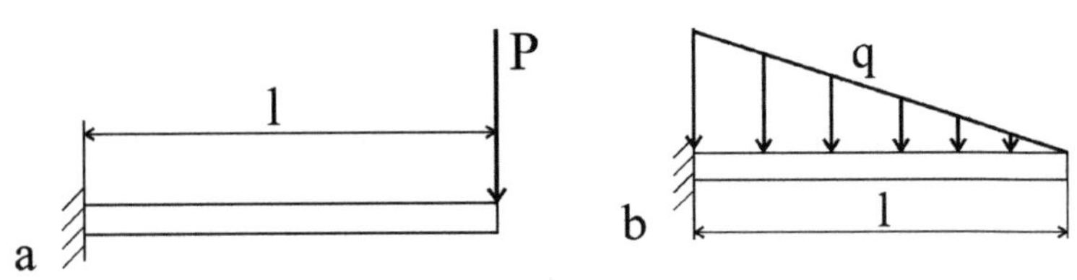

Fig. 5.6 Beam bending induced by a concentrated force (**a**) and a distributed force (**b**)

$$\delta = \frac{P \times l^3}{3 \times E \times I}, \tag{5.7}$$

$$\delta = \frac{q \times l^4}{30 \times E \times I}, \tag{5.8}$$

where P is the total vertical concentrated force, q is the total vertical distributed load, l is the length of the beam (span of the wing section or blade), E is the Young's modulus and I is the moment of inertia.

5.4.2 Load Path

In order to comprehend the significance of loads exerting forces on a specific structure, it is imperative to first delineate the load path. A load path is defined as a physical trajectory that establishes a linkage between the location of the applied force and the forces elsewhere that contribute to equilibrium with the aforementioned applied force. However, the structural components along this trajectory should be capable of carrying and transferring the loads. This phenomenon is illustrated by the action of a point force at a specific location in space, as depicted in Fig. 5.7 [3].

In instances where the force is applied in an upward vertical direction (see Fig. 5.7a), multiple solutions exist to ensure equilibrium. In some cases, a propeller shaft or strut may be adequate for establishing a connection between the point force and the base. In this scenario, the shaft or strut functions as the load path, and the load is applied in a tensile manner. In the second case illustrated in Fig. 5.7b, however, the shaft's stiffness may prove inadequate. The application of a point force on the beam could result in its deflection, thereby compromising the structural integrity of the system. This is particularly crucial for the movable components, such as the shaft. Consequently, equilibrium would not be achieved, and the shaft would be unable to provide the load path. It is imperative that the designer augment the cross-section.

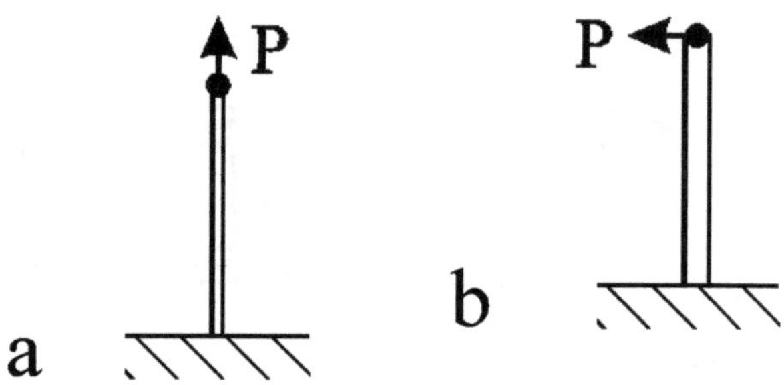

Fig. 5.7 Different load paths for a point force: (**a**) force along the beam axis, zero bending moment; (**b**) force perpendicular to the beam, bending moment arm equal to the beam length

5.4.3 Bending of Beam Structures

It is imperative to consider the deflection of a beam, as illustrated in Fig. 5.8 [3]. The vertical load applied at the free tip of the beam will cause a deflection. Upon analysis of the left-hand side illustration in the provided figure, it becomes evident that the load path comprises normal forces at the upper and lower surfaces (or girders) and shear forces within the web plate.

For both cases 1 and 3 in Fig. 5.8, the load path is in equilibrium with the applied force in the center of the cross-section. However, in the case of (2), the shear forces in the web plate are not aligned with the applied force. The displacement between the center of the web plate (see Fig. 5.8b) and the center of the cross-section will induce a resulting moment. This particular moment is classified as a torsional moment, which results in the beam being twisted together with the vertical deflection.

With regard to the load in the elements of the beam, it has previously been noted that the forces at the clamped side of the beam are substantial. The magnitude of these forces is contingent upon the applied load at the opposing tip and the length L of the beam. The magnitude of the normal forces in the girders at the root is directly proportional to the length of the beam.

The aforementioned equilibrium can be illustrated with Fig. 5.8b case 2. The product of the applied load and the length must be in equilibrium with the product of the normal forces and the height H. That is to say, at the extremity of the beam, the normal forces are zero, and they increase linearly towards the root. By extracting a segment of the girder from any specific point on the beam, it can be deduced that the stability of the system is ensured by the combined effects of the normal forces and shear forces exerted by the web plate on the girder.

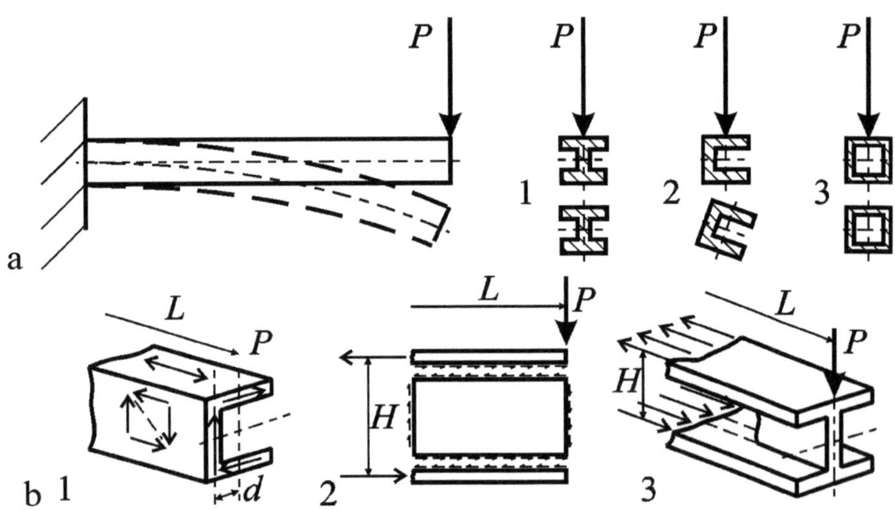

Fig. 5.8 Bending of beams with different cross sections: (**a**) point force applied to the beam; (**b**) internal force distribution for different cross-sectional shapes

Here, it is evidently that the shear forces at the web are constant throughout the web plate. Cutting away a part of the web plate, see Fig. 5.8b case 3, which means that the normal forces in the girders are not increasing in the area where the web plate is removed. Only due to the constant shear forces that the web plate applies to the girders, the normal forces increase. Without the shear forces, the normal forces will be constant.

5.4.4 Loads and Load Paths in a Frame of an UAV

As has been previously documented, the lift generated by the wing will result in an upward bending of the wing. It has been determined that all lift generated by the wings will be in equilibrium with the weight of the wing and fuselage. However, if the UAV is regarded as two intersecting beams, the lift will result in an upward bending of the wing beam, while the weight of the fuselage beam will induce a downward bending over the wing. The deformation is illustrated in Fig. 5.9a [3]. The lift generated by the wings and the weight of the wing and fuselage are considered distributed forces that act on the structure.

Similarly, one can conceptualize the frame of a multicopter (Fig. 5.9b). The propellers generate upward thrust, resulting in an upward bending of the arm beams. Conventionally, the copter's center of mass, which includes its batteries and equipment, is situated at the intersection of the beams. The resultant vector from this weight is directed downwards, thereby causing a downward deflection of the copter arms' beams.

Examples of concentrated loads that are forces applied to drones at the landing gears can be observed in Fig. 5.10 [3]. In the event that the drone is positioned on the designated platform, the forces exerted at the landing gears are found to be in equilibrium with the total weight of the UAV.

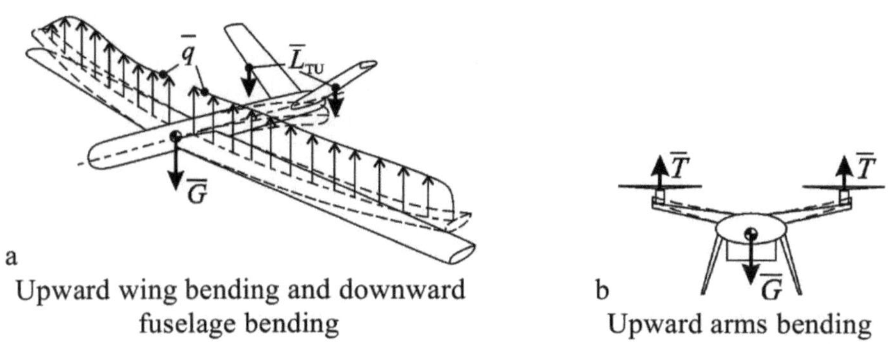

Fig. 5.9 Deformation of UAV structural components under loads: (**a**) fixed-wing UAV; (**b**) multicopter UAV

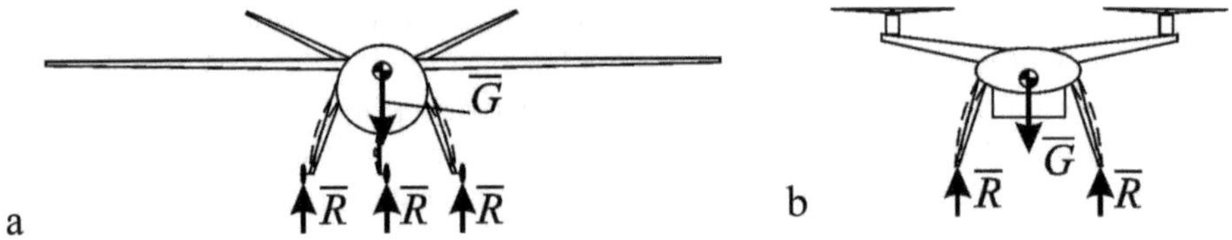

Fig. 5.10 Concentrated forces acting on the frame from the landing gear: (**a**) fixed-wing UAV; (**b**) multicopter UAV

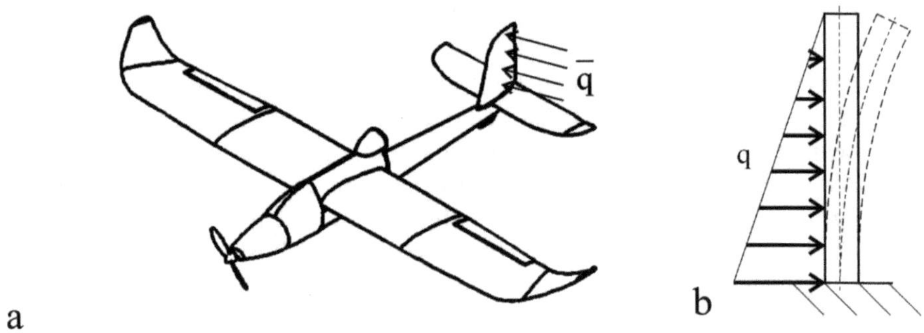

Fig. 5.11 Bending moment on the vertical tail induced by a gust: (**a**) general view of the UAV; (**b**) design scheme

The load path that can be identified in this case commences at the local load on the wheel (Fig. 5.10a), traverses the struts, and terminates at the reinforced bulkhead at that location, subsequently entering the fuselage structure.

As illustrated in Fig. 5.10b, the landing gear's skid type does not feature an additional component between the strut and the reaction force. The force passes directly from the strut to the fuselage.

In the following example, the focus will be on a fixed-wing drone in forward flight. As illustrated in Fig. 5.11 [3], gusts may exert a load on the vertical tail surface. In the event that the fin is considered to be clamped at the fuselage, the deformation may be represented by a beam under bending, as illustrated on the right-hand side of the figure. It is imperative to acknowledge that when the gust is applied to the full surface area of the fin, it is regarded as a distributed load. As the wetted surface decreases in size towards the fin tip, the forces exerted are likely to decrease concomitantly.

Therefore, the forces are in equilibrium by forces at the root of the vertical tail. However, if the fuselage is regarded as a cylinder "clamped" at the wing fittings, the bending moment on the vertical tail will cause the fuselage to turn, as demonstrated in Fig. 5.12 [3].

Upon further analysis of the fuselage structure, particularly with respect to its skin panels, it was observed that these panels would deform in a shearing manner due to torsional forces acting upon the fuselage.

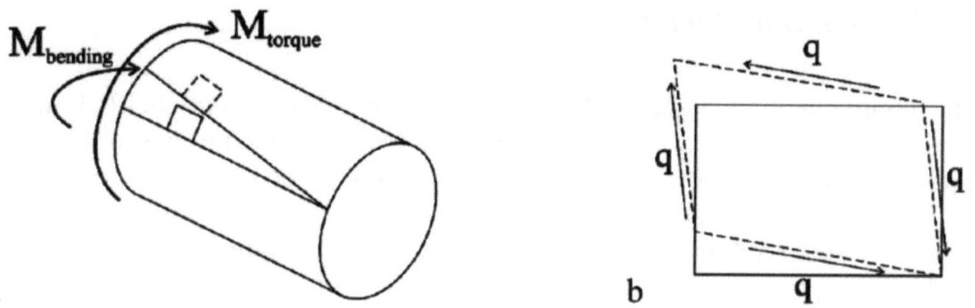

Fig. 5.12 Torsion of the fuselage (**a**) with as result shear deformation of the skin panels (**b**)

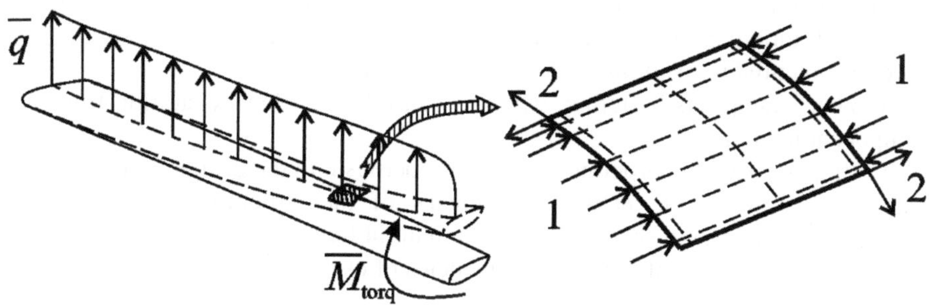

Fig. 5.13 Upward bending and torsion of the wing: 1—Compression; 2—Shear

A similar discussion may be applicable to the bending and torsion of the wing in flight (Fig. 5.13 [3]). Initially, the wing may be regarded as clamped at its root, with bending occurring through the actions of upper and lower wing skins, which are subjected to compression and tension forces, respectively. The torsional loading will be carried by shear in the torsion box, i.e., both wing skins and both web plates of the spars.

A similar approach can be employed in the analysis of blade loads in UAVs with propellers. The distribution of the air load is illustrated in Fig. 5.5b. In the wing, the bending and torsion moments, as well as the shearing force, must be perceived by the spars. In contrast to the configuration of a wing, the blade's limited space precludes the accommodation of multiple spars. Consequently, contemporary blades generally possess a single D-shaped spar or are simply composed of a solid structure. The components of this spar are designed to bear all the loads that are acting on the blade. Through this spar, the loads are transferred to the blade's fitting with the shaft. The design of this fitting may vary depending on the size of the propeller, and thus, on the drone model.

5.4.5 Dynamic Loads

The discussion of loads and load paths in this chapter has been implicitly limited to quasi-static loading. That is to say, an external load is applied to the structure, and equilibrium must be provided. However, it should be noted that not all loads within an UAV are quasi-static. It has been observed that a proportion of the loading appears to be of a dynamic nature. Despite the fact that fatigue pertains to cyclic loading, it is predominantly associated with quasi-static loading.

In certain instances, the interaction of aerodynamic forces, structural elastic reactions, and the inertia of components may induce oscillations of UAV components. These oscillations have the potential to be both significant and severe. As a case in point, these oscillations may manifest as flutter of the wing or blade. In the absence of anticipation regarding the methods of their dumping, these oscillations have the potential to result in the destruction of components.

Another example is buffeting, which is caused by airflow separation from one component to another.

It is imperative to incorporate damping properties into the design, as oscillations may amplify when the natural frequencies of the structure are approached. Furthermore, the design should account for natural frequencies and attempt to keep them outside the range of potential frequencies of oscillations.

5.5 Structural Materials Selection

Prior to the design stage, designers typically have numerous inquiries regarding structural materials, particularly with regard to their selection. Which material would be more rational for the specified purpose? The inquiry pertains to the presence of a particular material experience. The objective of this book section is to discuss the rational selection of materials from the point of view of their strength performance.

A series of inquiries must be addressed to ensure the adequacy of the design. Initially, the UAV design is distinguished by its adaptability and the utilization of a diverse array of materials. It is imperative to consider the factors previously discussed, including the element's shape and dimension, its value, and the type of loads it is subjected to.

However, it is imperative to acknowledge that the selection of structural material is not solely dictated by its maximal strength characteristics. The authors of the book frequently engaged in communication with aircraft manufacturers, including those specializing in UAVs. In numerous instances, we encountered circumstances in which extraneous factors, such as economic considerations, exerted a more substantial influence. However, the issue at hand lies beyond the scope of this particular book.

Figure 5.14 presents a common diagram that assists in comprehending the fundamental principles of structural materials selection [3].

5.5.1 Classes of Engineering Material

The conventional classification of materials of engineering is typically divided into six broad categories, as illustrated in Fig. 5.15 [3]: metals, polymers, elastomers, ceramics, glasses, and composites. The members of a class are defined by shared characteristics, properties, and processing routes. These characteristics and properties are often utilized in similar applications.

Structural metals possess properties that are, in comparison, high. The application of specific alloys, in conjunction with mechanical and heat treatment, has been demonstrated to enhance their strength. However, they maintain their own plasticity, a property that enables them to be shaped by deformation technology. Typically, high-strength alloys exhibit a plasticity level of approximately 2%. However, it is sufficient to provide the metal yields prior to the occurrence of breakage. This

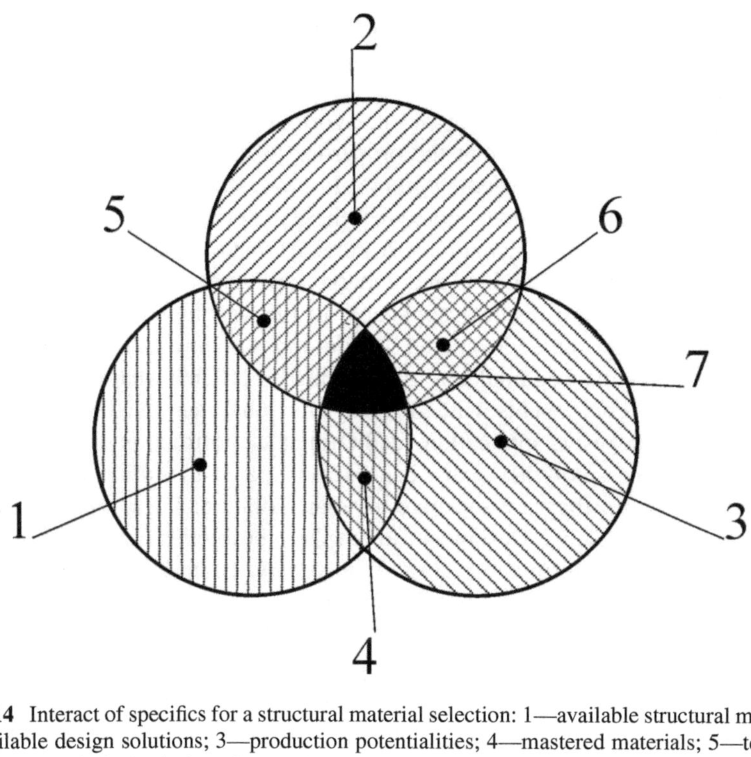

Fig. 5.14 Interact of specifics for a structural material selection: 1—available structural materials; 2—available design solutions; 3—production potentialities; 4—mastered materials; 5—technical solution; 6—technical solutions that can be manufactured; 7—reasonable design solutions

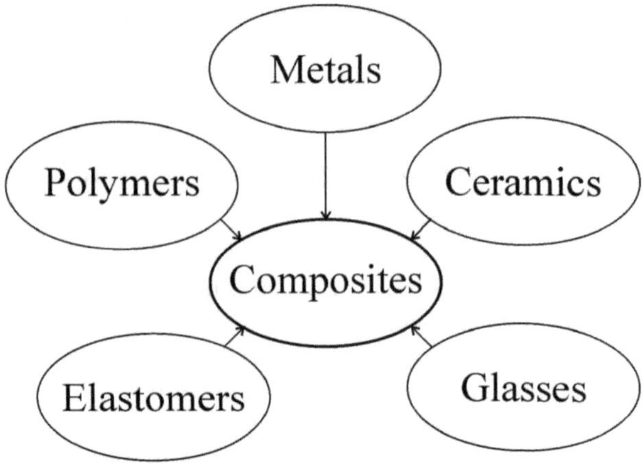

Fig. 5.15 Types of engineering materials

sensitivity to fatigue is partly attributable to their limited plasticity. In addition, all metals exhibit inadequate resistance to corrosion.

Ceramics and glasses possess certain advantageous properties; however, they are susceptible to breakage, a quality that differentiates them from metals. Furthermore, their low plasticity renders them vulnerable to stress concentrations, such as holes or cracks. In consideration of the aforementioned characteristics, in conjunction with the operational specifics of UAVs (a significant number of which are utilized in unprepared conditions), the application of these materials is inherently constrained.

Polymers and elastomers are distinct categories of structural materials [3, 14]. A comparative analysis reveals that the properties in question exhibit lower values than those observed in the preceding groups. In contrast, they exhibit optimal elastic properties. However, it should be noted that these materials exhibit a high degree of sensitivity to temperature fluctuations. A polymer that exhibits flexibility at 20 °C may demonstrate fragility at 4 °C and undergo rapid creep at 100 °C. In the event that these rationales are deemed applicable to the structural design, the benefits inherent to polymers can be leveraged. In addition to their elastic properties, these materials exhibit enhanced malleability, allowing for efficient shaping. The integration of polymer components with distinct functionalities can be accomplished through a single adhesive application, facilitating the construction of intricate three-dimensional shapes. Polymers have been demonstrated to exhibit high corrosion resistance; however, they are sensitive to solar radiation. These materials exhibit low frictional coefficients. Designers can use these properties rationally.

Composites are a formidable class of materials, comprising fibers and resin. These materials exhibit a combination of properties characteristic of diverse material groups. These materials possess low density, high rigidity, and strength. The composites are composed of a polymer matrix, which can include epoxy, polyester, and other materials. This matrix is reinforced by fibers from glass, boron, carbon, Kevlar, or metal wire. The cost of composite parts is attributable to the complexity of their fabrication and assembly. Engineers will apply them only when their properties can compensate their high cost.

5.5.2 *Common Requirements to Structural Materials*

The judicious selection of materials for members of structures can result in a substantial enhancement of the flight characteristics of UAVs and their profitability when in operation.

The basic requirements for structural materials:

1. The mechanical and physical properties of a structural material should provide strength and rigidity to a structure, while minimizing its mass, under both operational conditions and extreme temperatures.
2. The material should be adaptable for manufacturing.
3. It should have resistance to corrosion under atmospheric conditions.
4. It should have stable physical and mechanical properties.
5. The cost of the material should be relatively low.

5.5.3 *Several Structural Materials Characteristics*

5.5.3.1 Aluminum Alloys

The aerospace industry has long utilized alloys of aluminum [3]. The alloys exhibit a combination of qualities, including a low density, a high degree of durability, and a high degree of strength. Aluminum is often alloyed with other elements, including magnesium, coper, zinc, and silicon, to enhance its mechanical properties.

7075 aluminum alloys are among the most frequently utilized due to their high strength, which is attributed to the presence of zinc as the primary alloying element [15]. The material was utilized in the fabrication of the fuselage and wings.

Another frequently utilized alloy is 2024, which is renowned for its exceptional fatigue resistance, as evidenced by numerous studies [16].

Alloys such as 6061 and 6063 offer a superior balance of strength, corrosion resistance, and efficiency.

Some aluminum alloys properties are in the Table 5.2 [3, 14, 16].

5.5.3.2 Steel Alloys

Usually, steel alloys are not the main structural materials used in the construction of UAVs, as other materials, such as aluminum, titanium, and composites, are typically preferred. However, steel alloys still have an important place due to their strength,

Table 5.2 Some aluminum alloys and their strength indexes

Alloy	Additive	Ultimate strength (MPa)	Yield strength (MPa)	Modulus of elasticity (GPa)
2024	Copper	483	345	73
5052	Magnesium, chromium	262	214	70.3
6061	Magnesium, chromium	310	276	68.9
6063	Magnesium	290	269	69
7068	Zinc	683	655	73.1
7075	Zinc	559	496	73.1

Table 5.3 Some steel alloys and their strength indexes

Alloy	Additive	Ultimate strength (MPa)	Yield strength (MPa)	Modulus of elasticity (GPa)
301	Nickel, chromium	1275	965	193
410	Chromium	1393	1076	200
13–8	Nickel, chromium	1517	1414	200

durability, and cost-effectiveness. Some of the most commonly used steel alloys for producing UAVs are [3, 14]:

- 301 (stainless steel): A lower-cost alternative with nickel as the other element. This combination highlights properties such as formability and corrosion resistance, making it suitable for molding and trimming.
- 410 (Stainless Steel): These chromium stainless steel alloys have different carbon levels that influence properties such as high strength, good ductility, and heat resistance.
- 13-8 (stainless steel): These alloys have high strength, excellent fatigue properties, and corrosion-cracking resistance.

Some steel alloys properties are in the Table 5.3 [3, 14].

5.5.3.3 Titanium Alloys

Titanium alloys are an excellent choice for producing drones due to their exceptional properties, such as a high strength-to-weight ratio, superior corrosion resistance, and high temperature tolerance. Compared to aluminum and steel alloys, titanium alloys have higher strength and temperature stability, and are half the weight of most steels [3, 14]. However, they are much more expensive than steel and aluminum alloys. Some of the alloys used to construct UAV structures are [14]:

- Ti-6Al-4 V is the most widely used titanium alloy in the aviation industry. It has a good combination of mechanical properties and a wide operational temperature range. It is also highly weldable.
- Ti-54 M has superior superplastic forming properties compared to other titanium alloys, allowing it to be formed into complex shapes.
- Ti-6-2-4-2 is known for its strength and resistance to oxidation at high temperatures, making it ideal for applications like engine parts.

Some titanium alloys properties are in the Table 5.4 [3, 14].

5.5.3.4 Composite Materials

1. Carbon fibers

   Carbon fibers have perfect engineering properties, making them ideal for use in drone structures [17, 18]. When combined with resin to form a composite, the resulting structure is light and rigid, and is also known as a carbon fiber reinforced polymer (CFRP). CFRPs are advanced materials that are composed of thin, strong carbon crystalline fibers weaved together. One of the main advantages of CFRPs is their high strength-to-weight ratio. The most popular combination in industry is epoxy resin and woven carbon fiber. The crucial properties mentioned in Table 5.5 [17, 18] allow the structure to be light without compromising structural integrity for UAV producing. CFRPs also boast excellent corrosion resistance and can withstand extreme temperatures. CFRPs can be molded into complex shapes, offering great flexibility in creating aerodynamic and efficient structures. The main problem with all composite materials is their high cost. The production process is complex and dangerous to the environment.
2. Glass Fibers

   Glass fibers play an important role in the manufacturing of unmanned vehicles due to their high properties [17, 18]. The main advantage of using glass composite to manufacture drones is its perfect strength-to-weight ratio. Glass fiber shows high resistance to wear and corrosion, ensuring the longevity and performance of UAVs and reducing maintenance costs and extending service life. Another beneficial property of glass fiber is its thermal insulation, which allows the drone structure to withstand extreme temperature conditions without damaging sensitive components. The type of glass fiber most commonly used in UAV manufacturing is E-glass [17, 18]. E-glass is usually composed of

Table 5.4 Some titanium alloys and their strength indexes

Alloys	Additive	Ultimate strength (MPa)	Yield strength (MPa)	Modulus of elasticity (GPa)
Ti-6Al-4 V	Aluminum	1137	1068	122
Ti-54 M	Aluminum	925	845	105
Ti-6-2-4-2	Aluminum	1004	895	114

Table 5.5 Average properties of the CFRPs

Property	Value
Tensile strength	3000–6000 MPa
Shear strength	90 MPa
Compressive strength	570 MPa
Density	1600 kg/m^3
Shear modulus	10 GPa
Modulus of elasticity	230 GPa

Table 5.6 Average properties of the glass fibers

Property	Value
Tensile strength	2000–5000 MPa
Shear strength	27 MPa
Compressive strength	90 MPa
Density	2600 kg/m^3
Shear modulus	20 GPa
Modulus of elasticity	72 GPa

silica-based glass with boron oxide and alumina, providing a combination of mechanical properties, as shown in Table 5.6 [17, 18] below.

The choice of material for the UAV's structure is a pivotal decision that affects the UAV's strength, rigidity, performance, efficiency, and durability. The most widely used materials are metal alloys (such as, aluminum or titanium alloys) and composite materials (carbon or glass fibers), each of which has its own list of advantages and disadvantages.

5.6 Conclusions

This chapter is devoted to an exposition of the base concepts and principles of drone strength analysis. A thorough demonstration of the drones and their components structural features, along with their bear-loading structural schemes, was conducted. The chapter also discussed the basic principles for determining the external loads acting on UAVs. Examples of structural schemes that confirm the described patterns are provided. Structural materials are the basis for providing strength and rigidity. This chapter outlines the fundamental principles for selecting structural materials. It also provides brief characteristics of materials most applicable to UAV design.

References

1. Fomichev P., Zarutskii A., Mandzyuk S. The strength analysis of the aircraft, Kharkov: National Aerospace University, 2017.
2. Pogorelov V. Loadings and heating of UAVs, St. Petersburg: BSTU. 2009.
3. Tiniakov D, Li L. Airworthiness engineering basis. Datian Aviation Beijing Beixue Publishing House; 2020.
4. Motvila S. Strength analysis of rotor blades and hubs. Kharkov: Kharkov Aviation Engineering Institute; 1987.
5. Kumar A. Structural optimization of frame of the multi-rotor unmanned aerial vehicle through computational structural analysis. J Phys. 2021;1849:012004. https://doi.org/10.1088/1742-6596/1849/1/012004.
6. EASA Delegated Regulation (EU) 2019/945. [Internet]. https://www.easa.europa.eu/en/document-library/regulations/commission-delegated-regulation-eu-2019945. Accessed 24 Sept 2025
7. EASA Implementing Regulation (EU) 2019/947. [Internet]. https://www.easa.europa.eu/en/document-library/regulations/commission-implementing-regulation-eu-2019947. Accessed 24 Sept 2025
8. Smeur E, Yue N, Liu R. Peregrine racing drone. Final report. Delft: TUDelft; 2021.
9. Oliver D, Money A. Unmanned areal vehicles roadmap 2000–2025. Department of Defense's; 2001. [Internet]. https://www.govinfo.gov/content/pkg/GOVPUB-D-PURL-gpo140963/pdf/GOVPUB-D-PURL-gpo140963.pdf. Accessed 24 Sept 2025
10. Govindarajan B, Sridharan A. Evaluation of UAV configurations for package delivery missions though conceptual design. Warsaw: European Rotorcraft Forum; 2019.
11. Leishman G. Introduction to aerospace flight vehicles. Daytona Beach: Embry-Riddle Aeronautical University; 2025. [Internet]. https://eaglepubs.erau.edu/introductiontoaerospaceflightvehicles/. Accessed 19 Sept 2025
12. Tiniakov D, Li L, Su Y. Airworthiness design of civil aircraft system and structures. Datian Aviation Beijing Beixue Publishing House; 2018.
13. Wyllie T. Parachute recovery for UAV systems. Aircr Eng Aerosp Technol. 2001;73(6):542–51. https://doi.org/10.1108/00022660110696696.
14. Ismail A, Dahalan W, Ochsner A. Advanced structured materials. Cham: Springer; 2021. https://doi.org/10.1007/978-3-030-89988-2.
15. Merati A, Eastaugh G. Determination of fatigue related discontinuity state of 7000 series of aerospace aluminum alloys. Eng Fail Anal. 2007;14(4):673–85. https://doi.org/10.1016/j.engfailanal.2006.02.016.
16. Banerjee S, Robi P, Srinivasan A, Lakavath P. Effect of trace additions of Sn on microstructure and mechanical properties of Al-Cu-Mg alloys. Mater Des. 2010;31(8):4007–15. https://doi.org/10.1016/j.matdes.2010.03.012.
17. Li L., Tiniakov D. Airworthiness design of composite structures, Datian Aviation Beijing Beixue Publishing House. 2021.
18. Chen C, Randall R, Tiniakov D. Fiber-reinforced composites—recent advances, new perspectives and applications. In: Composite materials: application specifics for transport category aircraft wings. London: IntechOpen Limited; 2024. https://doi.org/10.5772/intechopen.1004217.

Chapter 6
Take-off and Landing

6.1 Introduction

Takeoff and landing are critical and complex phases of flight for all aircraft, including UAVs [1]. The specific problems depend heavily on the UAV's aerodynamic configuration [1–3].

Modern UAVs come in a variety of sizes and configurations [4]. However, most UAVs are small. This feature, along with other drone properties, creates certain problems during takeoff and landing. This type of flying vehicle can use non-traditional takeoff and landing methods because of its small size and lack of a human on board [3, 5]. For instance, a special acceleration catapult can be used for takeoff [6]. Similarly, traditional landing gear is not necessary for landing; a parachute landing method can be used instead [7]. This chapter will discuss these issues and other features of UAV takeoff and landing.

6.2 Basic Concept of Takeoff and Landing

From an engineering standpoint, takeoff and landing are the processes by which a UAV transitions from one dynamic state to another.

Takeoff is the point at which an aircraft transitions from a state of zero motion relative to the ground—that is, a state of zero kinetic and certain potential energies—to a state of steady flight with a given velocity and altitude. Landing is the reverse process, transitioning from steady flight to zero motion on the ground [1, 3].

Both processes are characterized by continuous changes in motion parameters (e.g., velocity, altitude, and angle of attack), operation of drone systems in transient modes, and high control accuracy requirements [2, 3].

L. Li, D. Tiniakov, *Unmanned Aircraft Design*,
https://doi.org/10.1007/978-981-95-7167-3_6

6.3 UAVs Take-off Process

The main goal of takeoff analysis is to ensure the UAV accelerates safely and effectively from velocity $V = 0$ to a velocity exceeding the stall speed (V_{stall}) by a given amount, followed by a climb [3].

6.3.1 Fixed-Wing Drones

For fixed-wing UAVs with traditional landing gear, the equation of motion during takeoff (when projected onto the horizontal axis) is as follows [2, 3]:

$$m \times \frac{dV}{dt} = T - D - F_{fr}, \tag{6.1}$$

where m is a mass of the drone; dV/dt is an acceleration; T is a powerplant thrust; D is a drag; F_{fr} is a friction force.

The lift L acts on the run-up, which increases with speed. This decreases the friction force, facilitating takeoff.

$$F_{fr} = f \times (m \times g - L), \tag{6.2}$$

where f is a friction coefficient.

Classical airworthiness rules do not apply to UAVs. However, the logic and principles behind these rules are often adapted and applied to the design and operation of UAVs. Therefore, it is possible to draw a parallel between the takeoff process of commercial aircraft and that of UAVs. Based on this, the following key parameters can be identified during a UAV takeoff [3]:

1. The takeoff distance is the total distance from the start of takeoff until the front wheels separate from the ground or climb 10.7 m, whichever comes first, for certification purposes. It depends on:
   - The mass of the UAV; the greater the mass, the longer the takeoff.
   - The thrust-to-weight ratio; the higher the thrust-to-weight ratio, the shorter the takeoff.
   - The lift-to-drag ratio in takeoff mode; a higher C_L with a lower C_D reduces the takeoff.
   - Other external conditions include altitude, air density, temperature, wind, slope, and runway conditions. For example, a headwind increases the takeoff distance.
2. Takeoff speeds are classified by stages [2, 3]:

- The speed of decision-making V_1. For an aircraft, this is the maximum speed at which the pilot can abort takeoff and come to a stop within the remaining length of the runway (speed brake). There is no direct V_1 equivalent for UAVs. However, autopilot algorithms may include decision-making logic based on other parameters. For example: "If the speed is higher than X m/s and the engine/system is operating normally, continue takeoff," or "If the speed is higher than Y m/s and a failure is detected, perform an emergency procedure: stop acceleration and slow down if possible, or perform an emergency takeoff followed by an emergency landing." Therefore, the decision is not made by a pilot in real time, but rather by a developer during the programming stage.
- The rotation speed of the nose wheel V_R. For an aircraft, this is the speed at which the pilot begins to deflect the elevator to liftoff the nose strut. For UAVs, this parameter often exists but is implemented automatically. When an UAV takes off from a runway, the autopilot is programmed to perform the command "rotate" at a certain speed. This parameter is critical to preventing premature or delayed takeoff.
- Takeoff safety speed V_2. For an aircraft, this is the minimum speed at which it can safely take off from an altitude of 10.7 m. For UAVs, the concept of safe speed is paramount. It is the minimum speed at which a UAV can steadily fly and gain altitude with a given margin. It is determined during the design process and is one of the primary target speeds for the autopilot during takeoff. However, referencing an altitude of 10.7 m is not mandatory and depends on the particular UAV's requirements.

3. The altitude in 10.7 m. For a manned aircraft, this is the standard altitude at which the speed V_2 must be achieved. However, this is not a standard value for UAVs. The altitude at which a UAV transitions from takeoff to climb or flight along the route depends on its flight mission, the terrain, and safety requirements. For a small UAV, it can be 20–30 m; for a heavy one, it can be higher [3].
4. The angle of attack is a critical parameter. Once V_R is reached, the optimal angle of attack is achieved, which provides maximum lift with minimal resistance to liftoff [2, 3].

The takeoff of a fixed-wing UAV can be divided into the following stages:

1. The takeoff run involves accelerating along the runway while increasing lift.
2. Liftoff occurs when the lift exceeds the weight of the aircraft ($L > G$).
3. Holding and climbing involves flying at a constant speed V_2 until reaching a safe altitude.
4. High-lift device retraction involves sequentially retracting the flaps and landing gear to reduce drag and accelerate.

6.3.2 Propelled Drones

Vertical takeoff UAVs (VTOL UAVs) include helicopter, multicopter and tiltrotor configurations, the equations of motion differ greatly from those of fixed-wing drones [3].

Simplifications of this type of UAV are acceptable, including the following:

- Takeoff in calm conditions without wind is assumed.
- The device is considered to be symmetrical, with its center of mass coinciding with its geometric center.
- The propellers are rigidly fixed, and the thrust axis is always perpendicular to the plane of the device.

The drone's movement in the earth's inertial coordinate system is what we will consider. In this system, the z axis points upward. In these conditions, takeoff is mainly a vertical movement. The equation is written in projection on the z-axis [3]:

$$m \times \frac{\mathrm{d}^2 z}{\mathrm{d}t^2} = \Sigma T - m \times g - F_D, \tag{6.3}$$

where $\mathrm{d}^2z/\mathrm{d}t^2$ is a vertical acceleration (the second derivative of the z coordinate); ΣT is the total thrust generated by all propellers. This is a control force. F_D is the force of aerodynamic drag, which is low during the initial stage of vertical takeoff and increases with velocity. Often, at the takeoff stage, it is neglected or modeled as

$$F_D = 0.5 \times \rho \times C_D \times S \times (\mathrm{d}z / \mathrm{d}t)^2, \tag{6.4}$$

where ρ is the air density, C_D is the drag coefficient, and S is the characteristic area.

It is also necessary to consider the equations of rotational motion in a body-axis coordinate system. For the takeoff to be stable and prevent the UAV from flipping over, zero angular velocities must be ensured. Here is the Euler equation for this [3]:

$$\Sigma M = I \times (d\omega / \mathrm{d}t) + \omega \times (I \times \omega), \tag{6.5}$$

where ΣM is the total control torque created by the thrust difference on the propellers; I is the inertia tensor of the UAV; $\omega = [p, q, r]^{\mathrm{T}}$ is the angular velocity vector (roll rate, pitch rate, yaw rate).

These equations describe how the orientation of the UAV changes over time. The UAV's orientation is described by three angles: roll (φ), pitch (θ), and yaw (ψ).

In simple terms, the essence of takeoff control is as follows [3]:

1. Vertically. The flight controller smoothly increases the ΣT from a value slightly higher than $m \times g$ (to lift off) to a value much higher than $m \times g$ (to climb with acceleration).

2. Orientation: The stabilization system continuously adjusts the moments ΣM to maintain angles of roll and pitch of approximately 0°, ensuring a strictly vertical lift.

6.3.3 *Takeoff from a Ski Jump or Catapult*

A special device is often used for takeoff with fixed-wing UAVs. In this case, the drone does not require traditional landing gear. However, this method has its own peculiarities. Below is a detailed description of this process.

The purpose of takeoff from a special device is to enable takeoff in areas without long runways, such as on ships, in fields, or on small sites. This method enables the UAV to reach the necessary flight speed in a very short distance.

6.3.3.1 Features of the Takeoff from the Ski Jump

A ski jump is an inclined guide at the edge of which the UAV, having picked up speed, performs a zoom climb and transitions to horizontal flight [3, 5, 6].

The following phases of takeoff can be distinguished with this method [3, 5]:

1. Acceleration along the guide. The UAV is mounted on a guide bases and moves along the ski jump. The UAV's engine (propeller or jet) is put into takeoff mode. The drone can be accelerated with the help of an external force, such as a catapult, or due to the thrust of the UAV's own engine.
2. Turn-off the ski jump. The trolley of catapult stops at the edge of the guide bases, and the UAV continues moving up the arc of the ski jump by inertia before turning off from it.
3. Transition mode. After exiting the ski jump, the UAV enters a short-term descent phase along a ballistic trajectory. In order to increase altitude, the autopilot must actively counteract this descent by increasing lift, for example by increasing the angle of attack.
4. Enter the horizontal flight mode.

   The key features of a ski jump are as follows [3, 5]:

   - Energy efficiency. The ski jump uses potential energy (altitude). The slope helps convert some horizontal velocity into vertical velocity, reducing the speed required to turn off.
   - Less overloads. Compared to a catapult, acceleration is smoother because climbing up compensates for some of the energy.
   - The device is relatively simple. No complicated high-pressure or steam start-up systems are required.
   - Ballistic takeoff characteristics. Accurate calculations and reliable automation operation are required since, after turning off the ski jump, the drone has a small margin in terms of speed and altitude.

– Weight dependence. Each UAV requires an optimal angle of inclination and guide length. A vehicle that is too heavy can "fail" after descent with an insufficient angle of inclination.
– Limited launch mass. It is less effective for very heavy UAVs than powerful catapults.

For example: The Bayraktar TB3 military drone has completed a takeoff from a ski jump [8].

6.3.3.2 Features of the Takeoff from the Catapult

A catapult is a device that propels a UAV at high speeds over a short horizontal distance using an external energy source [3, 6].

The following phases of the process are distinguished when starting from a catapult [3, 6]:

1. Engagement and tension. The UAV is mounted on an acceleration trolley. The holdback pin on the UAV's landing gear is hooked to the catapult's launching device. The UAV engine is brought to takeoff thrust mode, creating pre-tension.
2. Active acceleration. Upon command, the pin releases and a powerful pulse (steam, pneumatic, or electromagnetic) propels the UAV truck along the rails in 2–3 s.
3. Dump the trolley and detach. The trolley stops at the edge of the guides, and the UAV continues moving by inertia. It lifts off the ground independently due to the speed it has already reached, which exceeds the stall speed.
4. Stabilization and climb.

The catapult takeoff process has the following key features [3, 6]:

– High accuracy and repeatability. The catapult provides a consistent speed for the UAV, regardless of weather conditions (e.g., headwinds), similar to a ski jump.
– The ability to launch heavy vehicles. This feature enables the launch of UAVs with a high takeoff weight, including those with limited engine thrust for independent takeoff.
– A more predictable trajectory. Takeoff is almost horizontal, which simplifies autopilot operation at the initial stage.
– Versatility. It can be used to launch different types of drones of various weights.
– High overloads. Sudden acceleration creates significant overloads (up to 4–5 g), placing high requirements on the structural strength of the UAV and its onboard equipment.
– Complexity and cost. Catapults, especially steam or electromagnetic ones, are expensive, bulky, energy-intensive systems that require complex maintenance.
– Dependence on an external energy source. A powerful source of steam, compressed air, or electricity is required.

One example is the catapult launch of the MQ-25 Stingray.

The choice between a ski jump and a catapult depends on the requirements for the UAV, the location, and economic considerations. A ski jump is a simpler, cheaper solution for light and medium-sized vehicles. A catapult is a powerful and versatile tool, but it is expensive and complex. It is indispensable for launching heavy UAVs in cramped conditions.

6.4 UAVs Landing Process

The main goal of landing analysis is to ensure the UAV descends safely from a given altitude to a complete stop on the landing point [1–3].

6.4.1 Fixed-Wing Drones

This process is the reverse of takeoff, but has one key difference: the need to dissipate a high amount of kinetic and potential energy. For fixed-wing UAVs, the equation of motion for descend is as follows [2, 3]:

$$m \times \frac{dV}{dt} = T - D - G \times \sin\theta, \tag{6.6}$$

where θ is the angle of the glide path. During landing, the idle power mode ($T \approx 0$) is typically engaged, allowing the UAV to move under the effect of gravity.

There is an important feature of energy dispersal during landing. Energy is extinguished due to the following [2, 3]:

- Aerodynamic drag, which is caused by the extension of flaps, slats, spoilers, and interceptors (full set of possible ways is depending on an UAV weight and specifics).
- Friction forces that are providing by wheel braking.
- Thrust reverse is turning engines thrust forward (application of this method is depending on an UAV weight and landing conditions).

The following stages of landing are distinguished [2, 3]:

1. Landing approach. A stable descent along the glide path at a constant speed.
2. Flare. This is a gentle transition from descent to horizontal movement immediately in front of the runway, which reduces vertical speed.
3. Main wheel contact. Contact of the main landing gear with the runway.
4. Landing roll. Braking to a complete stop using all means of energy dispersal.

The following key parameters are typical for landing fixed-wing drones [2, 3]:

1. Full landing distance. The distance from the start of a descent maneuver on a glide path (e.g., from an altitude of 100 m) to a full stop. This point is not standardized and is set by the developer. It includes the gliding descent distance, the flare distance, and the ground roll distance. In this case, the ground roll distance is the distance from when the main wheels (or the fuselage) contact the runway to when the drone comes to a complete stop.
2. Landing Speeds. The reference landing speed V_{ref} is the minimum design speed at the runway threshold. It is usually equal to 1.3 times the stall speed with the high-lift devices extended. This provides a margin for stalling out. The approach speed V_{app} is V_{ref} plus the wind correction.
3. The angle of the glide path θ. The standard angle is usually 3° for regular aircraft. This provides a safe descent and visual orientation. However, safety and avoiding obstacles are paramount for UAVs. The approach path must be clear of obstacles such as trees, buildings, and power lines. Depending on the drone's purpose, the flight's mission, and the surrounding conditions, one can distinguish between:

   – Steep glide path (6°–8° or more, reaching 20°). This is selected when there are high obstacles near the runway. A steeper angle enables the drone to "roll over" obstacles and descend safely to the landing point. This is typical for field conditions.
   – A shallow glide path (2°–4°) is used when there are low obstacles near the runway. This is only possible at equipped airfields with long runways and cleared approach areas. This provides a smoother and more predictable descent.

6.4.2 Propelled Drones

For drones with propellers, the equation of translational motion during a vertical landing is as follows [3]:

$$m \times \frac{\mathrm{d}^2 z}{\mathrm{d}t^2} = \Sigma T - m \times g + F_D, \tag{6.7}$$

where the sign of the component of the equation is $+F_D$, when a drone is descanting, the speed is directed downward and the resistance force is directed upward.

The landing of a propeller-driven UAV has its own specifics. During the descent stage, the flight controller reduces the thrust to a value slightly less than the weight. This creates a slight negative acceleration, and the UAV begins to descend at a constant rate.

Then, during the hovering and precise positioning stage, before touching down at an altitude of 0.5–1 m, the UAV often switches to short-term hover mode

($\Sigma T = m \times g$ and $d^2z/dt^2 = 0$). This allows the drone to be accurately positioned above the landing point and the horizontal speed to be reduced.

And then the final touchdown happens. The autopilot slowly and smoothly reduces the thrust to the minimum amount necessary to compensate for the weight. The drone then smoothly descends to the surface. As soon as the sensors on the landing gear detect contact with the ground, the thrust automatically resets to zero and the motors turn off.

At the same time, a very important feature occurs: the "ground effect." When landing on a rigid surface, the air stream from the propellers reflects off the surface, creating additional lifting force. This means that less thrust is required to maintain the same altitude above the surface as above altitude. Advanced landing algorithms should compensate for this effect.

6.4.3 Parachute Landing

The system consists of a parachute [3, 7], which is usually round or cruciform, a packing compartment, and a deployment device. The deployment device can be pyrotechnic, pneumatic, or mechanical. When activated by the operator, timer, or in case of loss of communication or critical failure, the device deploys the parachute. The drone then descends to the ground.

Parachuting has the following advantages [3, 7]:

1. Simplicity and reliability. The system is structurally simple with few mechanical parts, which increases its reliability. It is the "last frontier" for saving the drone.
2. It is weatherproof for rescue. It can be used in conditions where other types of landings are impossible, such as in the event of complete electronics or control failure, when the drone would simply fall.
3. No skills required. Landing does not require special piloting skills—just press the activation button.
4. It is suitable for unprepared places. It does not require a special landing area. In theory, it is possible to land the UAV in a forest or hilly area, but there is a risk of damage.
5. It is safe for people on the ground. A controlled parachute descent is safer than an uncontrolled fall because the trajectory is more predictable and the propellers usually stop turning.

However, this landing method has disadvantages [3, 7]:

1. Inaccurate landing. This is the main drawback. The parachute relies heavily on the wind. The drone could land hundreds of meters from the target point, for example, in a pond, on a tree, or on the roof of a building. For military missions, this makes stealthy return impossible.
2. High risk of damage to the UAV:

  - Dynamic impact. Even with an operating parachute, the rate of descent is 5–8 m/s. An impact with the ground can be strong enough to damage sensitive payloads, such as cameras and sensors, or the frame.
  - Wind. Not only does the drone blow away in the wind, but it can also drag and hit the surface after touching the ground, causing additional damage.
  - Damage when landing on obstacles. The parachute may catch on a tree or power line, leaving the vehicle hanging.

3. Deployment time. After the command is given, it takes several seconds for the parachute dome to fill completely. If the altitude is insufficient, the parachute will not open in time to slow the fall.
4. Additional weight and space. For heavy UAVs especially, the parachute system adds weight and takes up space that could be used for a payload or battery.
5. Difficulty at high altitudes and speeds. Deploying a parachute at high speeds (e.g., in a fixed-wing UAV) can cause the parachute to rupture or damage the drone's frame due to sudden dynamic impact. A two-stage system is often used: first, a braking parachute resets the speed, and then the main parachute deploys.

Parachute landing is widely used for these types of UAVs:

- Small and medium fixed-wing UAVs. This is the main type for which the parachute system is most applicable. This is especially true for vehicles without landing gear or that are designed to land on difficult terrain.
- Multicopters. They are used less frequently because they have their own emergency landing capability, even if several engines fail. However, parachute systems are beginning to be used for mission-critical and expensive multicopters, such as those used for transporting donated organs or for industrial inspections, where payload loss is unacceptable. For these multicopters, rapid parachute system is often used, which open in a fraction of a second.
- Disposable or dropable UAVs. For example, there are drones designed for reconnaissance that can be carried in a backpack. After completing a task, these drones land on a parachute, eliminating the need to look for a special landing area.

Parachute landing is a compromise. Simplicity, reliability, and the ability to use in an emergency case are prioritized over accuracy and softness of landing. These drones are not precise enough for regular operations; rather, they are a "life jacket" for the UAV, ensuring its safety (or minimizing damage) in critical cases.

6.4.4 Net Recovery

Net recovery is one of the most effective and popular methods for small and medium-sized fixed-wing UAVs, especially in confined spaces such as ships, forests, and cities [9].

In this case, the system should consist of two main parts [9]:

1. A drone with a high-strength frame, a pickup system, and an autopilot with precise tracking capabilities.
2. A net tool consisting of a net, a frame, and an energy dissipation system. The net must be made of strong yet elastic materials, such as nylon, that can absorb the drone's kinetic energy. The energy dissipation system is usually a drag reel or spring device that brakes the drone smoothly, stretching the net rather than causing a sharp impact. Sometimes, the net simply hangs freely from the frame to pick up the drone.

The key features of using a net for landing are as follows [9]:

- Short landing distance. This is the main advantage. The system takes up minimal space, making it ideal for ship decks and mobile land-based complexes.
- High reliability in difficult conditions. It enables landing in strong crosswinds and on a swinging deck, where conventional landing with rolling is impossible.
- No landing gear required. This saves weight and reduces the complexity of the drone design.
- Automation. The entire process, from approach to pickup, can be fully automated, which minimizes the human factor.
- Demanding accuracy. Even the slightest mistake in tracking can lead to an accident. This requires highly accurate navigation.
- Impact loads on the drone frame. Despite the shock absorption, the drone experiences significant overloads, especially at points of contact with the net.
- The net tool system is complex and costly. A braked equipment is complex and expensive.
- The risk of bouncing. If the net is stretched too tightly or the shock absorption is insufficient, the drone may bounce and fall.
- Operator assistance is necessary. In semi-automatic systems, the operator must give the pickup command precisely at the right moment.

The process of landing on the net can be divided into the following stages [9]:

1. Landing approach. The drone enters the designated approach trajectory via radio command or independently. This trajectory usually leads directly to the center of the net.
2. Precise tracking. The drone's autopilot uses GPS, optical tags, or radio beacons on a net tool to adjust the course and ensure an encounter. The accuracy should be within a few decimeters.
3. Pickup. The drone moves into the net. If there is a hook, the drone grabs a special purchase cable. If there is no hook, the drone sticks its nose or wings into the net.
4. Energy dissipation. The net stretches under the impact. The braking system smoothly absorbs the drone's kinetic energy until it stops.
5. Absorption. The elasticity of the net and the braking system's operation prevent damage to the drone from dynamic impact.
6. Extraction. After stopping, the drone either falls onto soft mats or is manually or automatically removed from the net to prepare for the next flight.

There are different systems for net recovery [9]:

- Horizontal net. This is the classic option for ground use. The net stands vertically, like a volleyball court.
- Inclined net. It is often used on ships. It is installed at an angle to redirect the drone's trajectory upward, which compensates for the pitching and provides smoother speed damping.
- A system with a skyhook is another option. It's not exactly a net, but conceptually close. The drone hooks onto a stretched cable, rather than a net. The energy is dissipated due to the cable's movement through the braking system.

Net recovery is a highly specialized yet extremely effective technology that transforms a disadvantage (limited space) into an advantage. Its special feature is shifting the emphasis from landing distance to tracking accuracy and structural strength, making it indispensable for modern tactical UAVs.

6.5 VTOL UAVs Specifics

These UAVs combine vertical takeoff and landing capabilities, like a multicopter, with energy-efficient flight, like an aircraft [1, 3].

They are characterized by the following stages of takeoff and landing: vertical takeoff; transition to horizontal flight; and reverse transition before vertical landing.

The takeoff and landing processes for these drones have the following features:

- The complexity and reliability of the transition algorithm is the critical weakness of hybrids. Devices for turning the wings, engines, or redistributing thrust add points of failure, as well as weight and cost. Failure during transition mode is almost always fatal.
- Complex aerodynamics and control systems. The drone must fly stably in two fundamentally different modes. The control algorithms must be highly sophisticated to ensure a smooth and safe transition, which is the most stressful stage of flight.
- Transition losses. Considerable energy is expended on the transition process itself.
- Structural compromises. The design must combine elements of both types, often leading to a decrease in efficiency compared to specialized vehicles in each mode. For example, a wing may be less aerodynamically efficient due to vertical takeoff propellers located on it.

The multicopter's unique combination of all-terrain capability and aircraft range makes it promising for many tasks, such as delivery and monitoring large territories.

6.6 Conclusions

Takeoff and landing are critical phases of flight for all aircraft. Understanding the specifics of these stages enables designers to optimize the performance of UAVs during these phases. In contrast to commercial aviation, drones do not carry humans. However, they are no less dangerous to people on the ground than conventional aircraft. Therefore, studying drone takeoff and landing features helps improve safety conditions. The above chapter provided an overview of the basic patterns of UAV movement. The main aspects of takeoff and landing for various types of drones were also discussed.

References

1. Tiniakov D., Li L. Airworthiness engineering basis. Datian Aviation Beijing Beixue Publishing House. 2020.
2. Guglieri G, Riboldi C. Introduction to flight dynamics. Torino: Celid; 2014.
3. Moiseev V. Unmanned aerial vehicles flight dynamics and control. Kazan: RIC Shkola; 2017.
4. Chen C, Randall R, Tiniakov D. Fiber-reinforced composites—recent advances, new perspectives and applications. In: Composite materials: application specifics for transport category aircraft wings. London: IntechOpen Limited; 2024. https://doi.org/10.5772/intechopen.1004217.
5. Li J, Liu T, Guo H. Dynamics analysis of carrier-based UAVs with ski-jump launch. J Phys Conf Ser. 2020;1626:012146. https://doi.org/10.1088/1742-6596/1626/1/012146.
6. Novaković Z, Vasić Z, Ilić I, Medar N, Stevanović D. Integration of тactical-medium range UAV and catapult launch system. Sci Tech Rev. 2016;66(4):22–8.
7. Wyllie T. Parachute recovery for UAV systems. Aircr Eng Aerosp Technol. 2001;73(6):542–51. https://doi.org/10.1108/00022660110696696.
8. Bayraktar. Bayraktar TB3 passes first ski jump test with flying colors. [Internet]. https://baykartech.com/en/press/bayraktar-tb3-passes-first-ski-jump-test-with-flying-colors/. Accessed 26 Sept 2025
9. Wei X, Wei D, Sui X, Cao L, Su Q. Design and simulation study of shipboard UAV net recovery device. J Phys Conf Ser. 2022;2403:012038. https://doi.org/10.1088/1742-6596/2403/1/012038.

Chapter 7
Control System Design

7.1 Introduction

Designing a UAV flight control system (FCS) requires balancing the device's tasks, weight, size, cost, and required autonomy. The key principle is that the FCS must adequately perform the UAV's functions [1].

A drone's FCS is a combination of technical and logical devices that provide control. Control is defined as the process of adjusting the parameters of the UAV's movement in a desired direction and purposefully influencing the operation of its systems to achieve a specific target.

In order to understand the features of FCS by drones, it is necessary to understand the stages of their development. In general, the following stages can be distinguished [2]:

1. The early stage (before the 1990s). UAV predecessors include targets and aircraft-shaped missiles. Control was primitive, either by radio command or by wire. Example: The German V-1. The USSR and the USA had unmanned reconnaissance aircraft, such as the DBR-1 "Hawk," which were difficult to control and often resulted in the loss of the vehicle.
2. The nascent stage (1990s–2000s): This period is associated with a technological boom. The advent of affordable global positioning system (GPS) technology and miniature microelectromechanical systems (MEMS) sensors revolutionized the industry. UAVs evolved from expensive toys into effective tools. Vehicles such as the Pioneer and the Predator (both USA) have demonstrated their effectiveness in real conflicts. Their FCSs allowed them to carry out semi-autonomous missions.
3. The stage of mass production and autonomy (2010s to present). The emergence of the civilian market, exemplified by the DJI company (China), has significantly reduced costs and simplified FCSs. Meanwhile, military and industrial UAVs are moving towards full autonomy with computer vision and AI systems.

L. Li, D. Tiniakov, *Unmanned Aircraft Design*,
https://doi.org/10.1007/978-981-95-7167-3_7

At the current stage of technological development, the following specifics of FCSs can be identified [3]:

- Intellectualization and AI. Machine learning and computer vision are implemented for navigation without GPS (BeiDou or GLONAS). This includes orientation based on terrain, maps, and objects; flying around obstacles in real time at various speeds; and autonomous decision-making when classifying objects and their condition.
- Swarm Interaction. The objective is to design a FCS that enables the control of a group (swarm) of drones as a single entity. This requires new communication protocols and algorithms for collective behavior.
- Increase stability. Developing alternative navigation methods (e.g., visual odometry and magnetic field) and systems that protect communication channels from intrusive noises (e.g., frequency hopping and cryptography).
- Standardization and interoperability. This involves creating common standards, such as the Micro Air Vehicle Link (MAVLink) communication protocol, so that vehicles and ground stations from different manufacturers can operate together.
- Democratization of access. Powerful and reliable FCSs that were previously only available to the military are now available in commercial products, opening up new opportunities for business, science, and logistics.

The modern UAV FCS has evolved from a simple radio command device into a complex, autonomous system. Key development vectors include intellectualization, resistance to counteraction, and operating in heterogeneous groups. These features make UAVs increasingly independent players in battlefields, economics, and scientific research.

Thus, a rather deeply developed theoretical framework exists today. This framework is based on the experience of operating UAVs and other aircraft. This chapter will cover the general principles of developing FCS for modern drones.

7.2 Basic Concept of Control

As mentioned above, the flying vehicle control system consists of technical and logical devices that control the aircraft. Based on experience with operating different aircraft types, several main functions of the FCS can be identified [4, 5].

- Controlling the trajectory of the center of mass involves changing the forces acting on the flying vehicle (e.g., aerodynamic forces and engine thrust when flying in the atmosphere).
- Motion control relative to the center of mass (angular position control). This involves a change in the moment vector relative to the center of mass.
- Improving stability and controllability performance involves adjusting the controls during perturbed movement to increase stability or reduce transient time.

The following components can be included in the FCS to perform all the functions listed above:

- Command control levers are technical devices that facilitate the perception and transmission of pilot-operator control actions.
- Control effectors are technical devices that control the magnitude and direction of the forces and moments acting on a flying vehicle. Modern flying vehicles use both aerodynamic and jet (gas-dynamic) control effectors.
- Logical and technical subsystems that improve stability and controllability usually operate independently of the pilot-operator on the control effectors during perturbed movement. These subsystems adjust the functions of the control effectors according to angular velocity, angle of attack, and overload.

The process of controlling a flying vehicle includes several stages [1, 3]:

- Obtaining data about the control tasks and the current state of the flying vehicle.
- Analyzing the received data and developing a control solution.
- Executing the decision by deflecting the control effectors, changing the engine thrust, or activating the braking devices.

There are three types of control depending on the degree to which the pilot-operator and automatic systems participate in the control process:

1. Non-automatic. The pilot-operator performs full vehicle control functions (receiving and analyzing data, generating and executing decisions).
2. Automatic. Total vehicle control functions are performed by the automatic control system.
3. Semi-automatic. Some functions are performed by the pilot-operator and some by the automatic control system.

When developing an FCS, it is important to consider that control systems for each type of flying vehicle are nearly unique since vehicles differ in purpose, layout, and technical design.

7.3 UAVs Control Specifics

As previously mentioned, the design of a UAV control system necessitates a tradeoff among its functions, characteristics, cost, and autonomy. The control system must be adequate for the purpose of the specific UAV under consideration.

In Chap. 2, the classification of UAVs by weight and size was considered. It is reasonable to become attached to this classification.

7.3.1 Micro-UAVs and Mini-UAVs

These drones are typically utilized for a variety of purposes, including exploration, surveillance, filming, and recreational activities.

The following design features will be characteristic of such drones [1, 3, 6]:

The concept of centralization is a key component. A single, highly capable flight controller is utilized, which integrates all functions—stabilization, navigation, engine control, and camera data processing.

The primary emphasis is to prioritize simplicity and reliability. Algorithms, frequently in the form of PID controllers, must be designed to be both simple and fast in order to ensure their execution on computing resources with limited capabilities.

The integration of sensors is a critical component of the system. The emphasis on miniaturization is a salient feature of this technology. A variety of sensors, including MEMS gyroscopes, accelerometers, barometers, as well as optical and ultrasonic sensors, are employed for the primary functions of position control and obstacle avoidance.

The user interface also is the primary emphasis for such drones. The design is executed in conjunction with a convenient mobile application or remote control.

7.3.2 Tactical UAVs

The functionality of such drones encompasses a range of missions, including reconnaissance, targeting, strike operations, and relay communications.

The design features of this type of drone are as follows [1, 3, 6]:

- Modularity. The control system is divided into various subsystems, including a flight controller, a payload management system (which incorporates a camera and a laser rangefinder), a communication system, and a cargo drop system (weapons).
- Fault tolerance. The redundancy of critical systems is being designed. Examples of such systems include dual flight controllers, dual communication channels, and sensors.
- The integration of the system with the ground control system is paramount. The significance of data exchange protocols, such as STANAG 4586, is rapidly escalating. The drop system must be capable of accepting high-level commands, such as "follow the point" and "follow the target."
- Navigation. Typically, these systems are integrated with other positioning technologies, such as GPS/BeiDou, in conjunction with an inertial navigation system. In the event of a GNPS signal interruption, the control system is required to maintain the vehicle's operation by relying exclusively on data from the inertial system. This necessitates the implementation of sophisticated filtering algorithms (for example, Kalman).

7.3.3 *Strategic UAVs*

These UAVs are equipped with a range of sensors and communication systems, enabling them to perform a variety of functions, including long-term patrolling, global reconnaissance, and monitoring.

The design features of this case are as follows [1, 3, 6]:

The first feature is that of high autonomy. The control system is designed as a complex on-board system capable of carrying out a mission for hours and days without operator intervention, including automatic takeoff and landing.

The utilization of satellite communication is imperative in this context. The integration of satellite communication channels (SATCOM) facilitates control that extends beyond the limits of direct radio visibility. It is imperative for designers to consider the substantial signal delay.

The planning of complex missions necessitates the implementation of advanced algorithms. The control system must consider several factors, including weather conditions, affected air defense zones, the amount of remaining fuel, and the state of the onboard systems.

The degree of reliability is a critical factor in this case. Reliability requirements for manned aircraft are imminent. Full redundancy and sophisticated diagnostic systems are utilized.

7.3.4 *Racing and Experimental UAVs*

These UAVs are distinguished by their exceptional speed and maneuverability.

In this case, the following factors must be taken into consideration [1, 3, 6]:

The priority of response speed is paramount. In most cases, the utilization of control algorithms in manual mode, commonly referred to as "Acro," results in the deactivation of automatic stabilization functions. Disabling of auto stabilization is a consequence of this mode of operation. The flight controller's mission is to execute the pilot's commands with the utmost accuracy and expediency.

Minimalism is a significant concept. The vehicle is equipped with a minimal set of sensors, including a gyroscope and an accelerometer. The absence of a GPS module is a deliberate design choice aimed at reducing the drone's weight and delay.

It is imperative that customization be prioritized as a prerequisite. The design process entails meticulous calibration of the PID controls to align with the operator's preferences and the characteristics of the drone model.

7.3.5 *General Estimation of the UAV Control System*

The UAV control system is a multi-level cyber-physical complex that functions as the "brain" of the device.

Contemporary control systems for UAVs exhibit common characteristics across various drone types.

The system is characterized by a high degree of automation. This configuration serves to alleviate the physical demands placed upon the operator.

Reliability is a critical component for UAVs. This is achieved through the implementation of redundancy measures and advanced algorithms.

The concept of flexibility is paramount in this case. The capacity to adapt to novel tasks is a critical feature in the modern time.

Integration constitutes an additional specific requirement. The capacity to integrate data from multiple sensors to facilitate decision-making.

The present drones are depending on communication channels. These vehicles are susceptible to interference and signal interception.

The UAVs demonstrate a high degree of dependence on GNSS (GPS, BeiDou, GLONASS). The system's susceptibility to disruption is amplified in scenarios where the signal is suppressed or spoofed.

Contemporary control systems for diverse drone categories are characterized by intricacy and complexity. The design and verification of complex autonomous systems necessitates substantial expenses and qualifications.

The present situation is marked by the existence of security issues. The potential for cyberattacks against the control system is a matter of concern.

7.4 Control System Analysis Process

The final stage of the process of organizing the operation of UAVs is the development of software control rules that ensure flights along the required trajectories at specified time intervals.

In this section, the focus will be on the indirect control of UAV aircraft scheme, a principle that remains constant irrespective of the various configurations. Furthermore, the methods of establishing indirect control of an unmanned helicopter of a single-rotor scheme will be considered. The methods of establishing such control are predicated on the well-established theory of inverse problems of dynamics of controlled motion of objects [6, 7].

7.4.1 Control of the Movement of Fixed-Wing UAVs Along Spatial and Planar Trajectories

In general, the definition of the control vector $u = u(t)$ is a dynamic object described by a system of differential equations in vector form:

$$\dot{x} = f(x,u), t \in [t_0, t_F], \tag{7.1}$$

employing the theory of inverse problems, the task is reduced to the solution of systems of nonlinear parametric equations of the following form:

$$f\left(x^*(t),u(t)\right)-\dot{x}^*=0, t\in\left[t_0,t_F\right], \tag{7.2}$$

where, $x^*(t)$ denotes the requisite alteration in the phase vector of the vehicle across the designated time interval.

It is imperative to acknowledge that the system (7.2) is employed in instances where the dimensions of the vectors $x(t)$ and $u(t)$ are congruent. Conversely, expressions (7.2) are employed as constraints within the auxiliary parametric optimization problem.

The dynamic equations of motion for fixed-wing type UAVs utilize a control vector:

$$u(t)=\left(P(t),\alpha(t),\gamma(t)\right), t\in\left[t_0,t_F\right], \tag{7.3}$$

where $P(t)$ is the thrust force of the UAV engines; $\alpha(t)$ is the angle of attack of the UAV; $\gamma(t)$ is the angle of roll of the UAV.

These equations are to be regarded as a particular instantiation of the system (7.1).

In order to calculate the required change in the components of the phase vector during the spatial motion of the UAV, the following functions must be used:

$$V_{\text{targ}}=V_{\text{targ}}(t);\theta_{\text{path}}=\theta_{\text{path}}(t);\Psi_{\text{path}}=\Psi_{\text{path}}(t), t\in\left[t_0,t_F\right], \tag{7.4}$$

here V_{targ} is the target value of the UAV's flight speed, the functions $\Theta_{\text{path}}(t)$ and $\Psi_{\text{path}}(t)$ reflect the time change in the angles of banking and turning of the UAV's trajectory as it moves along the required flight path.

Subsequently, the system of equations is to be written in the following form:

$$\begin{aligned}
&\frac{P(t)-D\left(\alpha(t),y_{\text{path}}(t),V_{\text{targ}}(t)\right)}{m}-g\sin\theta_{\text{path}}(t)-\dot{V}_{\text{targ}}(t)=0;\\
&\frac{\left[P(t)\left(\alpha(t)+\varphi_{\text{eng}}\right)+L\left(\alpha(t),y_{\text{path}}(t),V_{\text{targ}}(t)\right)\right]\cos\gamma(t)}{mV_{\text{targ}}(t)}\\
&-\frac{g\cos\theta_{\text{path}}(t)}{V_{\text{targ}}(t)}-\dot{\theta}_{\text{path}}(t)=0;\\
&\frac{\left[P(t)\left(\alpha(t)+\varphi_{\text{eng}}\right)+L\left(\alpha(t),y_{\text{path}}(t),V_{\text{targ}}(t)\right)\right]\sin\gamma(t)}{mV_{\text{targ}}(t)\cos\theta_{\text{path}}(t)}\\
&-\dot{\Psi}_{\text{path}}(t)=0, t\in\left[t_0,t_F\right],
\end{aligned} \tag{7.5}$$

The system of nonlinear parametric Eq. (7.3) that is employed to determine the components of the control vector is a specification of the system (7.2).

In the event that the system is configured with specification of (7.5), two expressions are substituted that are based on (4.2)–(4.4), which are written as follows:

$$\begin{aligned} D\left(\alpha(t), y_{\text{path}}(t), V_{\text{targ}}(t)\right) &= 0.5C_{\text{D}}\left(\alpha(t), y_{\text{path}}(t), V_{\text{targ}}(t)\right)\times \\ &\times \rho\left(y_{\text{path}}(t)\right)V_{\text{targ}}^2(t)A_{\text{ref}}; \\ L\left(\alpha(t), y_{\text{path}}(t), V_{\text{targ}}(t)\right) &= 0.5C_{\text{L}}\left(\alpha(t), y_{\text{path}}(t), V_{\text{targ}}(t)\right)\times \\ &\times \rho\left(y_{\text{path}}(t)\right)V_{\text{targ}}^2(t)A_{\text{ref}}. \end{aligned} \tag{7.6}$$

The coefficients of drag and lift, which are integral to these calculations, can be expressed as follows:

$$\begin{aligned} C_{\text{D}}\left(\alpha(t), y_{\text{path}}(t), V_{\text{targ}}(t)\right) &= \left(c_0 + \frac{c_1 V_{\text{targ}}(t)}{a\left(y_{\text{path}}(t)\right)}\right) + \\ & K_{\text{DLF}}\left(d_0 + \frac{d_1 V_{\text{targ}}(t)}{a\left(y_{\text{path}}(t)\right)}\right)^2 \left(\alpha(t) + \alpha_0\right)^2; \\ C_{\text{L}}\left(\alpha(t), y_{\text{path}}(t), V_{\text{targ}}(t)\right) &= \left(b_0 + \frac{b_1 V_{\text{targ}}(t)}{a\left(y_{\text{path}}(t)\right)}\right) + \\ & \left(d_0 + \frac{d_1 V_{\text{targ}}(t)}{a\left(y_{\text{path}}(t)\right)}\right)\alpha(t). \end{aligned} \tag{7.7}$$

In this equation, a denotes the sound speed, and K_{DLF} represents the drag-due-to-lift factor.

As per the stipulated Eq. (7.5), the angular quantities $\alpha_{\text{path}}(t)$, $\Psi_{\text{path}}(t)$, along with their respective derivatives, have been computed through application of the following equations:

$$\begin{aligned} \theta_{\text{path}}(t) &= \arctan y'\left(x_{\text{path}}(t)\right); \\ \dot{\theta}_{\text{path}}(t) &= \frac{V_{\text{targ}}(t)\, y''\left(x_{\text{path}}(t)\right)}{\sqrt{\left(1 + y'^2\left(x_{\text{path}}(t)\right)\right)^3 \left(1 + z'^2\left(x_{\text{path}}(t)\right)\right)}}; \\ \Psi_{\text{path}}(t) &= \arctan z'\left(x_{\text{path}}(t)\right); \\ \dot{\Psi}_{\text{path}}(t) &= \frac{V_{\text{targ}}(t)\, z''\left(x_{\text{path}}(t)\right)}{\sqrt{\left(1 + y'^2\left(x_{\text{path}}(t)\right)\right)\left(1 + z'^2\left(x_{\text{path}}(t)\right)\right)^3}}. \end{aligned} \tag{7.8}$$

In this context, y', y'', z', z'' denote the first and second derivatives of the functions $y(x)$, $z(x)$, and $x_{path}(t)$ with initial conditions $x(t_0) = x_0$ and $z(t_0) = z_0$ over the time interval $[t_0, t_F]$.

When the required UAV trajectory is directly constructed in the form of functions $x_0(t)$, $y_0(t)$, $z_0(t)$, the dynamics of these angles and angular velocities are described by the following expressions:

$$\begin{gathered}
\theta_{path}(t) = \arctan\left(\frac{\dot{y}_0(t)}{\dot{x}_0(t)}\right);\\
\dot{\theta}_{path}(t) = \frac{\dot{x}_0(t)\ddot{y}_0(t) - \dot{y}_0(t)\ddot{x}_0(t)}{\dot{x}_0^2(t) + \dot{y}_0^2(t)};\\
\Psi_{path}(t) = \arctan\left(\frac{\dot{z}_0(t)}{\dot{x}_0(t)}\right);\\
\dot{\Psi}_{path}(t) = \frac{\dot{x}_0(t)\ddot{z}_0(t) - \dot{z}_0(t)\ddot{x}_0(t)}{\dot{x}_0^2(t) + \dot{z}_0^2(t)}, \quad t \in [t_0, t_F].
\end{gathered} \tag{7.9}$$

It should be noted that the dependencies $y_{path}(t)$ employed in the Eqs. (7.6) and (7.7) correspond to the solution of the aforementioned system of differential equations, or are designated as a function $y_0(t)$, $t \in [t_0, t_F]$.

The solution to the system of nonlinear parametric Eq. (7.5) is obtained through the application of numerical methods over the specified time interval $[t_0, t_F]$. The resulting solutions of this system are required to meet the operational limitations for the designated type of UAV:

$$\begin{gathered}
P_{min} \le P(t) \le P_{max};\\
\alpha_{min} \le \alpha(t) \le \alpha_{max};\\
\gamma_{min} \le \gamma(t) \le \gamma_{max}, \quad t \in [t_0, t_F].
\end{gathered} \tag{7.10}$$

Example 7.1 A mid-range UAV equipped with a turbojet engine, the parameters of which are shown in Table 7.1 [6].

The formation of a control vector of the form (7.3) is necessary to ensure the UAV's flight in normal atmospheric conditions to the designated zone of subsequent movement in a horizontal plane at an altitude of 3000 m along a circular trajectory:

$$z_{path}(x) = \sqrt{80000^2 - x^2} \text{ with the velocity } V_{targ} = 100\,\text{m/s}.$$

The final position of the UAV at $t = t_F$ can be described by the following parameters:

Table 7.1 Initial data of the UAV

UAV's parameter	Value
Mass of the UAV, m (kg)	350
Area of the wing, A_{ref} (m^2)	1.4
Wing swept angle, χ (°)	0
Wing span, l (m)	2.64
Mean aerodynamic chord, MAC (m)	0.546
Engine's setting angle, φ (°)	3.5
Thrust range of the engine, P_{min}/P_{max} (N)	58.86/1208.65
Acceptable angles of attack range, α_{min}/α_{max} (°)	–6/+14
Trim angle of attack, α_0 (°)	–0.4
Banking angle range, γ (°)	±65
Sliding angle range (°)	±10
Maximum operational overload	–3 to +9
Range of flight velocities, V_{min}/V_{max} (m/s)	80.55/200
Flight altitude range, H_{min}/H_{max} (m)	300/9000
Central moments of inertia (kg·m/s)	$Jx = 2.1$, $Jy = 31$, $Jz = 30$

$$\left(x_F;y_F;z_F\right)=\left(56568;3000;56568\right)\ \text{m};$$
$$\theta_F=0°;\ \Psi_F=-45°;\ V_{targ}=100\ \text{m/s}.$$

It is hypothesized that $t_0 = 0$ s and that the UAV initiates its flight into the zone at a velocity of $V = 100$ meters per second, according to the following values for the initial coordinates and trajectory angles:

$$x_0=20000\,\text{m},y_0=600\,\text{m},z_0=10000\,\text{m},\theta_0=20°;\Psi_0=40°.$$

Preliminary data will be used to develop polynomials that describe the projections $\overline{y}\,(x)$ and $\overline{z}\,(x)$ of the spatial trajectory of the UAV under consideration.

In general, these polynomials are expressed as follows:

$$\begin{aligned}\overline{y}(x)&=\overline{C}_1x^3+\overline{C}_2x^2+\overline{C}_3x+\overline{C}_4;\\ \overline{z}(x)&=\widehat{C}_1x^3+\widehat{C}_2x^2+\widehat{C}_3x+\widehat{C}_4.\end{aligned}\tag{7.11}$$

The system of equations that has been employed to determine the coefficients $\overline{C}_i$ for $i = (1, 4)$, based on the initial data, is expressed as follows:

$$\begin{aligned}
&\bar{C}_1 20000^3 + \bar{C}_2 20000^2 + \bar{C}_3 20000 + \bar{C}_4 = 600;\\
&3\bar{C}_1 20000^2 + 2\bar{C}_2 20000 + \bar{C}_3 = \tan 20°;\\
&\bar{C}_1 56568^3 + \bar{C}_2 56568^2 + \bar{C}_3 56568 + \bar{C}_4 = 3000;\\
&3\bar{C}_1 56568^2 + 2\bar{C}_2 56568 + \bar{C}_3 = \tan 0°.
\end{aligned}$$

The following values are obtained through the utilization of a numerical method in the resolution of the system of these expressions:

$$\bar{C}_1 = 1.740{\bullet}10^{-10}; \bar{C}_2 = -2.496{\bullet}10^{-5}; \bar{C}_3 = 1.154; \bar{C}_4 = -1.388{\bullet}10^4.$$

The coefficients $\hat{C}_i$ for $i = (1, 4)$ are calculated from the system of expressions:

$$\begin{aligned}
&\hat{C}_1 20000^3 + \hat{C}_2 20000^2 + \hat{C}_3 20000 + \bar{C}_4 = 10000;\\
&3\hat{C}_1 20000^2 + 2\hat{C}_2 20000 + \hat{C}_3 = \tan 40°;\\
&\hat{C}_1 56568^3 + \hat{C}_2 56568^2 + \hat{C}_3 56568 + \hat{C}_4 = 56568;\\
&3\hat{C}_1 56568^2 + 2\hat{C}_2 56568 + \hat{C}_3 = \tan(-45°).
\end{aligned}$$

The roots of this system of expressions take the following form:

$$\hat{C}_1 = -2.025{\bullet}10^{-9}; \hat{C}_2 = 2.074{\bullet}10^{-4}; \hat{C}_3 = -5.028; \hat{C}_4 = 4.379{\bullet}10^4.$$

Consequently, the expressions (7.11) are reduced to the following form:

$$\begin{aligned}
&\bar{y}(x) = 1.740{\bullet}10^{-10}x^3 - 2.496{\bullet}10^{-5}x^2 + 1.154x - 1.388{\bullet}10^4;\\
&\bar{z}(x) = -2.025{\cdot}10^{-9}x^3 + 2.074{\cdot}10^{-4}x^2 - 5.028x + 4.379{\cdot}10^4.
\end{aligned} \tag{7.12}$$

The graphs of these functions for $x \in [20{,}000; 56{,}568]$ m are shown in Fig. 7.1.

The parametric representation, designated as $x(t)$, $y(t)$, and $z(t)$, of the UAV's spatial flight path has been obtained through the solution of a system of differential equations:

$$\begin{aligned}
&\dot{x} = \frac{V_{\text{targ}}(t)}{\sqrt{(1 + y'^2(x))(1 + z'^2(x))}};\\
&\dot{y} = \frac{V_{\text{targ}}(t)\, y'(x)}{\sqrt{(1 + y'^2(x))(1 + z'^2(x))}};\\
&\dot{z} = \frac{V_{\text{targ}}(t)\, z'(x)}{\sqrt{(1 + y'^2(x))(1 + z'^2(x))}}, \quad t \in [t_0, t_F].
\end{aligned} \tag{7.13}$$

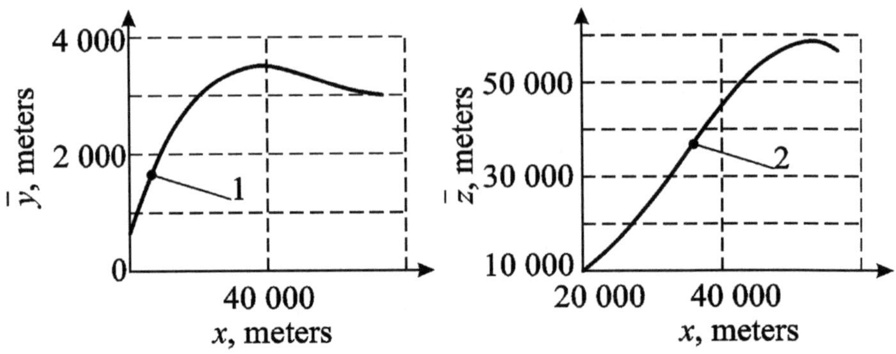

Fig. 7.1 Graphs of the functions (7.12): (1) $\overline{y}_1(x)$; (2) $\overline{z}_1(x)$

Table 7.2 The results of solving the expressions (7.13)

t (s)	x (m)	y (m)	z (m)
0	20,000	600	10,000
43.166	22,807	1511	12,988
86.332	25,206	2126	16,411
129.497	27,372	2564	20,044
172.663	29,399	2883	23,796
215.829	31,342	3114	27,616
258.995	33,241	3277	31,475
302.160	35,128	3385	35,349
345.326	37,035	3446	39,219
388.492	38,995	3467	43,064
431.658	41,054	3450	46,858
474.823	43,280	3398	50,554
517.989	45,791	3313	54,059
561.155	48,840	3193	57,095
604.321	52,820	3055	58,509
647.4865	56,568	3000	56,568

for the following initial conditions:

$$x(0) = 20000 \text{ m}, \; y_0 = 600 \text{ m}, \; z_0 = 10000 \text{ m}$$

is presented in increments of $\Delta t = 43.166$ s in Table 7.2.

As demonstrated in Table 7.2, the results indicate that the UAV attains a flight altitude that exceeds 3000 m at time $t \in [172.663; 215.829]$. Subsequently, commencing at time t $\in [388.492; 431.658]$, it undergoes a decline to the requisite altitude for the initiation of its horizontal movement, which is attained at time $t_F = 647.487$ s.

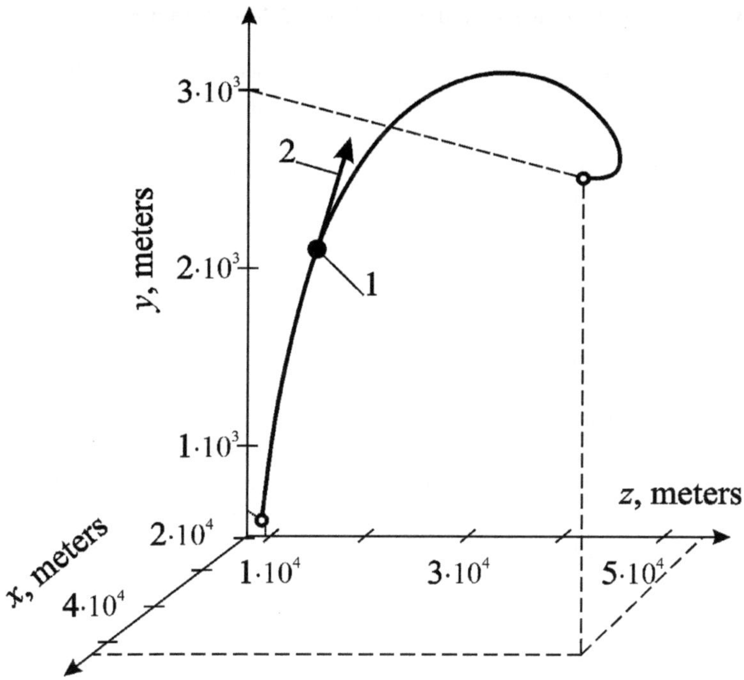

Fig. 7.2 The UAV's spatial trajectory: (1) UAV; (2) V_{targ}

It is important to note that, at this particular point in time, the remaining UAV coordinates are equivalent to the values established by the initial conditions.

As illustrated in Fig. 7.2, the UAV's spatial trajectory indicates its subsequent flight trajectory.

The subsequent step involves the formulation of the control functions $P(t)$, $\alpha(t)$, $\gamma(t)$, $t \in [0; 647.487]$ c, which are components of the vector (7.3).

In order to solve the system of Eq. (7.5), it is first necessary to calculate the angles of banking $\Theta(t)$ and turning $\Psi(t)$ changes of the UAV trajectory, as well as their derivatives (see Table 7.3) for the time interval $t \in [0; 647.487]$ using the Eq. (7.8).

As demonstrated in the foregoing table, the trajectory angles of the UAV at $t = t_0$ and $t = t_F$ meet the stipulated criteria outlined in the initial conditions, respectively.

The following assumptions are made about the UAV under consideration:

$$\alpha_0 = -0.401° = -0.007\text{rad}; K_{\text{DLF}} = 0.0759; A_0 = 0.017; A_1 = 0.025;$$
$$d_0 = 4.312; d_1 = 1.292.$$

Subsequently, the aerodynamic coefficients of the UAV will be expressed as follows:

Table 7.3 The angles of banking Θ(*t*) and turning Ψ(*t*) and their derivatives

t (s)	Θ (°)	$\dot{\theta}\cdot 10^{-4}$ (rad/s)	Ψ (°)	$\dot{\Psi}\cdot 10^{-4}$ (rad/s)
0	20	–18.46	40	72.60
43.166	15.990	–14.33	51.862	31.19
86.332	12.785	–11.77	57.490	16.44
129.497	10.106	–9.994	60.616	9.569
172.663	7.807	–8.647	62.460	5.647
215.829	5.808	–7.555	63.513	3.006
258.995	4.057	–6.623	63.992	0.9235
302.160	2.523	–5.795	63.982	–1.008
345.326	1.186	–5.024	63.478	–3.126
388.492	0.037	–4.269	62.385	–5.869
431.658	–0.923	–3.480	60.458	–10.06
474.823	–1.675	–2.568	57.130	–17.74
517.989	–2.167	–1.331	50.889	–35.47
561.155	–2.258	0.8558	36.599	–92.31
604.321	–1.542	5.208	–3.675	–230.0
647.487	0	6.461	–45	–96.32

$$C_D(t) = \left(0.017 + \frac{0.025V}{340 - 0.004y(t)}\right) + \left(1.188 + \frac{0.356V}{340 - 0.004y(t)}\right)\left(\alpha(t) + 0.007\right)^2;$$

$$C_L(t) = \left(4.312 + \frac{1.291V}{340 - 0.004y(t)}\right)\left(\alpha(t) + 0.007\right)^2.$$

The solution to the system of Eq. (7.5) was obtained by employing the advanced expressions (7.6) and (7.7) at V = 100 m/s and V = 0, with the aforementioned initial data spanning the time interval $t \in [0;\ 647.487]$ on a grid with a step of Δt_1 = 43.166 s. This calculation was performed with an initial approximation as follows:

$$_{\text{init}} = 1000\,\text{N}; \alpha_{\text{init}} = 1°; \gamma_{\text{init}} = 1°.$$

It was determined that the roots of this system of equations are independent of the initial approximations, thereby establishing the uniqueness of the solution.

The results of solving the system in the form of functions $P(t)$, $\alpha(t)$, $\gamma(t)$ are shown in Fig. 7.3 and in Table 7.4.

As illustrated in Fig. 7.3b, the banking angle curve $\gamma(t)$ exhibits a break. Within the specified neighborhood, the control functions $P(t)$ and $\alpha(t)$ demonstrate a monotonic behavior.

It is important to note that the constraints of the form (7.10) on the controls $P(t)$, $\alpha(t)$, $\gamma(t)$, as presented in Table 7.1, are satisfied at the considered flight time interval

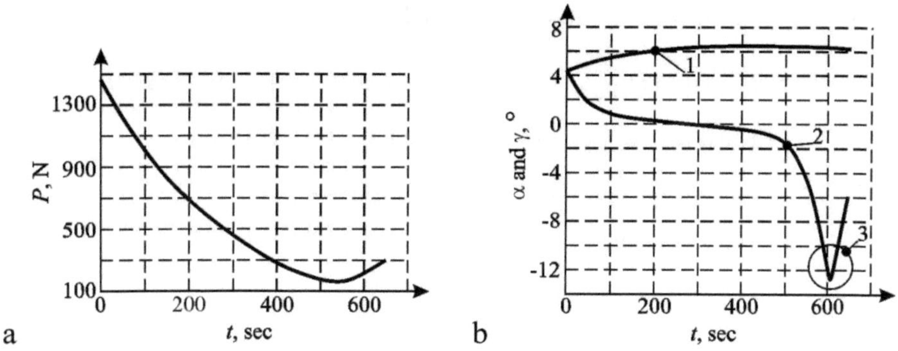

Fig. 7.3 Graphs of functions (**a**) $P(t)$, (**b**) $a(t)$ and $g(t)$: 1 – $a(t)$; 2 – $g(t)$; 3 – the break

Table 7.4 The results of solving the system in the form of functions $P(t)$, $\alpha(t)$, $\gamma(t)$

t (s)	P (N)	α (°)	γ (°)
0	1454.319	4.098	4.320
43.166	1222.471	4.687	1.850
86.332	1038.364	5.136	0.973
129.497	882.941	5.484	0.565
172.663	749.128	5.755	0.333
215.829	632.368	6.965	0.177
258.995	529.869	6.123	0.054
302.160	439.867	6.238	–0.059
345.326	361.297	6.314	–0.184
388.492	293.687	6.357	–0.344
431.658	237.221	6.371	–0.590
474.823	193.068	6.356	–1.039
517.989	164.452	6.319	–2.074
561.155	160.807	6.281	–5.373
604.321	210.621	6.348	–12.879
647.487	296.307	6.157	–5.573

of the UAV, with the exception of the engine thrust value at the time interval [0; 86.332].

In order to rectify the identified impediment that emerged during the process of calculating the UAV's control, it is hereby proposed that the subsequent values be utilized in replacement of the initial values of its trajectory. These values are derived from Tables 7.2 and 7.3 for $t = 86.332$ s:

$$x_0 = 25206\,\text{m};\, y_0 = 2126\,\text{m};\, z_0 = 16411\,\text{m};\, \theta_0 = 12.785°;\, \Psi_0 = 57.490°,$$

In the case at hand, when formulating the vector $u(t)$, all constraints of the form (7.10) for the UAV under consideration will be fulfilled.

The Eq. (7.5) will be rewritten in a revised form:

$$\begin{aligned}
&P(t)-D\big(\alpha(t),y_{\text{path}}(t),V_{\text{targ}}(t)\big)-mg\sin\theta_{\text{path}}(t)-m\dot{V}_{\text{targ}}(t)=0;\\
&\Big[P(t)\big(\alpha(t)+\varphi_{\text{eng}}\big)+L\big(\alpha(t),y_{\text{path}}(t),V_{\text{targ}}(t)\big)\Big]\cos\gamma(t)\\
&\qquad -mg\cos\theta_{\text{path}}(t)-mV_{\text{targ}}(t)\dot{\theta}_{\text{path}}(t)=0;\\
&\Big[P(t)\big(\alpha(t)+\varphi_{\text{eng}}\big)+L\big(\alpha(t),y_{\text{path}}(t),V_{\text{targ}}(t)\big)\Big]\sin\gamma(t)\\
&\qquad -mV_{\text{targ}}(t)\cos\theta_{\text{path}}(t)\dot{\Psi}_{\text{path}}(t)=0.
\end{aligned} \tag{7.14}$$

Subsequent analysis of these systems will yield the equations for determining the control vectors $u(t)$ of UAV flights in the vertical and horizontal planes of movement.

The following conditions must be met in order to ensure the motion of UAV in a vertical plane:

$$\gamma(t)=0;\Psi(t)=0;\dot{\Psi}(t)=0$$

and the control vectors will look like:

$$u(t)=\big(P(t),\alpha(t)\big). \tag{7.15}$$

Substituting the aforementioned conditions into expressions (7.14), it becomes evident that the latter equation is equivalent to an identity, thus rendering it superfluous.

Subsequently, the components of the control vector (7.15) are defined as solutions to the following system of nonlinear parametric equations of the second order:

$$\begin{aligned}
&P(t)-D\big(\alpha(t),y_{\text{path}}(t),V_{\text{targ}}(t)\big)-mg\sin\theta_{\text{path}}(t)-m\dot{V}_{\text{targ}}(t)=0;\\
&P(t)\big(\alpha(t)+\varphi_{\text{eng}}\big)+L\big(\alpha(t),y_{\text{path}}(t),V_{\text{targ}}(t)\big)-mg\cos\theta_{\text{path}}(t)\\
&\qquad -mV_{\text{targ}}(t)\dot{\theta}_{\text{path}}(t)=0,\ t\in[t_0,t_F].
\end{aligned} \tag{7.16}$$

In the process of solving these equations, the previously mentioned expressions are employed:

$$\begin{aligned}
&\theta_{\text{path}}(t)=\arctan y'\big(x(t)\big);\\
&\dot{\theta}_{\text{path}}(t)=\frac{V(t)y''\big(x(t)\big)}{\sqrt{1+y'^2\big(x(t)\big)}},\quad t\in[t_0,t_{\text{F}}].
\end{aligned} \tag{7.17}$$

The functions $x(t)$ and $y(t)$ presented in expressions (7.16) and (7.17), respectively, are determined through the resolution of a system of differential equations over the time interval $[t_0, t_{\text{F}}]$:

$$\dot{x} = \frac{V(t)}{\sqrt{1+y'^2(x)}};$$
$$\dot{y} = \frac{V(t)y''(x)}{\sqrt{1+y'^2(x)}}, \tag{7.18}$$

for the initial conditions:

$$x(0) = x_0; y(0) = y_0.$$

Example 7.2 It is needed to formulate the control functions of a UAV that is navigating at a constant altitude of 2000 m [6]. It is presenting in the vertical plane, with the time interval spanning from 0 to 400 s. The velocity of the UAV is constant and set at 97.50 meters per second.

The ensuing results are derived from the integration of the system of Eq. (7.18) for $y(x) = 2000$ and $t \in [0; 400]$ in increments of $t = 40$ s. These results are presented in the initial three columns of Table 7.5.

Expressions (7.17) are specified for conditions:

$$\theta(t) = 0\,\text{rad};\ \dot{\theta}(t) = 0\,\text{rad/s}, t \in [0{,}400]\,\text{s}.$$

The solutions for $P(t)$ and $\alpha(t)$ of the system of parametric Eq. (7.16) in increments of 40 s (Δt) and with the calculation of the parameters included in them, as utilized in Example 7.1, are exhibited in the final two columns of Table 7.5.

The problem of determining the controls functions $P(t)$ and $\alpha(t)$, which ensure the process of descent of the UAV along the required trajectory of the form:

Table 7.5 The results for the Example 7.2

t (s)	x (m)	y (m)	P (Newton)	α (°)
0	1000	2000	292.782	5.793
40	4900	2000	292.782	5.793
80	8800	2000	292.782	5.793
120	12,700	2000	292.782	5.793
160	16,600	2000	292.782	5.793
200	20,500	2000	292.782	5.793
240	24,400	2000	292.782	5.793
280	28,300	2000	292.782	5.793
320	32,200	2000	292.782	5.793
360	36,100	2000	292.782	5.793
400	40,000	2000	292.782	5.793

Table 7.6 The results for the Example 7.2

t (s)	θ (°)	$\dot{\theta}$ ·10⁻³ (degrees per second)	P (N)	α (°)
0	0	−1.852	295.445	5.428
18.033	−1.721	−1.479	193.118	5.438
36.067	−3.056	−1.105	114.100	5.415
54.100	−4.006	−0.7344	58.372	5.370
72.133	−4.574	−0.3663	25.752	5.312
90.166	−4.764	0	16.050	5.250
108.200	−4.574	0.3663	29.151	5.189
126.233	−4.006	0.7344	65.044	5.138
144.266	−3.056	1.105	123.797	5.100
162.300	−1.721	1.473	205.463	5.081
180.333	0	1.852	309.930	5.084

$$y(x) = C_1x^3 + C_2x^2 + C_3x + C_4, \tag{7.19}$$

where C_i are the integration constants, $i = (1, 4)$.

The initial data indicates that the equations for $y(x)$ is of the following form:

$$y(x) = 3.429 \cdot 10^{-10}x^3 - 1.132 \cdot 10^{-5}x^2 + 0.041x - 1.96 \cdot 10^3, \tag{7.20}$$

The second and third derivatives of the function $y(x)$ have the form:

$$y'(x) = 10.287 \cdot 10^{-10}x^2 - 2.264 \cdot 10^{-5}x + 0.041. \tag{7.21}$$

$$y''(x) = 20.574 \cdot 10^{-10}x - 2.264 \cdot 10^{-5}. \tag{7.22}$$

The alterations in the functions θ(t) and $\dot{\theta}$ (t), $t \in [0; 180.333]$, have been determined through the utilization of Eq. (7.17), employing the established relations for the derivatives $y'(x)$ (7.21) and $y''(x)$ (7.22), in conjunction with the values $x = x(t)$ derived from Table 7.5. The detailed results are presented in Table 7.6.

The solution to the system of parametric Eq. (7.16) with a step of $\Delta t = 18.033$ s is presented in the final two columns of Table 7.6.

The graphs of the obtained dependences $P(t)$ and $\alpha(t)$ are shown in Fig. 7.4.

When the UAV is moving in a horizontal plane at an altitude of h, the following flight conditions are observed:

$$\theta(t) = 0; \; \dot{\theta}(t) = 0; y(t) = h$$

and the control vectors are described by the expression (7.3).

In this case, Eq. (7.14) is written as:

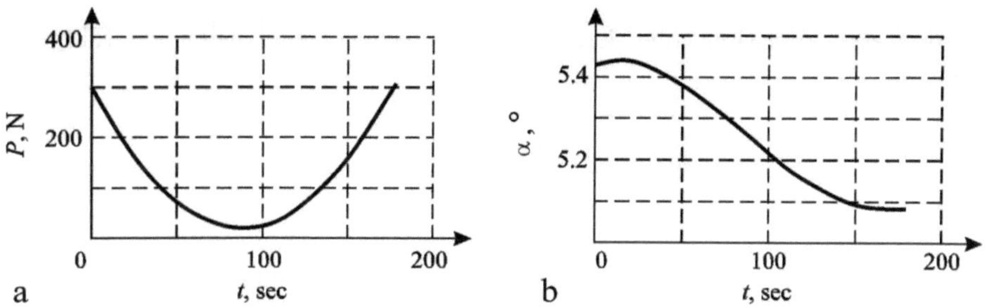

Fig. 7.4 Graphs for functions (**a**) $P(t)$ and (**b**) $a(t)$ Example 7.2

$$
\begin{aligned}
&P(t)-D(\alpha(t),h,V(t))-m\dot{V}(t)=0;\\
&\left[P(t)(\alpha(t)+\varphi_{\text{eng}})+L(\alpha(t),h,V(t))\right]\cos\gamma(t)=mg;\\
&\left[P(t)(\alpha(t)+\varphi_{\text{eng}})+L(\alpha(t),h,V(t))\right]\sin\gamma(t)=mV(t)\dot{\Psi}(t).
\end{aligned}
\tag{7.23}
$$

Subsequent to dividing the second of these equations by the first, the following expression is obtained:

$$
\tan\gamma(t)=\frac{V(t)}{g}\dot{\Psi}(t).
\tag{7.24}
$$

The law of changing the angle of roll of the UAV is defined as follows:

$$
\gamma(t)=\arctan\left(\frac{V(t)}{g}\dot{\Psi}(t)\right),t=[t_0,t_F].
\tag{7.25}
$$

Using the expression (7.23) and the expression:

$$
\cos\gamma(t)=\frac{1}{\sqrt{1+\tan^2\gamma(t)}}=\frac{g}{\sqrt{g^2+V^2(t)\dot{\Psi}^2(t)}}
\tag{7.26}
$$

the second equation from (7.23) can be rewritten as follows:

$$
\left[P(t)(\alpha(t)+\varphi_{\text{eng}})+L(\alpha(t),h,V(t))\right]-m\sqrt{g^2+V^2(t)\dot{\Psi}^2(t)}=0.
\tag{7.27}
$$

Therefore, in order to control the UAV's horizontal flight, it is necessary to determine the control functions $P(t)$ and $\alpha(t)$ by solving a system of nonlinear parametric equations of the second order (7.23), (7.27) over the time interval $[t_0, t_F]$. In this particular instance, the control function $\gamma(t)$ is determined by the Eq. (7.25).

In these expressions, the derivative $\dot{\Psi}(t)$ is calculated using the following equation:

$$\dot{\Psi}(t)=\frac{V(t)z''(x(t))}{\sqrt{1+z'^2(x(t))}},t=[t_0,t_F],\tag{7.28}$$

where $x(t)$ are the results of integrating of the system of equations over the time interval $[t_0, t_F]$:

$$\dot{x}=\frac{V(t)}{\sqrt{1+z'^2(x)}};$$
$$\dot{z}=\frac{V(t)z'(x)}{\sqrt{1+z'^2(x)}}\tag{7.29}$$

with initial conditions:

$$x(0)=x_0;z(0)=z_0.$$

Example 7.3 The formation of the control vector of the UAV, situated at the initial point with coordinates from Example 7.1, is to be accomplished through the utilization of the control function that was previously generated in Example 7.1.

In subsequent analyses, the UAV should be modeled as flying at $t_0 = 0$ at a speed of $V(t) = 100$ m/s at an altitude of $h = 3000$ m along a circular trajectory of the form:

$$z(x)=\sqrt{80000^2-x^2}.\tag{7.30}$$

The results of integrating the system of Eq. (7.29) with the initial conditions ($x(0) = 56{,}568$ m; $z = 56{,}568$ m) are shown in Table 7.7.

Numerical integration was performed up to the time t_F, at which the conditions for the UAV to reach a given endpoint of its movement with coordinates ($x(t_F) = 35{,}777$ m; $z(t_F) = 71{,}554$ m) were met.

The following value of flight time t_F is obtained from this table when the UAV is moving in a horizontal plane:

$$t_F=257.39\,\text{s}.$$

In defining the control functions $P(t)$, $\alpha(t)$, and $\gamma(t)$, the nature of the change in the derivative of $\dot{\Psi}(t)$ plays an essential role. This is reflected in the expressions (7.25) and (7.27) that are utilized.

Table 7.7 Numerical integration for Example 7.3

t (s)	x (m)	z (m)	P (Newton)	α (°)	γ (°)
0	56,568.00	56,568.00	296.102	6.152	–0.730
25.74	54,718.97	58,358.44	296.102	6.152	–0.730
51.48	52,813.30	60,088.47	296.102	6.152	–0.730
77.22	50,852.97	61,756.31	296.102	6.152	–0.730
102.96	48,840.00	63,360.22	296.102	6.152	–0.730
128.70	46,776.47	64,898.54	296.102	6.152	–0.730
154.44	44,664.53	66,369.69	296.102	6.152	–0.730
180.18	42,506.36	67,772.14	296.102	6.152	–0.730
205.92	40,304.18	69,104.44	296.102	6.152	–0.730
231.65	38,060.29	70,365.20	296.102	6.152	–0.730
257.3937	35,777.00	71,553.13	296.102	6.152	–0.730

It is evident that the UAV is operating in steady-state mode along a circular trajectory $z(x)$ with a radius $R = 80{,}000$ m and a velocity $V(t) = 100$ meters per second. Consequently, the following condition is satisfied:

$$\dot{\Psi}(t) = -\frac{V}{R} = -0.0125\,\text{rad/s}; t \in [t_0, t_F]. \tag{7.31}$$

In this context, the minus sign indicates that the UAV is moving in the direction of the origin of the selected coordinate system, as illustrated in Fig. 7.2.

It is noteworthy that the same expression can be obtained by employing the derivatives $z'(x)$ and $z''(x)$ of the function in consideration in relation to (7.28).

The law of variation of the derivative of the trajectory rotation angle should be substituted into the system of Eq. (7.23), (7.25), and Eq. (7.27). The numerical solution of the resulting system yields the following components of the control vector (see Table 7.7):

$$\text{P}(t) = 296.102\,\text{N};\ \alpha(t) = 6.152°; \gamma(t) = -0.730°\text{for time } t \in [0; 257.39].$$

A comparison of these data with the results of Example 7.1 will provide a valuable insight. According to the data presented in Table 7.4, the values of the thrust force, angles of attack, and roll at the final moment of the UAV's initial state for subsequent movement will be equal:

$$(647.49) = 296.307\,\text{N};\ \alpha(647.49) = 6.157°;\ \gamma(647.49) = -5.573°.$$

The transition from these values to the obtained control functions will require, if necessary, minor adjustments to the thrust force and angle of attack of the UAV and a change in its roll angle, which is used to fly along a given circular trajectory $z(x)$ by 4.843°.

7.4.2 Control of the Movement of Helicopter Type UAVs Along Spatial and Planar Trajectories

The movement of an unmanned helicopter over the time interval $[t_0, t_F]$ is generally described by the vector:

$$u(t)=\left(T_{MR}(t),\chi(t),T_{TR}(t)\right),t\in\left[t_0,t_F\right]. \tag{7.32}$$

The vector will be defined through the theoretical framework of inverse problems in the dynamics of controlled objects [6], leveraging the established dependencies: $V_{\text{targ}}(t)$, $\Theta_{\text{path}}(t)$, $\dot{\theta}_{\text{path}}(t)$, $\Psi_{\text{path}}(t)$, and $\dot{\Psi}_{\text{path}}(t)$, $t \in [t_0, t_F]$.

The dynamic equations of motion of the drone are expressed as follows:

$$\begin{aligned}
&T_{\text{mr}}(t)\cos\chi(t)-0.5C_{\text{D}}\rho(t)V_{\text{targ}}^2(t)S_{\text{mr}}-mg\theta_{\text{path}}(t)-m\dot{V}_{\text{targ}}(t)=0;\\
&T_{\text{mr}}(t)\sin\chi(t)-mg\cos\theta_{\text{path}}(t)-m\dot{V}_{\text{targ}}(t)\dot{\theta}_{\text{path}}(t)=0;\\
&T_{\text{tr}}(t)l_{\text{tr}}-M_{\text{t}}-J_y\dot{\Psi}_{\text{path}}(t)=0,t\in\left[t_0,t_F\right],
\end{aligned} \tag{7.33}$$

where T_{mr} denotes the thrust of the main rotor, l_{tr} signifies the offset of the tail rotor from the drone's gravity center, M_{t} represents the torque moment of the main rotor, S_{mr} designates the area of the main rotor disk, χ is the tilt angle of the main rotor axis, and J_y is the central moment of inertia of the drone about the vertical axis of the body axis coordinate system.

The variation in air density experienced by the UAV during its flight can be expressed as follows:

$$\rho(t)=\rho\left(y_{\text{path}}(t)\right),t\in\left[t_0,t_F\right],$$

where $y_{\text{path}}(t)$ is defined as the current value of the drone's flight altitude along the trajectory that is required.

In a similar manner, the variation in the torque of the main rotor over a given interval of time is described as follows:

$$M_{\text{t}}(t)=M_{\text{t}}\left(y_{\text{path}}(t)\right),t\in\left[t_0,t_F\right],$$

The right-hand side of this relation can be determined by the expression for $y = y_{\text{path}}(t)$, $t \in [t_0, t_F]$:

$$M_{\mathrm{t}}(y)=\frac{75N_{\mathrm{eng}}\eta_{\mathrm{mr}}g}{\omega_{\mathrm{mr}}}$$
$$=\frac{75N_{\mathrm{eng}}(0)\eta_{\mathrm{mr}}g\left(18.848e^{-0.129\left(\frac{y(t)}{1000}\right)}-31.696-0.715\left(\frac{y(t)}{1000}\right)\right)}{\omega_{\mathrm{mr}}\sqrt{288.15-6.50\left(\frac{y(t)}{1000}\right)}}, \tag{7.34}$$

where $N_{\mathrm{eng}}(H)$ denotes the engine power at an altitude of H; η_{mr} is the Figure of Merit of the main rotor; ω_{mr} is the rotational speed of the main rotor.

Expressions (7.33) represent a system of nonlinear parametric equations of the third order with respect to unknown functions $T_{\mathrm{mr}}(t)$, $\chi(t)$, $T_{\mathrm{tr}}(t)$, $t \in [t_0, t_{\mathrm{F}}]$.

The analytical simplicity of the system's configuration facilitates the determination of its solution.

The following auxiliary functions will be introduced:

$$\begin{aligned}
&A(t)=m\dot{V}_{\mathrm{targ}}(t)+mg\sin\theta_{\mathrm{path}}(t)+0.5C_{\mathrm{D}}\rho(t)V_{\mathrm{targ}}^{2}(t)S_{\mathrm{mr}};\\
&B(t)=mV_{\mathrm{targ}}(t)\dot{\theta}_{\mathrm{path}}(t)+mg\cos\theta_{\mathrm{path}}(t);\\
&C(t)=J_{y}\dot{\Psi}_{\mathrm{path}}(t)+M_{\mathrm{t}},t\in[t_0,t_{\mathrm{F}}].
\end{aligned} \tag{7.35}$$

Then, the system of Eq. (7.33) can be written as:

$$\begin{aligned}
&T_{\mathrm{mr}}(t)\cos\chi(t)=A(t);\\
&T_{\mathrm{mr}}(t)\sin\chi(t)=B(t);\\
&T_{\mathrm{tr}}(t)l_{\mathrm{tr}}=C(t),t\in[t_0,t_{\mathrm{F}}].
\end{aligned} \tag{7.36}$$

Performing the operation of squaring the right and left sides of the first two Eq. (7.36) and subsequently summing the results using the well-known trigonometric equality yields the following expression:

$$T_{\mathrm{mr}}^{2}(t)=A^{2}(t)+B^{2}(t). \tag{7.37}$$

The following inquiry is posited:

$$T_{\mathrm{mr}}(t)=\sqrt{A^{2}(t)+B^{2}(t)},t\in[t_0,t_{\mathrm{F}}]. \tag{7.38}$$

Employing second equation from (7.36), the following equation is obtained:

$$\sin\chi(t)=\frac{B(t)}{T_{\mathrm{mr}}(t)}. \tag{7.39}$$

The calculation equation is derived from the relations (7.38) and (7.39) as follows:

$$\chi(t) = \arcsin\left(\frac{B(t)}{\sqrt{A^2(t)+B^2(t)}}\right), t \in [t_0, t_F]. \tag{7.40}$$

The solution to the third equation from (7.36) is expressed as follows:

$$T_{\mathrm{tr}}(t) = \frac{C(t)}{l_{\mathrm{tr}}}, t \in [t_0, t_F]. \tag{7.41}$$

Accordingly, the components of the unmanned helicopter control vector (7.32), which ensure its movement along the required spatial trajectory with a given speed, are determined by the Eqs. (7.38), (7.40), (7.41) using expressions (7.35).

The expression (7.8) utilizes the first and second derivatives of the functions $y(x)$ and $z(x)$, which delineate the desired spatial trajectory of the drone in coordinate form.

The argument $x_{\mathrm{path}}(t)$ of these derivatives and the argument $y_{\mathrm{path}}(t)$ of the functions ρ and M_{t} are determined by solving the system of differential Eq. (7.13) with the given initial conditions over the time interval $[t_0, t_{\mathrm{F}}]$.

According to the general approach adopted in the field of flying vehicles control theory [6], the components of the vector (7.32) calculated over a given time interval must satisfy constraints of the form:

$$\begin{aligned} &T_{\mathrm{mr}}^{\min} \le T_{\mathrm{mr}}(t) \le T_{\mathrm{mr}}^{\max};\\ &\chi^{\min} \le \chi(t) \le \chi^{\max};\\ &T_{\mathrm{tr}}^{\min} \le T_{\mathrm{tr}}(t) \le T_{\mathrm{tr}}^{\max}, \quad t \in [t_0, t_F]. \end{aligned}$$

The left and right values of these inequalities are determined by the design characteristics of a particular UAV utilized.

Example 7.4 The initial data for the unmanned helicopter under consideration is presented in Table 7.8 [6].

The UAV with the parameters from Table 7.4 will carry out a descent in a given maneuverable coordinate system with a spatial turn at a speed of $V_{\mathrm{targ}} = 30$ m/s from the starting point with coordinates:

$$(x_0; y_0; z_0) = (2000; 3000; 400)\,\mathrm{m}.$$

The final destination is located at the following coordinates:

$$(x_F; y_F; z_F) = (4000; 1000; 1200)\,\mathrm{m}.$$

Table 7.8 The initial data for the Example 7.4

UAV's parameter	Value
Mass of the UAV, m (kg)	1962
Main rotor disk loading, p_{mr} (N/m^2)	245.25
Radius of the main rotor, R_{mr} (m)	5.0
Main rotor disk area, S_{mr} (m^2)	78.5
Radius of the tail rotor, R_{tr} (m)	0.8
Tail rotor disk area, S_{tr} (m^2)	2.01
Main rotor solidity ratio (σ_{mr})	0.053
Tail rotor solidity ratio (σ_{tr})	0.106
Main rotor figure of merit (η_{mr})	0.675
Tail rotor arm, l_{tr} (m)	6.0
Engine effective power in ground regime, $N_{eng}(0)$ (hp)	290
Maximum engine rotor speed, n_{max} (rpm)	2400
Mean power utilization factor (ξ)	0.835
Engine seizure angle, φ_{eng} (°)	6
Main rotor speed, n_{mr} (rpm)	300
Angular velocity of the main rotor, ω_{mr} (s^{-1})	31.4
Tail rotor speed, n_{tr} (rpm)	1672
Angular velocity of the tail rotor, ω_{tr} (s^{-1})	175
Parasitic drag coefficients:	
(a) vertical take-off/landing (b) other flight modes	$C_D = 0.09554$ $C_D = 0.03185$
The central moment of inertia of the drone about the vertical axis of the body axis coordinate system, J_y (kg·m^2·s^{-1})	300
Minimal flight speed, V_{min} (m/s)	30
Cruise flight speed, V_{cr} (m/s)	72.2
Maximal flight speed, V_{max} (m/s)	100

In this particular instance, the initial and final values of the trajectory angles are equal, respectively:

$$\theta_0 = 0°; \theta_F = 0°; \Psi_0 = 45°; \Psi_F = 80°.$$

Projections of the required flight trajectory will be formed on the xy and xz coordinate planes, which will take the following form:

$$y(x) = 5{\cdot}10^{-7}x^3 - 4.5{\cdot}10^{-3}x^2 + 12x - 7{\cdot}10^3;$$
$$z(x) = 1.468{\cdot}10^{-6}x^3 - 1.2{\cdot}10^{-2}x^2 + 31.556x - 2.629{\cdot}10^4.$$

The graphs of these functions for $x \in [2000; 4000]$ are shown in Fig. 7.5.

The application of the obtained projections $y(x)$ and $z(x)$, along with their respective derivatives, to the solution of the system of Eq. (7.13) with initial conditions over the time interval $[0, t_F]$ results in a parametric representation of the required trajectory of the drone descent, as presented in Table 7.9.

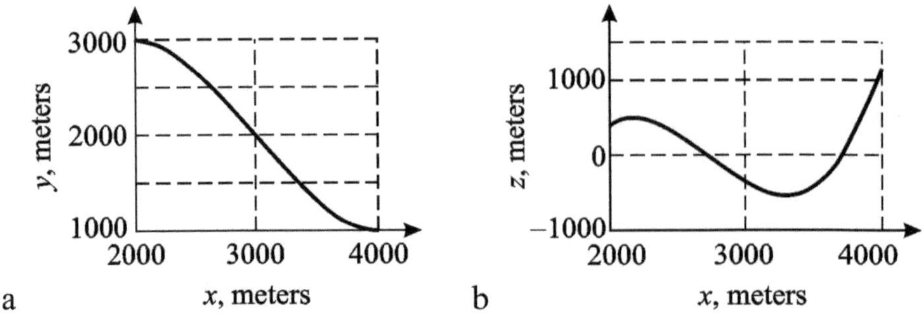

Fig. 7.5 The graphs of functions (**a**) $y(x)$ and (**b**) $z(x)$ for $x \in [2000; 4000]$"

Table 7.9 The solution of the system of Eq. (7.13) for the Example 7.4

t (s)	*x* (m)	*y* (m)	*z* (m)
0	2000	3000	400
17.054	2392	2799	382.830
34.108	2617	2547	130.240
51.162	2796	2302	–113.246
68.216	2973	2040	–338.501
85.270	3192	1716	–519.39
102.324	3472	1344	–457.179
119.377	3657	1156	–149.363
136.431	3791	1061	246.093
153.485	3904	1013	703.850
170.539	4000	1000	1200

The time point t_F of completion of the considered flight mode of the UAV is determined by the fulfillment of the initial conditions, and is the value $t_F = 170.539$ s.

A visual representation of the spatial trajectory of the drone, derived from the data contained within the final three columns of the resulting Table 7.5, is presented in Fig. 7.6.

The dynamics of changes in the angles $\Theta_{path}(t)$, $\Psi_{path}(t)$, and their derivatives at $t \in [0; 170.539]$ are determined by Eq. (7.8) and presented in Table 7.10.

As demonstrated in the foregoing Table 7.10, the angular boundary conditions for the formed trajectory of the drone are fully satisfied.

The UAV control vector, as delineated in Eq. (7.32), is to be formulated in accordance with the methodologies outlined in the subsequent Eqs. (7.38), (7.40), and (7.41). The values of its parameters are exhibited in Table 7.8.

In this case, the change in air density during the drone's flight will be described by the following equation:

$$\rho(t) = \rho_0 e^{-0.0001 y(t)}, \tag{7.42}$$

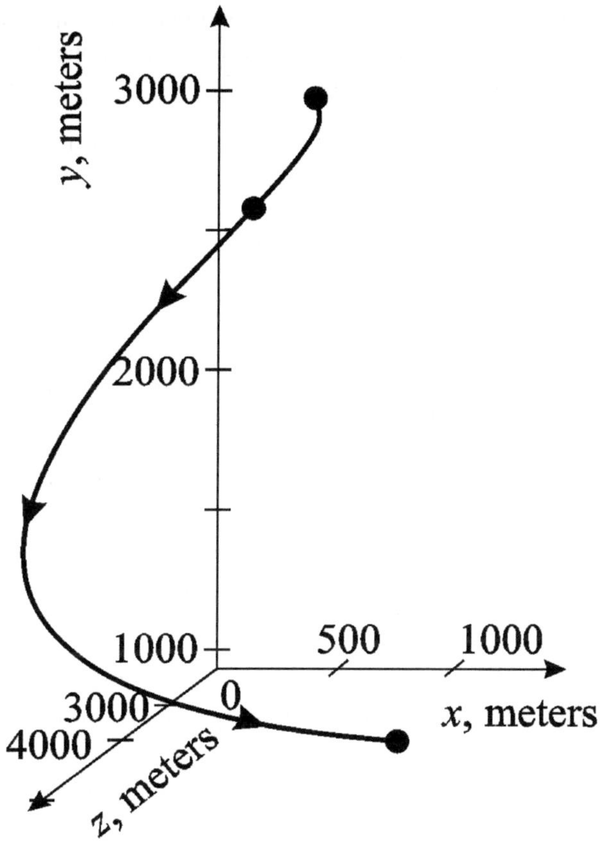

Fig. 7.6 The UAV's spatial trajectory

Table 7.10 The dynamics of changes in the angles $\Theta_{path}(t)$ and $\Psi_{path}(t)$

t (s)	θ (°)	$\dot{\theta}$ (radian per second)	Ψ (°)	$\dot{\Psi}$ (radian per second)
0	0	–0.064	45	–0.069
17.054	–43.420	–0.016	–40.743	–0.029
34.108	–51.996	$-4.870 \cdot 10^{-3}$	–52.775	$-4.249 \cdot 10^{-3}$
51.162	–55.172	$-2.029 \cdot 10^{-3}$	–53.699	$1.903 \cdot 10^{-3}$
68.216	–56.291	$-2.726 \cdot 10^{-3}$	–48.418	0.010
85.270	–55.310	$2.897 \cdot 10^{-3}$	–24.568	0.052
102.324	–49.374	$8.226 \cdot 10^{-3}$	45.468	0.044
119.377	–40.458	0.010	67.091	0.011
136.431	–29.271	0.013	74.220	$4.898 \cdot 10^{-3}$
153.485	–15.387	0.015	77.833	$2.787 \cdot 10^{-3}$
170.539	$-9.901 \cdot 10^{-7}$	0.016	80	$1.750 \cdot 10^{-3}$

where $y(t)$ is the altitude of the flight in the time interval $t \in [0; 170.539]$.

The results of the calculations of the components of the UAV control vector, as determined using the Eqs. (7.35), (7.38), (7.40), and (7.41), are presented in Table 7.11.

As illustrated in Fig. 7.7, the graphs depict the obtained dependences of $T_{mr}(t)$, $\chi(t)$, and $T_{tr}(t)$ at $t \in [0; 170.539]$.

When the UAV is flying in a vertical plane, we have that $\Psi(t) = 0$ and $\dot{\Psi}(t) = 0$, $t \in [t_0, t_F]$. In this particular instance, it can be deduced from expressions (7.35) and (7.41) that $T_{tr}(t) = M_t(t)/l_{tr}$, for all values of t that fall within the interval $[t_0, t_F]$.

The remaining components of the control vector (7.32) are calculated using the formulas (7.38) and (7.40).

Example 7.5 The formation of the control vector (7.32) for the unmanned helicopter is initiated [6], with the objective of executing an altitude climb within the time interval $t \in [0; 60]$ at a velocity of $V_{targ} = 72.2$ meter per second, while maintaining a trajectory angle of $\theta(t) = 15°$ from the initial point, as delineated by the following coordinates:

$$(E_0, C_0) = (500{,}100)\,\text{m}.$$

Table 7.11 The results of the calculations of the UAV control vector Example 7.4

t (s)	T_{mr} (Newton)	χ (°)	T_{tr} (Newton)
0	15,528.418	86.230	630.801
17.054	17,846.377	46.947	650.342
34.108	18,227.499	39.362	674.223
51.162	18,281.426	36.481	697.111
68.216	18,307.062	35.619	722.524
85.270	18,400.612	37.185	756.621
102.324	18,677.095	44.162	794.409
119.377	18,944.191	53.542	812.768
136.431	19,349.269	65.030	822.819
153.485	19,834.406	78.778	827.870
170.539	20,199.059	86.461	829.295

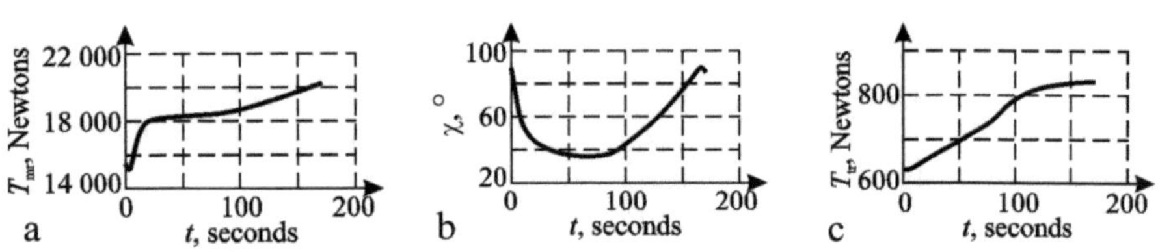

Fig. 7.7 The graphs of (**a**) $T_{mr}(t)$, (**b**) $c(t)$, and (c) T_{tr}(t) for Example 7.4

Table 7.12 The results of the calculations of the UAV control vector Example 7.5

t (s)	*x* (m)	*y* (m)	T_{mr} (Newton)	χ (°)	T_{tr} (Newton)
0	500	100	22,613.074	55.274	925.114
6	918.4	212.101	22,563.006	55.457	911.004
12	1337	324.202	22,513.725	55.640	897.044
18	1755	436.304	22,465.219	55.821	883.233
24	2174	548.405	22,417.477	56.002	869.571
30	2592	660.506	22,370.485	56.180	856.057
36	3011	772.607	22,324.233	56.358	842.696
42	3429	884.708	22,278.707	56.534	829.491
48	3848	996.810	22,233.898	56.709	816.451
54	4266	1109.911	22,189.794	56.883	803.473
60	4684	1221.012	22,146.382	57.056	790.903

The standard formulation of the rectilinear flight path required for the UAV, in its standard coordinate form, is expressed as follows:

$$y(x) - y_0 = \tan 15° (x - x_0).$$

Subsequent to the implementation of elementary transformations, the following equation is obtained:

$$\mathrm{C}(\mathrm{E}) = 0.2679\mathrm{E} - 33.9745.$$

The parametric representation of the required UAV climb trajectory, obtained by integrating the system of Eq. (7.18) with the specified initial conditions, is presented in the first three columns of Table 7.12.

The numerical values of the UAV parameters from Table 7.8 with additional initial data will be used to determine the components of the control vector $u(t)$, for $t \in [0; 60]$.

$$\theta(t) = 15° = 0.2618\,\text{rad}, \dot{\theta}(t) = 0.$$

When calculations are performed, the above dependence of air density on height (7.42) will be applied for altitudes $y(t)$ from Table 7.12.

The values of the moment $M_t(t) = M_t(y(t))$ are calculated using the Eq. (7.34).

The calculation results using expressions (7.38), (7.40), and (7.41) are shown in the last three columns of Table 7.12 and Fig. 7.8.

The UAV control is formed during flights in the horizontal plane at an altitude of h, with the vector $u(t)$ also expressed in the form of Eq. 7.32.

Equations (7.38), (7.40), and (7.41) apply under the following conditions:

$$\theta(t) = 0; \dot{\theta}(t) = 0, t \in [t_0, t_F]$$

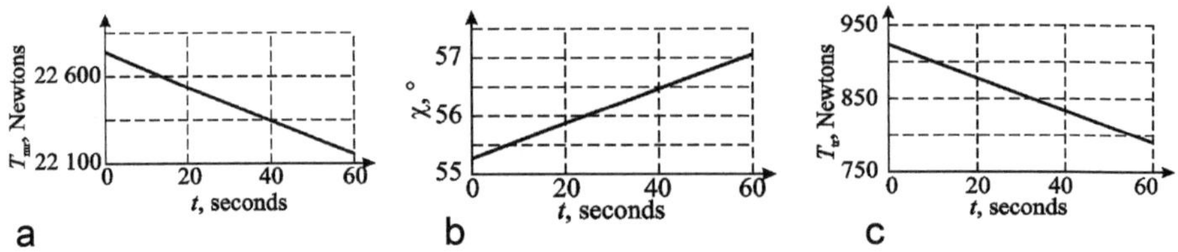

Fig. 7.8 The graphs of (**a**) $T_{mr}(t)$, (**b**) $c(t)$, and (**c**) $T_{tr}(t)$ for Example 7.5

Then, from expressions (7.35), the following has been derived:

$$A(t) = m\dot{V}_{\text{targ}}(t) + 0.5C_{\text{D}}\rho(h)V_{\text{targ}}^{2}(t)S_{\text{mr}};$$
$$B(t) = mg, t \in [t_0; t_{\text{F}}]. \tag{7.43}$$

When $V_{\text{targ}}(t)$ = const and $\dot{V}_{\text{targ}}(t) = 0$, the first of these relations takes the following form:

$$A(t) = 0.5C_{\text{D}}\rho(h)V_{\text{targ}}^{2}(t)S_{\text{mr}}. \tag{7.44}$$

In this equation, the value $\rho(h)$ is obtained by substituting the altitude $y(t) = h$, where $t \in [t_0; t_F]$, into expression (7.42).

The function $C(t)$ is calculated using Eq. (7.35), where the dependence $\dot{\Psi}_{\text{path}}(t)$ is formed using the coordinate $z = z(x)$ and the parametric representations $x(t)$ and $z(t)$ of the required flight path of the drone, with t belonging to the interval $[t_0, t_F]$.

Example 7.6 Consider a UAV performing a turn with a radius of $R = 2000$ m at an altitude of $h = 100$ m with different velocity modes $V_{\text{targ}}(t)$, where t belongs to the interval $[0; t_F]$ [6].

It will be assumed that, as in Example 7.4, a maneuverable system of coordinate is used with the beginning in the center of the circular trajectory of the turn (see Fig. 7.6). The initial and final flight conditions of the drone will be set to the following values:

$$(x_0; y_0; z_0) = (0; 100; 2000) = (x_{\text{F}}; y_{\text{F}}; z_{\text{F}}); \Psi_0 = 0°.$$

Let's assume the helicopter is turning clockwise.

In this case, the required change in the derivative of the UAV's heading angle is calculated as follows:

$$\dot{\Psi}_{\text{path}}(t) = -\frac{V_{\text{targ}}(t)}{R}, t \in [0; t_{\text{F}}]. \tag{7.45}$$

The control vectors of the drone are to be defined for various laws $V_{\text{targ}}(t)$, $t \in [0; t_F]$:

$$u(t)=\left(T_{\text{mr}}(t),\chi(t),T_{\text{tr}}(t)\right),t\in\left[0;t_{\text{F}}\right].$$

The formation of these laws will be guided by a linear relationship, expressed as follows:

$$V_{\text{targ}}(t)=a+bt,t\in\left[0;t_{\text{F}}\right]. \tag{7.46}$$

For the case under consideration, it is necessary to determine the values of the parameters a, b, and t_F for the following modes of UAV movement along a closed trajectory:

Case 1: a turn with constant velocity $V_{targ}(t) = 72.2$ m/s;
Case 2: a braking turn from $V_{targ}(0) = 72.2$ m/s to $V_{targ}(t_F) = 40$ m/s;
Case 3: a turn with acceleration from $V_{targ}(0) = 72.2$ m/s to $V_{targ}(t_F) = 100$ m/s.

When the UAV flies at a constant speed (Case 1), the values in expression (7.46) are as follows:

$$a=72.2\text{m}/\text{s};b=0;t_{\text{F}}=\left(2\pi R\right)/V_{\text{targ}}=174.09\text{s}.$$

The following values for the components of the desired control vector are obtained by applying the Eqs. (7.43), (7.44), (7.45), (7.38), (7.40) and (7.41):

$$T_{\text{mr}}(t)=20{,}800\text{N};\chi(t)=67.671^{\circ};T_{\text{tr}}(t)=933.373\text{N};t\in\left[0;174.049\right].$$

When the UAV flies with a braking (Case 2), the value a in expression (7.46) is 72.2 meter per second. The values of the remaining parameters, b and t_F, must be determined using the following system of nonlinear algebraic equations:

$$\begin{aligned}&\textstyle\int_0^{t_{\text{F}}}\left(72.2-bt\right)dt=2\pi R;\\&72.2-bt_{\text{F}}=40.\end{aligned}$$

Calculating the integral and substituting the values of $R = 2000$ m allows us to define this system as follows:

$$\begin{aligned}&72.2t_{\text{F}}-0.5bt_{\text{F}}^2-12560=0;\\&32.2-bt_{\text{F}}=0.\end{aligned} \tag{7.47}$$

The values of the desired parameters are found by solving this system of equations numerically:

$$b=0.144;t_{\text{F}}=223.999\text{s}.$$

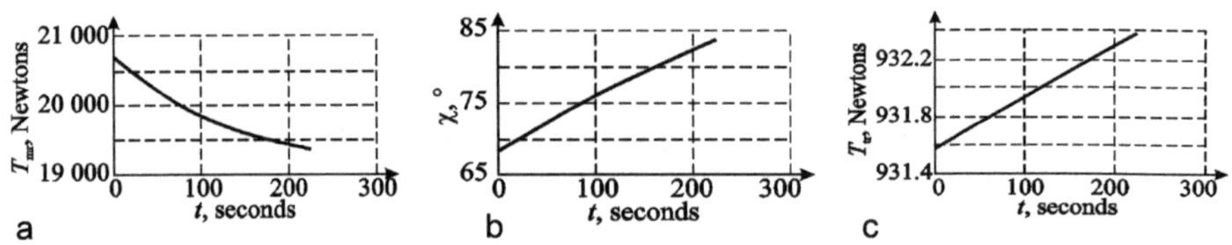

Fig. 7.9 The graphs of (**a**) $T_{mr}(t)$, (**b**) $c(t)$, and (**c**) $T_{tr}(t)$ for Example 7.6 Case 2

Table 7.13 The results of the calculations of the UAV control vector Example 7.6 Case 2

t (s)	T_{mr} (Newton)	χ (°)	T_{tr} (Newton)
0	20,694.688	68.393	931.568
22.4	20,450.562	70.193	931.649
44.8	20,236.816	71.948	931.729
67.2	20,051.311	73.652	931.810
89.6	19,891.853	75.299	931.891
112.0	19,756.219	76.882	931.971
134.4	19,642.185	78.395	932.052
156.8	19,547.551	79.834	932.133
179.2	19,470.163	81.194	932.213
201.6	19,407.939	82.472	932.294
224.0	19,358.884	83.664	932.375

Therefore, the law of drone velocity variation (7.46) is defined as follows:

$$V_{\text{targ}}(t) = 72.2 - 0.144t, t \in [0;223.999].$$

Figure 7.9 and Table 7.13 show the results of the calculations for UAV control during a braking turn.

When a drone turns with acceleration (Case 3), the law of speed change is of the form:

$$V_{\text{targ}}(t) = 72.2 + bt, t \in [0;t_{\text{F}}].$$

The system of nonlinear equations used to determine the parameters b and t_{F} is written in a manner similar to the expressions in (7.47):

$$72.2t_{\text{F}} - 0.5bt_{\text{F}}^2 - 12560 = 0;$$
$$bt_{\text{F}} - 27.8 = 0.$$

The roots of this system are the following values:

$$b = 0.190; t_{\text{F}} = 145.951\text{s}.$$

Thus, the control vector for the third flight mode of the UAV is calculated using a law of the following form:

$$V_{\text{targ}}(t) = 72.2 + 0.190t, t \in [0;145.951].$$

The results of this calculation are shown in Table 7.14 and Fig. 7.10.

Note that the computational procedures used to solve this example with different $V_{\text{targ}}(t)$ laws, where t belongs to the interval $[0; t_F]$, showed that the obtained control vectors ensure the unmanned helicopter moves along a closed circular trajectory with a radius of $R = 2000$ m.

Now, let's examine the features of forming control vectors for executing specific flight modes of helicopter-type UAVs.

When implementing the UAV braking mode with hovering, the control vector $u(t)$ is determined using the expressions (7.35), (7.38), (7.40), and (7.41), and when substituting the laws of velocity change $V_{\text{targ}}(t)$ and acceleration $\dot{V}_{\text{targ}}(t)$ are used.

Note that when the drone decelerates at a constant altitude, $T_{\text{tr}}(t)$ is constant for all t in the interval $[t_0, t_F]$.

For the rearward flight mode of the UAV along the trajectory of a previously performed direct flight, the following transformations are performed in expressions (7.35):

Table 7.14 The graphs of $T_{\text{mr}}(t)$, $\chi(t)$, and $T_{\text{tr}}(t)$ for Example 7.6 Case 3

t (s)	T_{mr} (Newton)	χ (°)	T_{tr} (Newton)
0	20,944.849	66.727	931.568
14.595	21,196.932	65.191	931.499
29.190	21,474.238	63.636	931.429
43.785	21,777.867	62.067	931.360
58.380	22,108.839	60.490	931.291
72.975	22,468.089	58.910	931.221
87.571	22,856.463	57.331	931.152
102.166	23,274.713	55.759	931.083
116.761	23,723.496	54.197	931.014
131.356	24,203.373	52.651	930.944
145.951	24,714.811	51.124	930.875

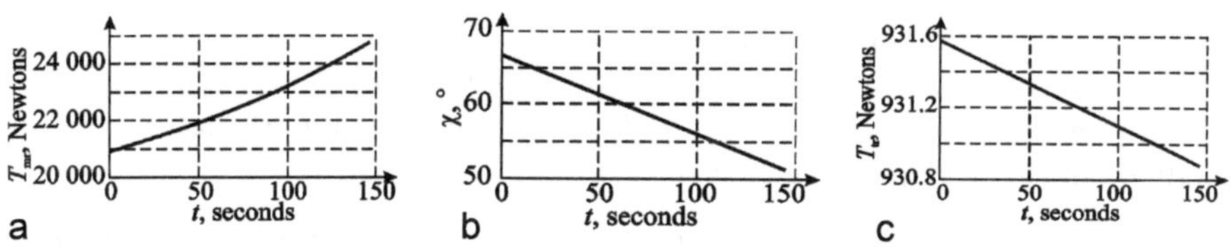

Fig. 7.10 The graphs of (**a**) $T_{mr}(t)$, (**b**) $c(t)$, and (**c**) $T_{tr}(t)$ for Example 7.6 Case 3

$$\sin\theta_{\text{path}} = \sin\left(\theta_{\text{path}}^{\text{dir}} + \pi\right) = -\sin\theta_{\text{path}}^{\text{dir}};$$
$$\cos\theta_{\text{path}} = \cos\left(\theta_{\text{path}}^{\text{dir}} + \pi\right) = -\cos\theta_{\text{path}}^{\text{dir}},$$

where $\theta_{\text{path}}^{\text{dir}}$ is the angle of turning of the drone's trajectory in direct flight.

In the process of turning the UAV while it is hovering, the control vectors implemented in the time interval $t \in [t_0, t_F]$ will be equal to:

$$T_{\text{mr}}(t) = mg; \chi(t) = 0; T_{\text{tr}}(t) = \frac{\Psi_2 - \Psi_1}{(t_2 - t_1)l_{\text{tr}}} + \frac{M_{\text{t}}(y_{\text{hov}})}{l_{\text{tr}}}, t \in [t_0, t_F]. \tag{7.48}$$

Here, Ψ_1 and Ψ_2 are the initial and final heading angle values when the drone turns; h_{hov} is the hovering and turning altitude of the unmanned helicopter.

The UAV's hovering mode at altitude h_{hov} with a heading angle of Ψ_{hov} is performed at time t with a control vector of the form:

$$T_{\text{mr}}(t) = mg; \chi(t) = 0; T_{\text{tr}}(t) = \frac{M_{\text{t}}(y_{\text{hov}})}{l_{\text{tr}}}, t \in [t_0, t_F]. \tag{7.49}$$

The value of the tail rotor thrust, which is used only to balance for the reactive torque of the main rotor, is obtained from the expression (7.48) for $\Psi_1 = \Psi_2 = \Psi_{\text{hov}}$.

The following are the calculated ratios for controlling an unmanned helicopter in vertical takeoff and landing modes.

In these modes, the drone is controlled by changing the $T_{\text{mr}}(t)$ and $T_{\text{tr}}(t)$ forces over time at $\chi(t) = 0$.

For a vertical takeoff of the UAV, expression (7.37) yields the following calculation formula when $B(t) = 0$:

$$T_{\text{mr}}(t) = m\dot{V}_{\text{targ}}(t) + 0.5C_{\text{D}}\rho_0 V_{\text{targ}}(t) S_{\text{mr}} + mg, t \in [t_0, t_F]. \tag{7.50}$$

Similarly, the expression can be written to describe the change in thrust of the main rotor of the UAV during its vertical landing mode:

$$T_{\text{mr}}(t) = mg - m\dot{V}_{\text{targ}}(t) - 0.5C_{\text{D}}\rho_0 V_{\text{targ}}(t) S_{\text{mr}}, t \in [t_0, t_F]. \tag{7.51}$$

Due to the low take-off and landing altitudes in expressions (7.50) and (7.51), the air density value at ground level is used.

The value of $T_{\text{tr}}(t)$ can be determined from expression (7.49) for $y_{\text{hov}} = 0$.

Additionally, the laws of change in the control action of $T_{\text{mr}}(t)$ depend significantly on the given dependencies of $V_{\text{targ}}(t)$.

Example 7.7 The drone's control vector during its vertical takeoff to an altitude of $h = 500$ m must be determined [6].

The following laws govern the drone's speed and acceleration:

$$V_{\text{targ}}(t) = 0.075t^2 - 2.545 \cdot 10^{-3} t^3 + 2.172 \cdot 10^{-5} t^4;$$
$$\dot{V}_{\text{targ}}(t) = 0.15t - 7.635 \cdot 10^{-3} t^2 + 8.688 \cdot 10^{-5} t^3, t \in [0;58.594]. \tag{7.52}$$

Figure 7.11 and Table 7.15 show the graph and values of these dependencies.

Substitution of the corresponding parameters of the UAV under consideration from Table 7.8 into expression (7.50) and utilization of the value $\rho_0 = 1.225$ kg/m^3 results in the following form of the desired control vectors:

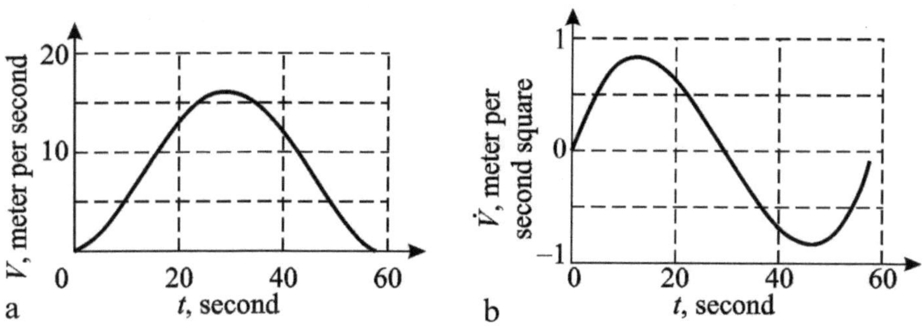

Fig. 7.11 Graphs of (**a**) $V_{\text{targ}}(t)$ and (**b**) $\dot{V}_{\text{targ}}(t)$ for Example 7.7

Table 7.15 Values of $V_{\text{targ}}(t)$ and $\dot{V}_{\text{targ}}(t)$ for Example 7.7

t (s)	V_{targ} (meter per second)	$\dot{V}_{\text{targ}}$ (meters per second square)
0	0	0
4	1.036	0.480
8	3.558	0.749
12	6.790	0.840
16	10.087	0.787
20	12.940	0.623
24	14.971	0.382
28	15.937	0.097
32	15.729	–0.200
36	14.369	–0.474
40	12.014	–0.692
44	8.955	–0.820
48	5.616	–0.826
52	2.553	–0.676
56	0.458	–0.337
68.594	0	0

$$T_{mr}(t) = [1962\left(0.15t - 7.635{\bullet}10^{-3}t^2 + 8.688{\bullet}10^{-5}t^3\right) +$$
$$68.906\left(0.075t^2 - 2.545{\bullet}10^{-3}t^3 + 2.172{\bullet}10^{-5}t^4\right) + 19240.647];$$
$$\chi(t) = 90°; T_{tr} = 945.656, t \in [0;48.925]. \quad (7.53)$$

The results of calculations employing the first expression (7.53) are presented in Table 7.16 and Fig. 7.12.

It is imperative to acknowledge the necessity of mandatory verification of the formed vectors $u(t)$ of indirect UAV control. This verification is achieved by integrating the corresponding equations of motion on the interval $[t_0, t_F]$ and subsequently comparing the obtained functions $V(t)$, $\Psi(t)$, $x(t)$, $y(t)$, $z(t)$ with the given and required dependencies $V_{targ}(t)$, $\Psi_{path}(t)$, $t \in [t_0, t_F]$.

Table 7.16 Results of calculations for Example 7.7

t (s)	T_{mr} (Newton)
0	19,240.647
4.077	20,361.579
8.154	20,959.784
12.231	21,230.328
16.308	21,232.968
20.385	20,933.780
24.462	20,311.137
28.540	19,445.537
32.617	18,546.849
36.694	17,906.755
40.771	17,798.224
44.848	18,378.001
48.925	19,240.647

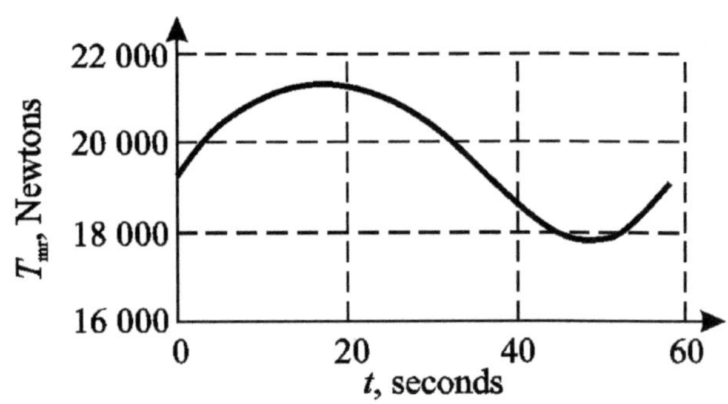

Fig. 7.12 Graph of $T_{mr}(t)$ for Example 7.7

7.5 Conclusions

This chapter has examined the fundamental principles of the control of flying vehicles. Examples of various types of UAVs and the features of their control systems were provided. The present chapter analyzes the indirect control of UAVs of fixed-wing, which is invariant for all their design schemes, based on the calculated examples. Furthermore, the methods of establishing indirect control of a single-rotor unmanned helicopter were analyzed.

References

1. Klenke R, Sleeman W, Motter M. A high-throughput processor for flight control research using small UAVs. In: 25th AIAA Aerodynamic Measurement Technology and Ground Testing Conference, 5–8 June 2006. San Francisco, CA: American Institute of Aeronautics and Astronautics. p. 2006.
2. Li L, Tiniakov D. Aeronautics—Characteristics and emerging technologies. In: Chapter: UAVs and hydrogen technologies are the future of civil aviation [Internet]. IntechOpen; 2025. https://doi.org/10.5772/intechopen.1008352.
3. Ducard G, Allenspach M. Review of designs and flight control techniques of hybrid and convertible VTOL UAVs. Aerosp Sci Technol. 2021;118:107035. https://doi.org/10.1016/j.ast.2021.107035.
4. Tiniakov D, Li L. Airworthiness engineering basis. Datian Aviation Beijing Beixue Publishing House; 2020.
5. Tiniakov D, Li L, Su Y. Airworthiness design of civil aircraft systems and structure. Datian Aviation Beijing Beixue Publishing House; 2018.
6. Moiseev V. Unmanned aerial vehicles flight dynamics and control. Kazan: RIC Shkola; 2017.
7. Sebbane Y. Smart autonomous aircraft. NW: CRC Press is an imprint of Taylor & Francis Group; 2016.

Chapter 8
Navigation System

8.1 Introduction

For the UAV, the navigation system is the basis of autonomy and the "sense of space." The primary objective of the aforementioned entity is to methodically and precisely address three fundamental inquiries: Where is the drone located? What is its intended flight path? How will it get there? [1] The absence of resolution to these issues renders the UAV a financially unviable and, in its practical applications, ineffectual device. Navigation entails more than merely determining coordinates; it integrates position, speed, and orientation data, thereby enabling the autopilot to stabilize the vehicle and adhere to a predetermined route while accomplishing its mission, whether it pertains to aerial photography, cargo delivery, or monitoring.

The evolution of UAV navigation is inextricably intertwined with the advancement of communication technologies, automation, and, subsequently, electronics [1–3]. The initial iterations of UAVs were manually operated via radio communication by a pilot-operator, thereby resembling remote piloting rather than autonomous navigation. The advent of Global Satellite Navigation Systems (GNSS), exemplified by GPS, Galileo, GLONASS, and Beidou, marked a pivotal breakthrough, ushering in a new era of technological sophistication. This innovation empowered the drones to ascertain their coordinates with an accuracy of up to meters, irrespective of their location on the planet.

Concurrently, Inertial Navigation Systems (INS) underwent development, employing accelerometer and gyroscope data to determine movement absent external landmarks. However, INS are susceptible to the accumulation of errors over time. The synthesis of GNSS and INS has emerged as the prevailing solution: satellites rectify the "drifting" values of inertial sensors, while INS facilitates uninterrupted navigation during brief periods of satellite signal interruption. This combination has led to the remarkable proliferation of commercial and civilian UAVs in the twenty-first century.

L. Li, D. Tiniakov, *Unmanned Aircraft Design*,
https://doi.org/10.1007/978-981-95-7167-3_8

Notwithstanding the remarkable strides made, contemporary UAV navigation systems confront a multitude of substantial challenges.

The primary concern pertains to the vulnerability of GNSS. The transmission of satellite signals is susceptible to interference, including jamming or spoofing. In situations involving conflict or the engagement with UAVs, the utilization of this technology renders the device ineffective or prompts it to pursue an erroneous trajectory.

The employment of drones in environments characterized by limited satellite signal constitutes a challenging undertaking. Flights within buildings, subterranean tunnels, and dense forest areas or urban canyons where GNSS signals are not available to necessitate the implementation of alternative solutions. In such cases, the utilization of computer vision-based technologies, such as correlation with the map and optical flow analysis, along with lidar and ultrasonic sensors, can prove instrumental in addressing the challenges posed by these dynamic environments. However, the implementation of these systems often necessitates substantial computing capabilities and the prior preparation of maps.

The accuracy and reliability of navigation systems are paramount [4]. In the context of precision farming, construction, and autonomous parcel delivery, achieving accuracy within a few meters is no longer sufficient. Achieving centimeter-level accuracy and ensuring reliable performance are paramount, driving the development of differential correction systems (Real-Time Kinematic, Plan Position Presentation) and sophisticated sensor fusion algorithms.

It is imperative to consider regulatory barriers when assessing the utilization of UAVs for civilian applications [4]. The issue is further complicated by the fact that the laws pertaining to the use of drones vary significantly from one country to another. Moreover, they are continually updated, ensuring their currency and relevance. The integration of UAVs into the existing common airspace necessitates the assurance of navigation reliability, comparable to that of manned aircraft, for operations such as delivery or taxiing. This necessitates the establishment of redundancy and backup systems that are capable of ensuring safety even in the case of partial failures.

Consequently, the progression of navigation systems for UAVs is shifting from a mere reliance on satellite coordinates to the development of an integrated, sustainable, and intelligent "sense of space." [3] The future of drone technology lies in hybrid systems that seamlessly integrate GNSS, INS, video camera, lidar, and radio beacon data, thereby creating a reliable global vision for the drone, even in the most challenging conditions. The successful navigation of contemporary challenges will serve as a catalyst for the substantial and pervasive integration of drones into our daily lives.

8.2 Basic Concept of Navigation

Navigation can be defined as the process of determining the location, direction, and control of the movement of a vehicle from one point to another [1–5]. It is the cornerstone of transportation in general and, in particular, aviation. In the context of manned aviation, safety is paramount for passengers and crew. In contrast, for

UAVs, safety is not only a priority but also a critical factor in determining their effectiveness and the safety of others. The concepts of navigation for UAVs have been largely inherited from classical aviation; however, they possess unique features due to their autonomy, scale, and tasks.

At the core of aviation navigation are several fundamental concepts.

The location of a specific point in space is determined by a system of coordinates that comprises latitude, longitude, and altitude. In the context of UAVs, altitude is frequently gauged not from sea level but from the takeoff point (Fig. 8.1a) [6]. This distinction is of paramount importance when operating UAVs in close proximity to the ground.

The trajectory of a vehicle is defined as a sequence of points (waypoints) that the vehicle must successfully navigate. In contrast to the trajectory of an airliner, the route of a UAV can be significantly more complex, encompassing maneuvers such as hovering, navigating around obstacles, and following non-linear trajectories (Fig. 8.1b) [3].

The orientation and heading of a vehicle in space are determined by roll, pitch, and yaw, which are defined as the angular velocities about the longitudinal, lateral, and vertical axes, respectively [6]. For UAVs, precise orientation detection is paramount for ensuring stability, particularly in windy conditions.

The navigation (wind) triangle is defined as the ratio between ground speed (relative to the ground), airspeed (relative to the air), and the wind vector (Fig. 8.1c) [3]. This concept directly impacts the accuracy of the exit trajectory and the UAV's flight duration.

As previously mentioned, navigation tasks can be distilled into three fundamental inquiries: Where is the drone located? What is its intended flight path? How will

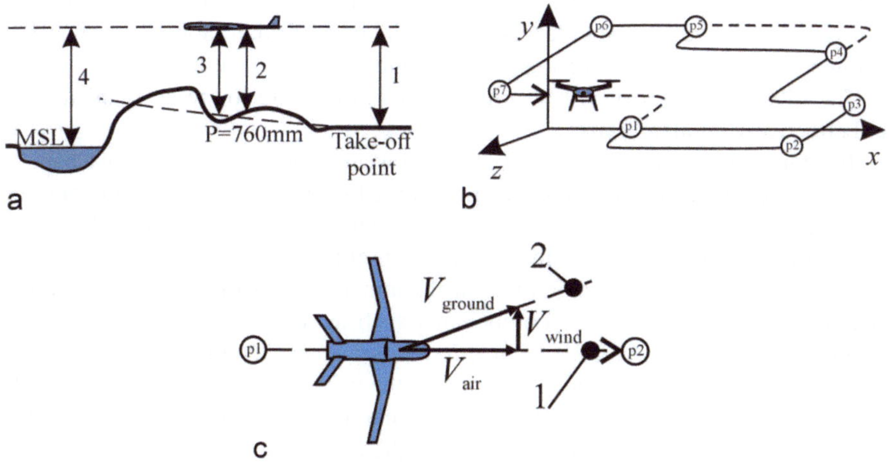

Fig. 8.1 The fundamental concepts of the navigation of UAVs. (**a**) 1—altitude above take-off point; 2—true flight altitude; 3—barometric altitude; 4—altitude relative to mean sea level. (**b**) Drone flight by waypoints. (**c**) The navigation (wind) triangle: 1—assigned track; 2—instrument course

it get there? [1] The implementation of solutions to these problems is achieved through a series of systems.

A thorough analysis of these tasks is warranted.

The determination of location and orientation is of paramount importance.

Contemporary UAVs address this issue not through a single method, but rather through a multifaceted approach that incorporates redundancy and accuracy.

The utilization of satellite navigation systems, such as GPS, GLONASS, BeiDou, and Galileo, constitutes the foundation for the majority of UAV flights [1–5]. The GNSS receiver provides precise geographical coordinates. However, the system is vulnerable to signal loss in urban areas, suppression, and spoofing. RTK or PPK (Post-Processed Kinematic) technologies are employed to achieve centimeter-level accuracy, a critical factor in cartography and precision farming.

The INS comprises accelerometers and gyroscopes that quantitive accelerations and angular velocities [3]. The system integrates this data to determine the position, speed, and orientation of the vehicle. The primary disadvantage associated with this approach is the gradual accumulation of errors over time. Consequently, the INS is best utilized in conjunction with GNSS. GNSS periodically corrects the drifting readings of the INS, thereby enabling the vehicle to maintain its course and position in the event of a short-term loss of satellite signal.

The capacity to perform visual odometry and Simultaneous Localization and Mapping (SLAM) distinguishes advanced UAVs. Utilizing cameras and lidar, the vehicle generates a map of the environment and concurrently ascertains its position on that map. This is imperative for autonomous flights in indoor environments, among forests, or amidst buildings where GNSS is either unavailable or has a weak signal. For instance, a copter utilizing SLAM to inspect the interior of a pipe employs this technology [4].

The measurement of altitude is achieved through the use of two instruments: a barometric altimeter and a sonar. These instruments serve to determine the altitude of a vehicle relative to ground level or the initial takeoff point. Sonar is particularly advantageous for precise hovering at low altitudes and automatic landing.

The subsequent task of utmost importance is to adhere to the designated route and effectively regulate traffic.

To address this task, it is imperative to employ an autopilot and FCS (see Chap. 7).

Subsequent to the reception of data pertaining to the current location and orientation, the UAV autopilot undertakes continuous calculations to determine the deviation from the designated route. In consideration of the data presented, the autopilot is responsible for the generation of control commands for various aircraft systems, including engines and control components such as aileron servos and elevators. Algorithms (e.g., PID controllers) function in a manner that aims to minimize this deviation by providing a smooth guidance to the vehicle along the trajectory. In the context of commercial aviation, the pilot frequently utilizes autopilot functionality to navigate the aircraft, thereby establishing the desired course and altitude. In the context of a UAV, the pilot-operator's role is typically limited to the delineation of the route between designated points, with the autopilot system assuming responsibility for all piloting functions.

The subsequent task is to circumvent potential impediments and ensure safety. To achieve this aim, the development of obstacle detection and avoidance systems is imperative.

This is a matter of paramount importance for the integration of UAVs into the existing common airspace. There are several types of collision avoidance approaches:

Reactive avoidance is the utilization of ultrasonic sensors, lidar, and stereo cameras to detect sudden obstacles (e.g., tree branches or buildings). The vehicle's response to the presence of the subjects is characterized by either cessation of movement or the initiation of aerial navigation.

Proactive flight planning entails the utilization of preloaded maps and terrain data, in addition to the acquisition of information regarding other flying vehicles through systems such as Automatic Dependent Surveillance-Broadcast (ADS-B), which transmits location data. Advanced delivery UAVs are currently equipped with ADS-B receivers, which facilitate the detection of manned aircraft and enable the UAVs to make appropriate adjustments in their flight path to ensure safe operation.

To illustrate this point, let us consider the following example: a comparative analysis of the process of cargo delivery from point A to point B [4].

In the context of commercial aviation, the designated pilots are the ones who receive the approved route from the relevant air traffic control authorities. This aircraft employs ground-based radio beacons (VOR, DME), inertial systems, and primarily GNSS to navigate air routes. The aircraft's position is monitored by radars on the ground.

In the context of UAVs, the operator then uploads the flight task, which includes the coordinates of the points, to the tablet. The UAV's takeoff and subsequent autonomous navigation along the designated route are facilitated by a combination of GNSS and INS technologies. In scenarios where precise landing is paramount, such as on the rooftop of a building, computer vision technology is employed to identify the designated landing marker. In the event of GNSS failure within an urban environment, the device transitions to camera and INS navigation, thereby completing the mission or returning to base.

The evolution of navigation for UAVs has been a progression from rudimentary remote control to sophisticated, multi-level, and intelligent autonomy [1]. Classical aviation concepts, such as the utilization of INS and GNSS, continue to serve as the foundational framework. However, these concepts are now augmented by technologies that have emerged in the contemporary era of robotics and artificial intelligence. These include computer vision, lidar, and sophisticated algorithms capable of performing SLAM. The future of UAV navigation is contingent upon the creation of a unified, stable, and interconnected system in which the vehicle is capable of not only accurately determining its position but also anticipating changes in the environment, interacting with other participants in air traffic, and autonomously making decisions, thereby ensuring the safe and efficient performance of the most challenging tasks.

8.3 Determination of Position, Velocity, and Orientation

The process of determining the position, velocity, and orientation of an UAV based on data supplied by a multi-channel GNSS receiver is comprised of two fundamentally different tasks. The first of these tasks is navigation proper, which is usually solved by processing so-called code measurements (pseudorange and pseudovelocity), determined based on the receiver's navigation message. This phenomenon has been thoroughly examined and documented in the extant references [7]. The other, namely the determination of the angular position and angular velocities of the drone in a particular coordinate system, is solved based on the processing of so-called phase measurements, the receipt of which is associated with the need to calculate the phase difference of the carrier frequency on different receiver antennas. It is important to note that the solution to the secondary problem is generally inaccessible without first addressing the primary problem. Consequently, this chapter delves into the solution to both of these problems, primarily from the vantage point of analyzing the potential accuracy of determining the position, velocity, and orientation of flying vehicles in specific conditions.

It is imperative to acknowledge that the issue of ascertaining the position, velocity, and orientation of the UAV is addressed by incorporating the impact of the following uncontrolled factors [3]:

1. Errors in the determination of the ephemeris of navigation satellites GNSS have been identified as resulting from the determination of the ephemerides of the navigation satellite by means of the ground-based navigation and control complex of these satellite systems.
2. Errors in pseudorange and pseudovelocity measurements can be categorized as either systematic or random. These errors can be attributed to various factors, including ionospheric and tropospheric delays, the lag of the receiver clock, and internal noise.
3. Systematic and random errors in measuring the phase difference of the carrier frequency may be attributed to the multipath effect, the lag of the receiver clock, and its internal noise.
4. Errors in system initialization may be classified as either systematic or accidental. Accidental errors may be attributed to inaccurate knowledge of the initial motion conditions.

The following algorithms can be used to process the information received from the receiving equipment [3, 7]:

- The recurrent Bayesian algorithm is a modification of the Kalman filter.
- The method of least squares is employed for a complete sample.

As previously stated, in the most general case, the task of determining the position, velocity, and orientation of an UAV using a multi-channel GNSS receiver can be solved only if the flying vehicle is equipped with an antenna system consisting, at a minimum, of four antennas arranged symmetrically in the drone's horizontal plane of symmetry (Fig. 8.2).

The determination of the UAV velocity vector's coordinates and components is achieved through the utilization of pseudoranges and pseudovelocities, which are received at the receiver's input from the currently visible navigation satellites and the available astronomical almanac. Consequently, the prevailing approach to problem resolution involves the implementation of either the ordinary least squares method (OLS) [1, 3] or the recurrent Bayesian estimation algorithm [4, 5], contingent upon the specific characteristics of the receiver model in question.

In the following section, we will assume that the task of determining the UAV's orientation is solved on the basis of an OLS based on a complete sample, using as measurements the phase difference of the carrier frequency from each navigation satellite formed on the two main bases of the antenna system. This assumption enables the elimination of the necessity to employ a mathematical model of the UAV's angular motion when solving this problem on board the flying vehicle. Alternatively, the model can be employed in an extremely simplified form.

The formation of an on-board integrated system that utilizes GNSS technologies entails not only the determination of the drone's position and velocity but also its orientation. This process necessitates the selection of coordinate systems and the delineation of all potential connections between them, which must be stored in the memory of the on-board computer.

The analysis indicates that the minimum set required for on-board implementation should contain the following coordinate systems (Fig. 8.3) [3]:

1. Inertial Frame of 2000 (J2000.0 reference frame) [8]. The coordinate basic origin of the IF2000 is located at the center of the Earth's mass. The primary plane is defined as the equatorial plane at 0 h 00 m 00 s on January 1, 2000 (epoch J2000.0). The X_{IF} axis is oriented towards the epoch-average point of the vernal equinox. The Z_{IF} axis is oriented along the Earth's axis of spin, aligning with the International Convention of the Moon for the years 1900 to 1905. The Y_{IF} axis serves as a complementary element to the right-hand coordinate system.

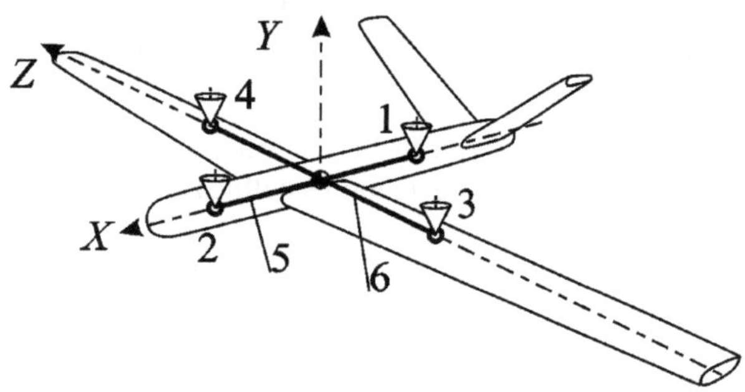

Fig. 8.2 Antennas arrangement: 1–4—Antennas; 5—First main base; 6—Second main base

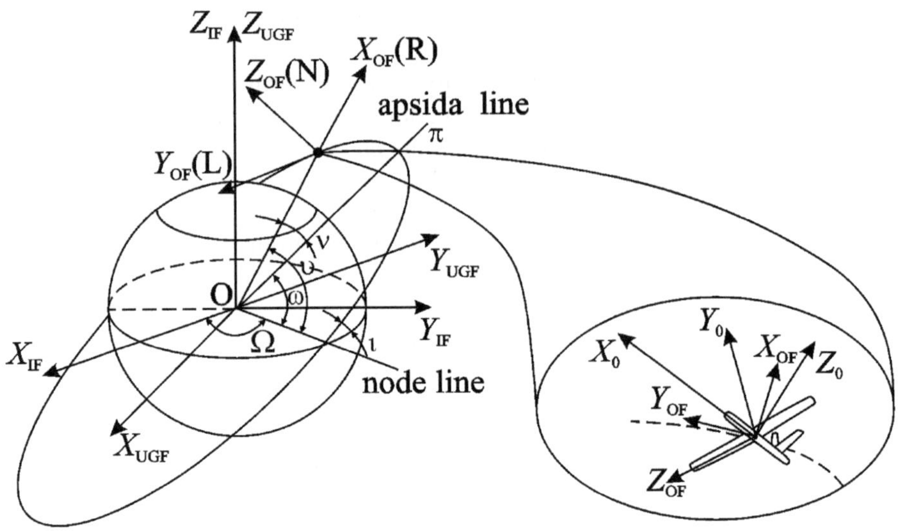

Fig. 8.3 Set of the coordinate systems

2. Earth-fixed coordinate system (Universal Greenwich Frame—UGF) [2]. The UGF coordinate basic origin is located at the center of the Earth's mass. The Z_{UGF} axis is aligned along the axis of spin of the Earth, corresponding to the Conventional International Origin (CIO) for 1900–1905. The X_{UGF} axis is aligned with the Greenwich Meridian, which serves as the CIO. The Y_{UGF} axis serves to complement the coordinate system located to the right-hand coordinate system.
3. Orbital coordinate system (Orbital Frame—OF) [3]. The reference point OF is situated at the center of the UAV mass. The X_{OF} axis is directed along the radius vector of the UAV (referred to as R bar). The Z_{OF} axis is directed along the vector of the moment of the amount of UAV movement (the so-called *N*-axis). The Y_{OF} axis functions as a complement to the coordinate system to the right-hand coordinate system (referred to as the *L*-axis).
4. Body axis coordinate system (Body frame—BF) [3]. The reference point BF is situated at the center of the drone mass. The axes BF (X_0, Y_0, Z_0) are defined as the axes of symmetry of the UAV.

The antenna coordinates are set in the body axis coordinate system. In order to ascertain the visible navigation satellites and calculate the range vectors and range derivatives, it is necessary to recalculate the coordinates and components of the velocity vector of each antenna into an inertial coordinate system. In order to accomplish this objective, the following relations are herein delineated. These relations describe the transition between the coordinate systems utilized.

The standard matrix operators of rotation around each of the axes by a certain angle *a* have the following form [3]:

$$R_x(a)=\begin{bmatrix}1 & 0 & 0\\ 0 & \cos a & \sin a\\ 0 & -\sin a & \cos a\end{bmatrix},$$
$$R_y(a)=\begin{bmatrix}\cos a & 0 & -\sin a\\ 0 & 1 & 0\\ \sin a & 0 & \cos a\end{bmatrix}, \tag{8.1}$$
$$R_z(a)=\begin{bmatrix}\cos a & \sin a & 0\\ -\sin a & \cos a & 0\\ 0 & 0 & 1\end{bmatrix}.$$

The transition matrix from the inertial to the Earth-fixed UGF coordinate system $A_{\mathrm{IF}}^{\mathrm{UGF}}$ is expressed as follows:

$$A_{\mathrm{IF}}^{\mathrm{UGF}} = R_z(\mathrm{GST}), \tag{8.2}$$

where GST is the Greenwich sidereal time.

The transition matrix from the inertial to the orbital coordinate system $A_{\mathrm{IF}}^{\mathrm{OF}}$ is expressed as follows:

$$A_{\mathrm{IF}}^{\mathrm{OF}} = R_x(\mathrm{u})\times R_y(\mathrm{i})\times R_z(\Omega), \tag{8.3}$$

where Ω, i, and u are the longitude of the ascending node, the inclination and the argument of latitude, respectively.

The transition matrix from the orbital to the body fixed frame $A_{\mathrm{OF}}^{\mathrm{BF}}$ coordinate system is expressed as follows:

$$A_{\mathrm{OF}}^{\mathrm{BF}} = R_x(\gamma)\times R_y(\psi)\times R_z(\upsilon)\times R_x(\pi)\times R_z(\pi/2), \tag{8.4}$$

where ϑ, ψ, and γ are the Eulerian angles.

Subsequently, the coordinates of the antenna in the inertial coordinate system will be expressed as follows:

$$A_{\mathrm{ant}}^{\mathrm{IF}} = \left(A_{\mathrm{OF}}^{\mathrm{BF}}, A_{\mathrm{IF}}^{\mathrm{OF}}\right)^{\mathrm{T}} R_{\mathrm{ant}}^{0} + R_{\mathrm{mc}}, \tag{8.5}$$

where R_{ant}^{0} is the coordinate of the antenna in the body fixed coordinate system, and R_{mc} is the coordinate of the mass center.

The antenna velocity vector in the inertial coordinate system is expressed as follows:

$$V_{\mathrm{ant}}^{\mathrm{IF}} = \frac{d}{dt}\left(A_{\mathrm{OF}}^{\mathrm{BF}}, A_{\mathrm{IF}}^{\mathrm{OF}}\right)^{\mathrm{T}} R_{\mathrm{ant}}^{0} + V_{\mathrm{mc}}, \tag{8.6}$$

where V_{mc} is the velocity vector of the center of mass.

The complete state vector of the antenna in the inertial coordinate system is expressed as follows:

$$X_{ant}^{IF} = \left(R_{ant}^{IF}, V_{ant}^{IF}\right)^{T}. \tag{8.7}$$

In the following section, the mathematical measurement models implemented in this method of problem-solving navigation and orientation determination will be described.

Two categories of models are recognized: true measurement models and on-board models, which are employed directly for data processing.

The true measurement model, otherwise known as the simulation model, is implemented by the following approach.

8.3.1 True Measurement Model

The following relationship is employed to measure the range:

$$\rho^{tr} = \left(\left(R_{NS}^{tr} - R_{ant}^{tr}\right), \left(R_{NS}^{tr} - R_{ant}^{tr}\right)\right)^{1/2} + \delta_{\rho}^{chr} + \delta_{\rho}^{ion} + \eta_{\rho}, \tag{8.8}$$

where ρ^{tr} is the true value of the range between the drone's antenna and the navigation satellite; R_{NS}^{tr} is the true radius vector of a navigation satellite; R_{ant}^{tr} is the true radius vector of the antenna in the inertial coordinate system; δ_{ρ}^{chr} is a systematic error that caused by the difference in time scales between the navigation satellite and the antenna receiver; δ_{ρ}^{ion} is a systematic error caused by the ionospheric signal delay; η_{ρ} is a random additive error that attributed to the presence of internal antenna receiver noise.

The derivative of the range is determined by the equation:

$$\delta\rho^{tr} = \left(\left(V_{NS}^{tr} - V_{ant}^{tr}\right), \rho_0^{tr}\right) + \delta_{\delta\rho}^{sys} + \eta_{\delta\rho}, \tag{8.9}$$

where $\delta\rho^{tr}$ is the true value of the derivative of the range between the drone's antenna and the navigation satellite; V_{NS}^{tr} is the true velocity vector of a navigation satellite; V_{ant}^{tr} is the true velocity vector of the antenna in the inertial coordinate system; ρ_0^{tr} is a unit vector that points in the direction of the true range between the UAV's antenna and the navigation satellite; $\delta_{\delta\rho}^{sys}$ is a systematic error of the measurement of the derivative of the range; $\eta_{\delta\rho}$ is a random additive error that attributed to the presence of internal antenna receiver noise.

The phase difference will be determined according to the following algorithm. As illustrated in Fig. 8.4 [3], a schematic diagram is provided of the measurement of the phase difference of the carrier frequency of a navigation satellite signal

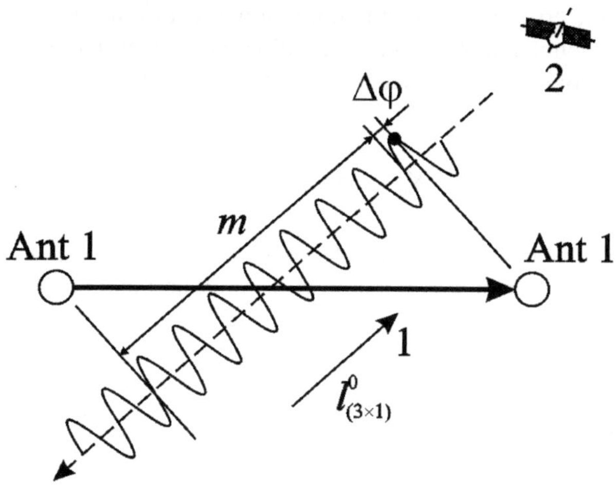

Fig. 8.4 Phase difference measurement scheme: 1—The direction to the navigation satellite; 2—The navigation satellite; Ant 1—The antenna 1; Ant 2—The antenna 2; m is the total number of carrier frequency wavelengths in the phase difference between the navigation satellite signals received by the first and second antennas.; $\Delta\varphi$ is the measurable phase difference; l_k^0 is a unit vector of the line of sight of the UAV–navigation satellite

located within the field of view of both antennas of the UAV navigation equipment. These measurements enable the precise determination of the drone's orientation in space, facilitated by the projection of the antenna base onto the viewing directions of visible navigation satellites.

The receiver does not directly measure the full values of the projections. It measures their fractions, denoted as $\Delta\varphi$. These fractions are obtained by subtracting an integer number of carrier frequency wavelengths. The full values of the projections are determined algorithmically, based on the use of information redundancy due to the large number of visible navigation satellites and/or a priori information supplied by the inertial UAV system.

Therefore, the equivalent linear value will be regarded as the measured value of the phase difference at the primary bases of the antenna system:

$$\Delta l_1^{\text{tr}} = \left(B_1^{\text{tr}}, \rho_1^{\text{tr0}}\right) + \delta_\varphi + \eta_\varphi, \quad \Delta l_2^{\text{tr}} = \left(B_2^{\text{tr}}, \rho_2^{\text{tr0}}\right) + \delta_\varphi + \eta_\varphi, \tag{8.10}$$

where $\varnothing_1^{\text{tr}}$ and $\varnothing_2^{\text{tr}}$ are the true phase differences at the first and second base stations, respectively; B_1^{tr} and B_2^{tr} are the true position vectors of the first and second stations in the inertial coordinate system, respectively; ρ_1^{tr0} and ρ_2^{tr0} are the unit vectors pointing in the direction of the true range between the first antenna and the second antenna, as well as the satellite, respectively; δ_φ is a systematic error caused by multipath effects, which is a phenomenon where the signal is reflected multiple times before reaching its destination; η_φ is a random additive error that attributed to the presence of internal antenna receiver noise.

The systematic error caused by the multipath effect of the signal received by the GNSS receiver, denoted as δ_φ, has a correlation coefficient that depends on the difference in angles of elevation above the local horizon of navigation satellites, which are measured consistently:

$$K_{\delta_{\varphi i}} = \cos\left(\varphi_i - \varphi_{i-1}\right), \tag{8.11}$$

where φ_i is the angle of climb of the UAV at which the measurement was taken, and $\varphi_{i\text{-}1}$ is the angle of climb at which the previous measurement occurred.

8.3.2 *On-Board Measurement Model*

The following relationship is employed to measure the range:

$$\rho^{\text{ref}} = \left(\left(R_{\text{NS}}^{\text{ref}} - R_{\text{ant}}^{\text{ref}}\right), \left(R_{\text{NS}}^{\text{ref}} - R_{\text{ant}}^{\text{ref}}\right)\right)^{1/2}, \tag{8.12}$$

where ρ^{tr} is the reference value of the range between the drone's antenna and the navigation satellite; $R_{\text{NS}}^{\text{ref}}$ is the reference radius vector of a navigation satellite; $R_{\text{ant}}^{\text{ref}}$ is the reference radius vector of the antenna in the inertial coordinate system.

The derivative of the range is determined by the equation:

$$\delta\rho^{\text{ref}} = \left(\left(V_{\text{NS}}^{\text{ref}} - V_{\text{ant}}^{\text{ref}}\right), \rho_0^{\text{tr}}\right), \tag{8.13}$$

where $\delta\rho^{\text{ref}}$ is the reference value of the derivative of the range between the drone's antenna and the navigation satellite; $V_{\text{NS}}^{\text{ref}}$ is the reference velocity vector of a navigation satellite; $V_{\text{ant}}^{\text{ref}}$ is the reference velocity vector of the antenna in the inertial coordinate system; ρ_0^{ref} is a unit vector that points in the direction of the reference range between the UAV's antenna and the navigation satellite.

The equivalent linear value is defined as the measured phase difference at the main bases of the antenna system, B_1 and B_2.

$$\Delta l_1^{\text{ref}} = \left(B_1^{\text{ref}}, \rho_1^{\text{ref0}}\right), \quad \Delta l_2^{\text{ref}} = \left(B_2^{\text{ref}}, \rho_2^{\text{ref0}}\right), \tag{8.14}$$

where $\varnothing_1^{\text{ref}}$ and $\varnothing_2^{\text{ref}}$ are the reference phase differences at the first and second base stations, respectively; B_1^{tr} and B_2^{tr} are the reference position vectors of the first and second stations in the inertial coordinate system, respectively; ρ_1^{tr0} and ρ_2^{tr0} are the unit vectors pointing in the direction of the reference range between the first antenna and the second antenna, as well as the satellite, respectively.

8.3.3 Data Processing

The selection of an appropriate algorithm for the processing of incoming data is contingent upon the type of receiver employed. Two algorithms are currently in vogue for this purpose: the recurrent Bayesian algorithm, which is based on a modification of the Kalman filter, and the traditional least squares method, which is based on a complete sample of measurements [5].

In the first case, it is advisable to use the so-called scalar modification of the Kalman filter [5], the peculiarity of which is that the components of the measurement vector are processed alternately. This process obviates the need for matrix inversion.

The following scheme presents a convenient representation of the process of utilizing the scalar modification of the Kalman filter:

1. The cycle formation for $j = 1, ..., N_c$, where N_c is the number of iterations used to determine the coordinates and components of the velocity vector.
2. Predicting the covariance matrix P_j

$$P_j^* = \Phi_{j,j-1} P_{j-1} \Phi_{j,j-1}^{\mathrm{T}} + K_{\Delta W}, \tag{8.15}$$

where $\Phi_{j,j-1}$ is the fundamental matrix for the system in the j-th iteration of the navigation process.
3. The prediction of the UAV's state vector, X_j^-, is the integration of the motion equations of the drone's center of mass up to time t_j.
4. The cycle formation for $i = 1, ..., N$, where N represents the number of visible navigation satellites.
5. Determination of the reference range ρ_i^{ref} for the i-th navigation satellite and the range observability matrix H_ρ^{ref}

$$H_\rho^{\mathrm{ref}} = \left(-\bar{\rho}_i^0 \middle| \bar{0} \middle| 1 \middle| 0\right), \tag{8.16}$$

where $\bar{\rho}_i^0$ is a unit vector that points in the direction of the range between to the i–th navigation satellite.
6. The calculation of the change in the state vector, denoted as ΔX:

$$\Delta X = \frac{P_j^* H_\rho^{\mathrm{ref}}}{\left(H_\rho^{\mathrm{ref\ T}} P_j^* H_\rho^{\mathrm{ref}}\right) + \sigma_\xi^2}. \tag{8.17}$$

7. The calculation of the a posteriori covariance matrix P_i:

$$P_i = \frac{P_j^* H_\rho^{\mathrm{ref}} \left(P_j^* H_\rho^{\mathrm{ref}}\right)^{\mathrm{T}}}{\left(H_\rho^{\mathrm{ref\ T}} P_j^* H_\rho^{\mathrm{ref}}\right) + \sigma_\xi^2}. \tag{8.18}$$

8. The determination of the state vector X_j:

$$X_j^+ = X_j^- + \Delta X. \tag{8.19}$$

9. The determination of the derivative of the range in the direction of $\delta\rho_i^{\text{ref}}$, the observability matrix for the velocities $H_{\delta\rho}^{\text{ref}}$

$$H_{\delta\rho}^{\text{ref}} = \left(\frac{V_i - V}{R_i - R} \left(\overline{\rho}_i^0 \times \overline{\rho}_i^{0\text{T}} - E \right) \middle| -\overline{\rho}_i^0 \middle| 0 \middle| 1 \right), \tag{8.20}$$

where R_i and V_i are the position and velocity of the i–th navigation satellite in the inertial coordinate frame, and R and V are the position and velocity of the UAV in the inertial coordinate frame.

Subsequently, analogous to the preceding instance, the augmentation of the state vector ΔX, the matrix P, and the novel state vector X_j are computed employing the aforementioned equations.

The second type of algorithm, the least squares method based on a complete sample, is a traditional method for processing trajectory measurements [4], where the array of measurements is a combination of ranges and derivatives of ranges in the direction to each visible navigation satellite. The observability matrices for each measurement are calculated using the equations given above.

To determine the orientation of the UAV using a multi-channel GNSS receiver, the least squares method for a complete sample can also be recommended [2]. It is essential to highlight the following aspects of the algorithm's implementation.

- A set of phase difference values at the two main antenna bases for each of the visible navigation satellites is utilized as an array of measurements.
- The observability matrices for each measurement are determined numerically due to the complexity of the relationships linking the measured and estimated parameters.

8.4 A Conceptual Diagram of the Navigation System

As is known [9], the search for navigation satellite signals is performed by so-called delay and carrier tracking systems. Carrier tracking systems have the capacity to function in two distinct modes: as a frequency auto-tuning circuit or as a phase auto-tuning circuit. The frequency auto-tuning circuit is utilized in signal capture, tracking, and measurement mode in navigation mode. The phase-locked frequency circuit is utilized in the information reception mode.

The receiver sequentially searches for and captures the signal of each navigation satellite. In this scenario, the C/A code, frequency, and phase of the carrier are initially captured for each signal. Subsequently, these parameters are synchronized in bits and bytes.

The search time is determined by both internal and external factors. The internal factors include the number of receiver channels, the characteristics of the element base used (mainly speed), the ability of the receiver software to effectively control it, and the external factors include the amount of a priori information available at the start of the search in the receiver, the probability of shadowing the antennas of the navigation user equipment (NUE) by the structural elements of the UAV due to its angular evolutions, and the maneuverability of the flying vehicle.

Furthermore, in the latter case, the primary factor is not the values of the overload or angular velocity of the drone, which it is capable of developing during movement, but the speed of their change. It is these parameters that ultimately determine the degree of unpredictability of movement. The measure of this unpredictability is the minimum time for predictable behavior of an UAV. This is generally recognized as the minimum time interval during which it can be assumed that the dynamic parameters of a drone's movement-overload and angular velocity-practically do not change their values or change in a known way. This interval is critical in determining the capacity of the UAV's delay and carrier tracking systems to effectively monitor the evolving situation during flight.

The most critical stage in terms of the possible loss of the navigation satellite signal is the stage of separation of an unmanned vehicle from the carrier, since it is the initial stage of movement that is characterized by high values of linear overloads and a high rate of their change. A similar pattern exists with respect to angular velocities and angular accelerations. Consequently, the maximum angular acceleration values of the *Y* and *Z* axes of the drone coordinate system can attain 1000 $°/s^2$, while the angular velocities along the same axes can reach values of 70 °/s and 60 °/s, respectively.

In the context of the aforementioned conditions, fluctuations are observed in the quartz oscillator. In the event that the overload exerts an influence in the direction of the quartz sensitivity axis, the transient process persists for a duration of approximately 1 s.

The reduction of the transient process in a quartz oscillator can be achieved by the following methods:

- The quartz installations on the UAV are to be conducted in accordance with the position of its sensitivity axis, oriented in the direction of maximum overload during the initial stage of flight.
- Additional measures are being taken to install quartz on specialized shock-absorbing supports within the receiver housing.

In such conditions, the primary factor determining the initiation of the search for a signal will be the mutual movement of the carrier and the drone, which dictates the moment when the latter emerges from the shadows.

Preliminary findings indicate that the following methodologies may be employed to mitigate the impact of an UAV's dynamics on the loss of communication with the NUE-navigation satellite:

- The utilization of spatial signal selection through the employment of specialized antennae constitutes a pivotal element in this field of study.
- The utilization of specialized signal processing, incorporating a priori knowledge of digital service data transmitted from a satellite, in conjunction with a meticulously arranged series of communication sessions with a navigation satellite, is employed to ensure redundancy in communication channels.
- The objective is to minimize the magnitude of dynamic impacts in order to reduce the range of tracking systems. This is achieved by leveraging data on the dynamics of the UAV obtained from the inertial system.

A detailed discussion of the aforementioned methods is warranted.

The conventional requirement for NUE antennas is to guarantee the uniform overlap of the upper hemi-plane, given that navigation satellites must be viewed from the horizon to the zenith.

However, the high dynamics of the drone and its rapid evolution during the initial seconds of flight suggest that installing a single antenna on the *Y* axis of the UAV coordinate system, or even two antennas located symmetrically on both sides on the *Z* axis of the UAV coordinate system, with a cone-shaped radiation pattern and an angle at the top of approximately 160°, may prove ineffective. Indeed, in the application conditions under consideration, the same satellite may fall into the field of view of one or the other antenna due to the evolution of the drone. Due to the non-identity of the frequency-phase characteristics of these antennas, the signals from them entering the receiver will have a strong disruptive effect on the delay and carrier tracking systems. This will lead to a sharp decrease in the signal-to-noise ratio, and, as a result, to the resumption of the search process from the initial state. It is evident that this may result in a discernible deceleration in the communication process.

Furthermore, it is imperative to acknowledge that the carrier's maneuvers prior to an UAV separation can result in UAVs manifesting at varying angles to the horizon at the moment of separation. It is important to note that the orientation of the antenna relative to the horizon will vary under different discharge conditions.

In such circumstances, the implementation of an annular antenna is a conceivable option, as it constitutes a constructive combination of multiple antennas exhibiting the aforementioned radiation patterns.

The accelerated signal search can be arranged by leveraging the redundancy in the number of measuring channels of the receiver, which exceeds the minimum required number of four. This value corresponds to the minimum number of navigation satellites whose signals must be obtained to form a set of measurements of the full volume. That is to say, this is the number of satellites required to obtain a set of measurements that can be used to solve the problem of navigation definitions based on a single sample.

Indeed, with regard to an unmanned vehicle, it is not necessary to achieve maximum accuracy from the first navigation determination and even from several subsequent ones with a period of 1 s. At the first stage of flight, it is only necessary to slightly reduce the area of uncertainty of knowledge of the kinematic parameters of motion caused by errors in knowledge of the initial conditions at the moment of separation of the UAV from the carrier and the influence of the error of the initial exhibition of the inertial system of the unmanned vehicle. It is evident that these factors exert a predominant influence on navigation errors during the initial stage. This narrowing of the area is necessary in order, as far as possible, to delay the moment on the trajectory when the errors of the free-form inertial system reach such a level that they cannot be compensated for by control on the remaining part of the trajectory due to the limited effectiveness of the latter.

A systematic approach to problem-solving facilitates the identification of additional resources that can contribute to the successful resolution of the issue.

In order to actualize these possibilities, during the prelaunch preparation of an UAV, information containing supplementary data on the GNSS almanacs and ephemeris of those navigation satellites located in the upper hemisphere at the time of launch is entered into its control system.

The availability of precise ephemeris data from all operational navigation satellites up to the moment of an UAV separation from the carrier facilitates the refusal of the UAV during flight.

The presence of this information facilitates the identification of four satellites from the totality of the navigation satellites located in the upper hemisphere. The probability of this occurrence is maximized at the first stage of the independent flight of the drone outside the area of possible shadowing of the antennas.

The processor of the UAV's NUE generates commands to organize the search for signals from only the selected four satellites using all measuring channels of the equipment. Consequently, a twelve-channel receiver will allocate three or four channels to the search for the signal of each satellite, whereas a two-channel receiver will dedicate six channels to this task. Consequently, it is feasible to reduce the examined area of uncertainty in coordinates by an equivalent amount.

As previously mentioned, it is imperative to input data concerning almanacs and ephemerides of navigation satellites into the on-board system.

The former are necessary for selecting the initial grouping of navigation satellites and organizing the optimal procedure for searching for their signals. In the same procedure, as well as signal tracking procedures designed to minimize the likelihood of tracking failure and, consequently, reduce the likelihood of resuming the search for a signal after it is lost, data about the current coordinates and velocities of the UAV is also involved.

In light of the aforementioned data, the receiver executes a preliminary adjustment of the signal search and tracking systems. This adjustment is performed prior to the screening process and is intended to align the signal search and tracking systems with the values of the current ranges from the UAV to the corresponding navigation satellites. It is important to note that this adjustment is performed separately for each receiver channel, and it is equally important to consider the rate of change associated with each individual range. This connection also provides more stable tracking of navigation satellite signals, reducing the likelihood of their loss. This is especially important in conditions of high drone dynamics and short flight duration.

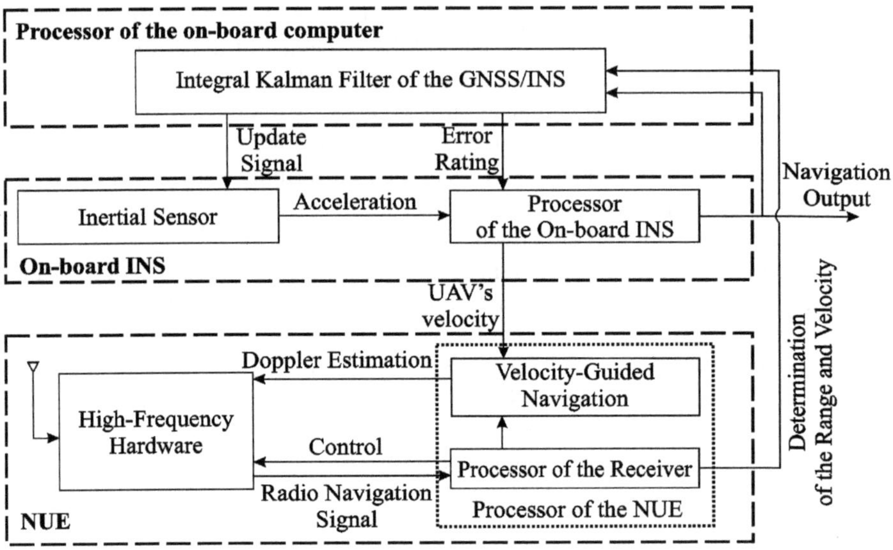

Fig. 8.5 A typical functional scheme of an integrated navigation system

A typical functional diagram of an integrated navigation system implementing the formulated approach is shown in Fig. 8.5 [3, 9].

The ephemerides of the navigation satellites engaged in the aforementioned sessions are utilized during the resolution of the navigation definition problem. Under standard conditions, ephemerides contained in the digital part of navigation satellite signals are received during sessions. However, even under optimal conditions, the temporal span for generating a comprehensive set of ephemerides is limited to tens of seconds. Conversely, the absence of a complete set of measurements for all navigation satellites involved in the session, from which pseudo-range and pseudo-speed measurements have already been obtained, renders the processing of these measurements unfeasible.

Consequently, the prelaunch input of precise ephemeris data from all operational navigation satellites enables the avoidance of such interference during flight.

In addition to the aforementioned coordinates and speed components, timestamps must be transmitted to the UAV's on-board computer on the State Primary Standard of Coordinated Universal Time UTC at the same frequency during the entire pre-launch period. Subsequently, this data is transferred to the NUE processor. In order to streamline this procedure, it is possible to restrict the transmission of digital information from the carrier in the form of the aforementioned digitization of second marks. Concurrently, it is imperative to guarantee that they are transmitted and, consequently, received into the NUE following a strictly deterministic and known time interval.

8.5 Conclusions

The contemporary UAV navigation system has undergone a substantial evolution, transitioning from a rudimentary remote control configuration to a sophisticated, integrated, and intelligent system. The system is predicated on a hybrid solution that integrates GNSS for global positioning and INS for short-term navigation in the event of signal loss, thereby providing fundamental autonomy. However, significant contemporary challenges, including the vulnerability of GNSS, the necessity of indoor work, and the demand for centimeter-level accuracy, necessitate further development. The future of UAV navigation is contingent upon the development of sustainable hybrid systems that seamlessly integrate data from cameras, lidars, and other sensors. These systems will create a reliable "sense of space," enabling the safe performance of complex tasks in any environment.

References

1. Huang H, Savkin A, Huang C. Autonomous navigation and deployment of UAVs for communication, surveillance and delivery. Canada: Wiley; 2023.
2. Panigrahi N, Tripathy S. Design principles of autonomous systems. New York, NY: CRC Press; 2025.
3. Krasilshchikov M, Sebryakov G. Control and guidance of unmanned maneuverable aircraft based on modern information technologies. Moscow: PhizMatLit; 2003.
4. L'Afflitto A, Inalhan G, Shin H. Control of autonomous aerial vehicles. Switzerland: Springer; 2024.
5. Shi H, Zhu J, Chen Z, Kuang M. Path planning, trajectory tracking and guidance for UAVs. Switzerland: MDPI; 2024.
6. Tiniakov D, Li L. Airworthiness engineering basis. Datian Aviation Beijing Beixue Publishing House; 2020.
7. Hu Y., Yuan X.,·Cao Y., Meng K. Reliable and efficient trajectory design in UAV-aided wireless networks. Singapore: Springer. 2025.
8. Capitaine N, Wallace P, Chapront J. Expressions for IAU 2000 precession quantities. J Astron Astrophys. 2003;412(2):567–86. https://doi.org/10.1051/0004-6361:20031539.
9. Hofmann-Wellenhof B, Lichtenegger H, Wasle E. GNSS—global navigation satellite systems. New York, NY: Springer; 2008.

Chapter 9
Specifics of UAVs Manufacturing

9.1 Introduction

The production of UAVs represents a distinctive convergence of the aviation industry, radio electronics, and contemporary digital technologies [1]. The evolution of this industry from artisanal enthusiast assemblies to full-scale industrial processes clearly demonstrates how the requirements for vehicles, materials, and methods of their creation have changed. The prevailing production challenges at present pertain to the optimization of designs for enhanced weight and cost reduction, ensuring the reliability of electronic components, the scalability of production processes, and the integration of intricate hardware and software systems.

The history of UAV production predates the digital era, dating back to the initial attempts to create remote-controlled targets and reconnaissance vehicles, such as the German Fi-103 (V-1) [2]. At that time, the production of drones was almost entirely analogous to the construction of classical aircraft. The fabrication of the UAVs involved the use of plywood, steel, and aluminum alloys. The assembly process was primarily manual, characterized by low volume and significant expense. The advent of compact electronics and satellite navigation systems at the close of the twentieth century signaled the onset of a genuine revolution [2]. The initial models of modern UAVs, such as the American Predator, exhibited characteristics of manned aviation in their design and production. However, it should be noted that the "stuffing" in question had already been digitalized.

The pivotal moment that signaled a paradigm shift in the industry was the meteoric rise in popularity of civilian multicopters, a development that was spearheaded by DJI [2]. Consequently, production encountered novel challenges, necessitating a transition from piece assembly to mass production, a shift that was imperative to achieve a substantial reduction in cost without compromising functionality. This necessitated a comprehensive revision of the underlying design principles, a transition to novel materials, and the automation of processes. While military UAVs

L. Li, D. Tiniakov, *Unmanned Aircraft Design*,
https://doi.org/10.1007/978-981-95-7167-3_9

frequently consist of discrete components assembled in equipment analogous to the assembly of airliners, civilian drones are developed on advanced assembly lines that resemble the production of smartphones.

The selection of materials for UAVs invariably entails a compromise among three pivotal parameters: strength, weight, and cost [3, 4]. The flight characteristics, durability, and price of the final product are contingent on this choice.

Polymer composites and plastics have become the fundamental materials in the modern engineering of both civilian and, to a lesser extent, military drones [4]. However, metals such as aluminum alloys, certain steels, and titanium alloys maintain their pertinence, particularly in the context of industrial UAVs and fixed-wing type frames [5]. The utilization of foamed materials (i.e., foams, styrofoam) in the context of budget- and amateur-type UAVs represents a novel development.

Consequently, the contemporary production of UAVs can be categorized into two primary components: the manufacturing of a frame (mechanical component) and the integration of electronic systems.

The manufacturing of UAV frame components employs conventional approaches, including traditional molding, stamping, and machining processes. New technologies such as injection molding, vacuum infusion, autoclave laying, and 3D printing are also employed in this process [6]. The technology employed is contingent upon the structural materials and the capabilities of the manufacturer.

The production and assembly of UAV electronic components is predicated on technologies that have already been extensively utilized in both the civil and military electronics industries.

Subsequent to mechanical assembly, each drone undergoes a calibration stage [1]. The precise positioning of gyroscopes, accelerometers, compasses, and cameras relative to each other is paramount for the effective functioning of the stabilization and navigation system. This process is frequently automated, but it necessitates specialized software and testing platform.

The primary challenge currently is the scaling up of production [7]. The transition from a prototype assembled by a laboratory equipment to a serial production of thousands of units per month is characterized by significant challenges. This transition necessitates the establishment of supply chains, the standardization of processes, the implementation of 100% quality control at every stage, and the automation of tasks that were previously performed manually [1].

Another issue that has been identified is the integration of UAV systems. A UAV is not merely a "frame with motors"; rather, it is a sophisticated cyber-mechanical system [8]. The assembly of a drone constitutes merely a component of the overall procedure. Ensuring the seamless interaction of hardware with firmware, the operating system, and the user application necessitates close collaboration between mechanical engineers, electronics engineers, and programmers.

Consequently, the contemporary production of UAVs constitutes a high-tech industry that effectively integrates the optimal practices of the aviation, electronics, and software industries. The progression from simple structural materials and handcrafted techniques to composite materials, additive manufacturing technologies, and fully automated electronics assembly has facilitated the transition from costly, individually assembled devices to mass-produced products. This transition has engendered novel prospects in the domains of private and industrial drone applications. The industry's future trajectory is inextricably linked to the ongoing pursuit of materials that are both exceptionally lightweight and resilient. This pursuit, in turn, is inextricably linked to the ongoing automation and miniaturization of electronic components. These developments are poised to propel the industry forward, catalyzing advancements in a multitude of sectors.

9.2 Basic Concepts of Manufacturing Process in Aviation Industry

The particulars of the aviation industry are contingent upon the characteristics of a given aircraft [3]. The aircraft must meet certain airworthiness requirements. In addition, its dimensions must align with prescribed parameters, while its intended use is another critical factor that necessitates consideration. The composition of any aircraft (including various types of drones) generally encompasses an airframe and its associated subsystems, such as the power plant, landing gear, control system, navigation system, and mission equipment, among others [3] (see Fig. 9.1).

A considerable proportion of the components and equipment that comprise an aircraft are intricate vehicles that exhibit marked distinctions from one another in terms of their structural design, functional requirements, and manufacturing technologies. The field of production technology, in particular, demands a distinct cadre of designers, manufacturing engineers, and laborers. Consequently, the design and manufacture of aircraft components, such as engines and electronics, is entrusted to specialized enterprises.

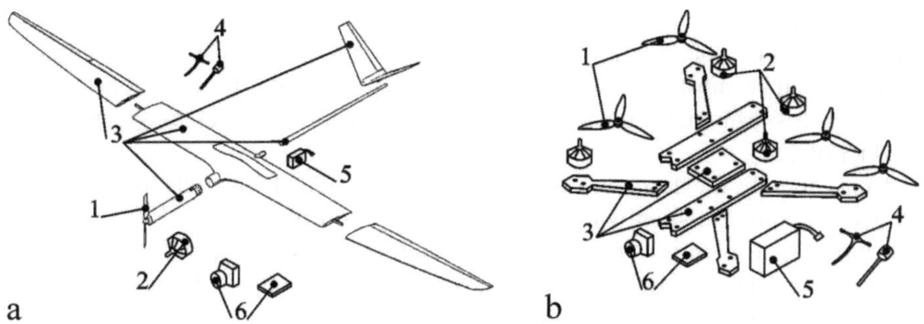

Fig. 9.1 Typical UAV Breakdown Structure (**a**) fixed-wing and (**b**) multicopter: 1—Propellers; 2—Motors; 3—Frame structural components; 4—Different types of antennas and receivers; 5—Batteries; 6—Electronic components, such as cameras, processors, controllers, etc.

9.2.1 Manufacturing Specifics of Aviation Industry [3]

9.2.1.1 Wide Assortment of Structural Components

Modern aircraft boast a plethora of systems, units, and equipment. The quantity of elementary components in a heavy UAV (excluding standard elements) can exceed 10,000 (see Fig. 9.1). The quantity of the aforementioned equipment and units is in the hundreds. The inherent property of an airplane necessitates the utilization of a multitude of manufacturing technologies and specialized tools, which complicates the planning, inspection, and monitoring of work-in-progress.

Contemporary manufacturing methodologies have emerged as a means to mitigate the volume of UAV parts. For instance, the processes of molding, casting, punching, and the utilization of composite material structures are pertinent to this field. The aforementioned entities are characterized by their ability to facilitate the formation of monolithic units and panels.

9.2.1.2 Wide Range of Structural Materials

The judicious selection of a structural material necessitates a trade-off among parameters such as strength, weight, corrosion resistance, and cost. It is imperative to acknowledge the significance of these factors, particularly in light of the recognized variability in performance requirements based on the specific application area of flying vehicles.

Historically, aluminum alloys have been utilized in various applications within the aviation industry, and more recently, they have also found application in the drone industry. Steel is a material of choice in the design of traditional aircraft due to its capacity to withstand substantial loads. For UAVs, such components can be fabricated from titanium alloys, despite their substantial cost. The contemporary era of aviation is distinguished by the substitution of numerous metal alloys with composite materials. This approach is also evident in the context of drones. However, in the context of civilian drones, where cost is a paramount consideration, composite materials can be substituted with plastics. This approach has been demonstrated to allow for a substantial reduction in vehicle cost while concomitantly increasing production rate.

Consequently, the designer's role entails the selection of materials from a diverse array of options that align with the customer's specifications (Figs. 5.14 and 5.15). Concurrently, it is imperative to maintain uniformity in the type of materials utilized, as substantial variations in materials can adversely impact the cost of production and the intricacy of assembling the drone's components.

However, a paucity of statistics exists in the public information regarding the ratio of the use of structural materials for UAVs. However, the authors of this book have visited numerous exhibitions and civilian drone manufacturing enterprises. It has been determined through the course of communication with production engineers that, in a multitude of instances, the cost of the material in question have proven to be a pivotal factor in the context of commercial UAVs.

9.2.1.3 Geometric Complexity of Frame Parts

The subsequent particulars pertain to dimensions, intricate forms (see Fig. 9.2), and minimal rigidity. The rigidity of the structure is contingent upon the imposed weight limitations. A significant proportion of airframe units is characterized by the implementation of distinctive manufacturing technologies.

The manufacturing of flying vehicles is a complex process, with numerous elements contributing to its intricacy. The dimensions of components vary significantly, ranging from millimeters (for example, the dimensions of fasteners and the thickness of the skin) to several meters (for instance, the length of skin panels and spars in the case of large UAVs). Elements with high dimensions exhibit low stiffness. This phenomenon engenders complications in the process of assembling pertinent units, thereby impeding the attainment of the requisite precision levels.

9.2.1.4 The High Labor-Intensiveness

The high labor intensity of manufacturing, adjusting, and testing operations is a significant factor in the industry.

A comprehensive overview of the tooling, machining, assembly, inspection, and testing processes in the aviation industry is provided. These processes are of paramount importance due to the stringent reliability requirements necessary for ensuring uninterrupted and efficient operation.

A considerable number of flying vehicle systems are subjected to rigorous inspection, entailing the execution of repeated tests and the implementation of adjustments that necessitate the utilization of specialized equipment.

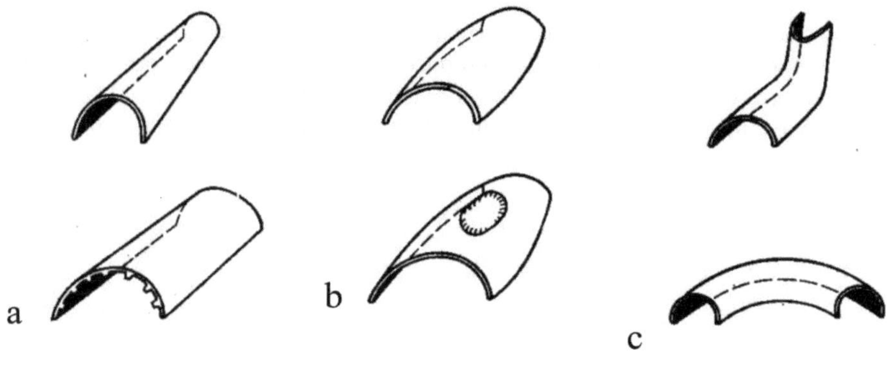

Fig. 9.2 External shapes of skins: (**a**) Single curvature; (**b**) Double curvature; (**c**) Triple curvature [3]

9.2.1.5 Multiple Joint Types

A considerable number of joints are present, including non-detachable types such as riveting, welding, and gluing (see Fig. 9.3).

The implementation of such jointing necessitates the utilization of specialized equipment, including riveting and welding machines operated in a neutral gas or vacuum envelope.

The implementation of specialized sealants and equipment is imperative for the execution of joint pressure-tightening procedures on vehicle compartments.

9.2.1.6 High Standards for Manufacturing Quality

The quality of the flying vehicle in terms of its manufacture is contingent upon the requirements and performance expectations that have been established. It is imperative to provide a rational structure for vehicles that is both manufacturable and capable of being produced with the requisite degree of accuracy. It is imperative to acknowledge that certain external surfaces necessitate meticulous accuracy. It is generally accepted that such elevated requirements invariably result in increased expenditures and heightened labor intensity.

The flying vehicle operational reliability requirements are stringent, necessitating the development of high fatigue-life joints, as well as specific manufacturing processes and equipment.

9.2.2 Vehicles Production Scales

The aviation industry is distinguished by its unique characteristics, which set it apart from other industries, such as the car industry. It is not feasible for the aviation industry to produce the same number of standard samples as the car industry, which typically produces hundreds of thousands [9]. For instance, the annual production of commercial airliners, such as the Boeing 737, can reach 600 samples [10]. In comparison, the Airbus A320 can produce up to 800 samples per year [10]. Furthermore, the most massive general aviation aircraft, the Cessna 172, which experienced its peak popularity in the 1970s, could produce up to 1900 units

Fig. 9.3 Multiple Joint Types: (**a**) Bolt joints; (**b**) Rivet joints; (**c**) Gluing joints [3]

annually [11]. The current state of affairs in the UAV industry bears a certain resemblance to that of the car industry. For instance, analysts have reported that DJI produces several million drones per year [12]. However, for the majority of companies specializing in drone production, the attainment of such figures remains unattainable. For such enterprises, the capacity to produce several thousand drones per annum is a standard benchmark [13].

In accordance with these specifications, the particulars of the production process within the aircraft industry must be taken into consideration.

Depending on the production scale, three primary categories of manufacturing are distinguished: batch, single, and pilot [3].

9.2.2.1 Batch Production

This production strategy is distinguished by a relatively limited range of vehicles manufactured or repaired for a cyclic batch of products, and it exhibits the following characteristics:

- The production process was developed in a step-by-step manner.
- A specific cyclic production operation is performed at each workplace;
- The equipment in the workshop is arranged in accordance with the sequence of operations in the production process.

The allocation of serial production into discrete categories is contingent upon the number of vehicles in a given batch. The categorization of serial production is as follows: small-scale, medium-scale, and large-scale.

Given the limited quantity of each product, the workshops have been configured to process multiple units of identical products in terms of size, configuration, and material. Consequently, these products undergo a uniform production process. This reduction in the variety of equipment in the workshop enables its utilization at a high intensity.

The execution of multiple operations on the same equipment necessitates its reconfiguration, thereby necessitating the implementation of a batch-based production methodology.

A production batch constitutes a group of products of the same type and nominal size that are sent for processing simultaneously or continuously for a certain period of time.

Subsequent to the processing of one batch of products, the equipment is set up for another operation. The duration of equipment operation between reconfigurations is determined by two factors: the number of products in the production batch and the complexity of the manufacturing.

In the context of batch production, there is a tendency to utilize universal equipment with greater frequency. In certain instances, the implementation of high-performance specialized equipment has been observed to result in a notable enhancement in overall productivity.

It is important to note that the frequent reconfiguration of universal equipment, in conjunction with the limited use of high-performance specialized equipment, has the potential to result in an increase in the cost of vehicles. However, when compared with other types of production, this approach provides the lowest cost of the vehicle.

9.2.2.2 Low-Rate Production of Vehicles

This production type is distinguished by a relatively extensive array of vehicles, which are manufactured or repaired in modest batches. The following characteristics are notable:

- The development of the production process at the macro level is of particular significance.
- The execution of distinct production operations is undertaken at each workplace, devoid of cyclical repetition.
- The workshop's equipment is methodically organized into groups according to their respective types.

A vast array of components is utilized within a single-unit production facility, with each component being fabricated in a distinct manner. Therefore, universal equipment, devices, and tools are extensively utilized in single-unit production. Highly skilled staff are employed, and the cost of production is elevated.

9.2.2.3 Prototype Aircraft Manufacturing

This production approach is typified by the fabrication of prototypes and the production of vehicle batches for research or the development of manufacturing documentation for serial production.

In terms of specifics, it is similar to a low-rate production, but differs in a more detailed approach to the development and implementation of improved production processes, taking into account the specifics of serial production.

9.3 UAV Production Specifics

The manufacturing processes employed in the production of drones are not monolithic. The design, engineering, and production of a UAV are profoundly dictated by its intended use case. A clear dichotomy exists between the manufacturing concepts for consumer-grade drones (for personal use) and commercial-grade drones (for industrial use) [7]. The latter represent a shift in requirements, complexity, and strategic importance. The production of commercial UAVs is characterized by an unwavering commitment to reliability, the integration of payload and sensors, adherence to regulatory compliance, data security, and operational durability. This commitment distinguishes them from consumer drones.

9.3.1 Consumer Drones

Consumer drones, exemplified by the prominent Mavic and Phantom series from DJI, represent a remarkable feat of contemporary electronics manufacturing. The production of these devices is a direct descendant of the smartphone industry, with optimization for scale, cost-efficiency, and rapid iteration [1, 6].

This variety of drone is characterized by its large-scale production and accessibility.

The design is intended to appeal to a wide audience while maintaining cost-effectiveness. The predominant factor is the consumer market's price sensitivity. This approach necessitates design-to-cost engineering, wherein each component is meticulously selected to align with a predetermined budgetary threshold. Manufacturing is contingent on global supply chains, with components such as plastic molding, brushed or low-end brushless motors, and standard cameras being sourced in the millions to achieve economies of scale. The assembly process is highly automated and often takes place in large-scale industrial facilities that resemble smartphone production lines.

A critical factor in achieving success with this product is the establishment of a cohesive, user-friendly platform. Consumer drones are generally characterized as closed systems. The airframe, software, camera, and controller are designed as an integrated, proprietary package. This "walled garden" approach has been shown to simplify the user experience, ensure compatibility, and allow the manufacturer to control the entire ecosystem, from flight controls to video editing applications. The manufacturing process has been streamlined to produce a single, cohesive product with minimal end-user configuration.

In the context of drones of this category, prioritizing user experience over durability emerges as a pivotal consideration. While durability is a salient feature, the primary design objectives are portability, ease of use, and camera quality. The materials utilized, including lightweight carbon-fiber composites and high-impact plastics, are selected to achieve a balance between strength and weight [3, 4]. However, these materials are not typically engineered to withstand extreme industrial environments. A crash is classified as either a user error or an accident, rather than an anticipated operational hazard [7]. The software has been meticulously engineered to facilitate intuitive flight operations, encompassing automated takeoff, landing, and obstacle avoidance functions. This sophisticated system serves to mitigate the user's exposure to the inherent intricacies of the underlying complexities.

In essence, the production of consumer drones is a matter of achieving maximum output. The objective is to optimize a singular, refined product and to replicate it in millions of samples with minimal variation, leveraging the efficiencies of high-volume electronics manufacturing.

9.3.2 Commercial Drones

For this particular drone category, the application-specific engineering and reliability of the drones is a critical factor [1, 6].

The objective of producing commercial UAVs is not to create a single, perfect product for millions of users, but rather to develop a highly adaptable and reliable platform for hundreds of critical applications [7]. The manufacturing philosophy undergoes a transition from a paradigm of serial production to a more adaptable, low-rate, high-value production model.

In this particular instance, the pivotal considerations are payload-centric design and modularity. The most significant manufacturing differentiator is the shift from a camera-centric to a payload-centric approach. A commercial drone is defined as a flying platform, the value of which is determined by the sensor or tool it can carry. These tools may include a multispectral camera for agricultural purposes, a LiDAR scanner for surveying, a thermal imager for infrastructure inspection, or a delivery mechanism for logistics. Consequently, the manufacturing process emphasizes the production of a robust airframe equipped with a standardized, frequently vibration-damped, payload mounting system and universal power and data interfaces. This necessitates a modular design approach from the ground up, entailing the development of the airframe, avionics, and payload bays as semi-independent systems that can be configured and assembled to order.

A salient factor in this matter is the considerable demand for reliability and redundancy. For industrial users, the failure of UAVs is not merely an inconvenience; it poses a substantial financial and safety hazard. The failure of a drone can result in the loss of data from a costly survey, the disruption of a supply chain, or a catastrophic accident in a sensitive environment, such as a power line or an oil rig. This necessitates a manufacturing process with rigorous quality control, akin to that employed in the aerospace industry. Commercial drones are equipped with redundant systems that are not commonly found in consumer models. These systems include dual GPS modules, redundant flight controllers, and often redundant power systems and communication links. The production line undergoes rigorous stress testing, encompassing vibration, temperature, and electromagnetic compatibility assessments, to ensure operational integrity in extreme conditions ranging from desert heat to arctic cold.

While consumer drones utilize advanced materials, commercial UAVs further refine this technology. The airframes are frequently composed of aluminum, titanium alloys, or custom-laid carbon fiber with higher tensile strength [5]. These materials are designed to withstand not just a crash, but also years of vibration, moisture, and chemical exposure. The manufacturing process for these components is more labor-intensive, involving precision CNC machining and advanced composite layup methodologies. These methodologies are slower and more expensive than the injection molding methodologies employed for consumer drones.

In the context of commercial drones, regulatory compliance and data security are identified as critical manufacturing imperatives. It is imperative that commercial drones be designed from the outset to meet the stringent aviation regulations established by federal agencies such as the FAA, EASA, or CAAC [3, 14]. These features include Remote ID, geofencing, and certified maintenance schedules. The manufacturing process now encompasses a multitude of activities that extend beyond the mere production of physical products. It involves the creation of comprehensive documentation, the compilation of certification dossiers, and the establishment of traceability for each critical component. Moreover, safeguarding data security is of the utmost importance. For enterprises leveraging drones in the context of mapping critical infrastructure or conducting surveillance operations, the secure transmission and storage of data is an imperative. This necessitates the fabrication of drones equipped with encrypted data links and secure data storage, frequently requiring specialized hardware that is integrated during the manufacturing process, rather than incorporated as an afterthought.

A relevant example is the distinction between a consumer drone inspecting a residential backyard and a commercial drone inspecting a wind turbine. The former constitutes a straightforward flight, whereas the latter necessitates a drone capable of withstanding high winds, operating reliably in proximity to substantial metal structures that interfere with GPS, acquiring high-resolution data suitable for engineering analysis, and doing so with a degree of safety that meets the standards of corporate insurers and national aviation authorities. The engineering challenge of manufacturing a drone for this purpose is comparable to that of manufacturing any other professional industrial tool.

9.4 Conclusions

The realm of UAV production is characterized by two distinct yet interconnected realities. The development of consumer drones represents a remarkable achievement in the fields of electronics miniaturization and serial-production efficiency, thereby making sophisticated flight technology accessible to the general public. Nevertheless, the drone industry's true economic impact lies in the commercial and industrial sectors. In this instance, the manufacturing specifics undergo a substantial shift. This environment is characterized by low production rates but high value, application-specific customization, and an uncompromising focus on reliability, regulatory compliance, and data security. The fabrication of a commercial UAV is less analogous to the assembly of a smartphone and more analogous to the construction of a specialized piece of aerospace machinery. As drones continue to become integrated into the modern industrial landscape, the sophistication and specificity of their manufacturing processes will continue to increase, thereby solidifying the distinction between the drone as a consumer gadget and the drone as an indispensable industrial platform.

References

1. Villacнs S, Burneo P. UAVs' efficient assembly: lean manufacturing implementation in a UAVs' assembly company. Int J Ind Eng Manag. 2020;11(4):237–52. https://doi.org/10.24867/IJIEM-2020-4-268.
2. Li L, Tiniakov D. Aeronautics—Characteristics and emerging technologies. In: Chapter: UAVs and hydrogen technologies are the future of civil aviation [Internet]. IntechOpen; 2025. https://doi.org/10.5772/intechopen.1008352.
3. Tiniakov D., Li L. Airworthiness engineering basis. Datian Aviation Beijing Beixue Publishing House. 2020.
4. Li L., Tiniakov D. Airworthiness design of composite structures, Datian Aviation Beijing Beixue Publishing House. 2021.
5. Kordos A, Kmiotek M, Żyłka W, Muszyński T. The problem of recycling of unmanned aerial vehicles. Adv Sci Technol Res J. 2024;18(7):277–88. https://doi.org/10.12913/22998624/193478.
6. Chan C. Drone manufacturing: types, process and uses. 2025. [Internet]. https://richconn.com/drone-manufacturing/. Accessed 19 Sept 2025
7. Harrison G. Unmanned aircraft systems (UAS): manufacturing trends. CRS Report for Congress. 2013. [Internet]. https://www.files.ethz.ch/isn/159662/203882.pdf. Accessed 19 Sept 2025
8. Marshall D, Barnhart R, Shappee E, Most M. Introduction to unmanned aircraft systems. New York, NY: CRC Press; 2016.
9. ACEA. Economic and market report global and EU auto industry: full year 2024. 2025. [Internet]. https://www.acea.auto/files/Economic_and_Market_Report-Full_year-2024.pdf. Accessed 19 Sept 2025
10. KPMG. The supply strain aviation leaders report 2025. 2025. [Internet]. https://assets.kpmg.com/content/dam/kpmg/ie/pdf/2025/01/ie-aviation-leaders-report-2025-2.pdf. Accessed 19 Sept 2025
11. Green P. Most successful aircraft in history: a look at the Cessna 172. 2024. [Internet]. https://simpleflying.com/cessna-172-guide-2024/. Accessed 19 Sept 2025
12. Dey M., Jambhale R. DJI statistics by usage, sales, revenue and facts. 2025. [Internet]. https://electroiq.com/stats/dji-statistics/. Accessed 19 Sept 2025
13. DJI. Agricultural drone industry insight report. 2025. [Internet]. https://www-cdn.djiits.com/cms_uploads/ckeditor/attachments/9191/508d08998272fa45f44460ab514195da.pdf. Accessed 19 Sept 2025
14. Thurman A. Navigating the skies: a guide to the world's leading aviation regulatory bodies. 2025. [Internet]. https://thurmanco.com/navigating-the-skies-a-guide-to-the-worlds-leading-aviation-regulatory-bodies/. Accessed 19 Sept 2025

Chapter 10
UAV Testing and Validation

10.1 Introduction

In the contemporary era, characterized by accelerated technological advancement, UAVs have become an indispensable component of contemporary society. These vehicles, initially developed for military purposes, have undergone a significant evolution, becoming a highly effective tool for civilian applications (see Chap. 1). These applications include amateur aerial photography, environmental monitoring, commercial cargo delivery operations, and infrastructure inspections. However, such expansion is not without its challenges, including ensuring flight safety, minimizing risks to the public and the environment, and adhering to stringent regulatory standards. This underscores the significance of civilian UAV testing, a comprehensive process entailing ground checks, flight tests, and the verification of results. This ensures the reliability and safety of the vehicles.

The advent of affordable quadrocopters from companies such as DJI and Parrot in the 2010s marked a significant turning point, leading to a substantial increase in the popularity of drones. As of the end of 2024, the number of registered UAVs in China surpassed 1.7 million, exhibiting a substantial increase from the previous year, with a recorded growth of approximately 1.87 million units by the close of 2025 [1]. In the United States, the FAA maintains records of over one million active drones [1]. By 2025, industry experts project that the number of registered civilian UAVs in the EU will exceed one million units [1]. This phenomenon is not only a catalyst for innovation; it also introduces potential risks. Incidents involving loss of control of UAVs, collisions with manned aircraft, and even cyber-attacks underscore the necessity for rigorous testing. For instance, in 2023, over 200 drone incidents were documented in Europe [2], which prompted the EASA to implement more stringent regulations.

The testing of civilian UAVs is not merely a technical routine; rather, it is a multilevel process regulated by national and international standards. In China, the

L. Li, D. Tiniakov, *Unmanned Aircraft Design*,
https://doi.org/10.1007/978-981-95-7167-3_10

primary document that establishes the regulatory framework for the UAV industry is the Interim Measures for the Management of Civil Unmanned Aerial Vehicles, which was published by the CAAC in 2018 [3]. In the United States, the FAA establishes regulations for commercial operations through Part 107 [3, 4], which necessitates pilot certification and preflight inspections. The European Union EASA has established a categorization system for aviation operations [3, 4]. The classification system is as follows: "Open" for low-risk flights, "Specific" for moderate risks, and "Certified" for high-risk flights, similar to manned aviation. These frameworks ensure that every drone, ranging from a toy to a professional inspection vehicle, undergoes a series of stability, autonomy, and safety tests.

Why are the tests so critical? Firstly, it is imperative to note that their primary function is to prevent accidents. According to data from the FAA, a preponderance of drone incidents (approximately 70%) are attributable to technical malfunctions that can be identified during the testing phase. Secondly, they foster innovation. Successful tests pave the way for new applications, such as delivering medicines to remote areas or monitoring forest fires. Thirdly, within the paradigm of globalization, the implementation of unified procedures has been shown to facilitate cross-border transactions. Nevertheless, discrepancies in regulatory frameworks—ranging from permissive policies concerning personal use in the United States to stringent certification procedures in the European Union or China—present substantial challenges for manufacturers.

For personal use, the tests center on fundamental assessments, including check-in, visual inspection, and basic flight tests within visual range. A more thorough examination of commercial drones is imperative, encompassing risk assessment (SORA—Specific Operations Risk Assessment) and certified field tests. The final step in the process is verification of results, which includes data auditing, modeling, and the consultation of external experts to confirm compliance with established standards.

The present chapter will examine the procedures for testing UAVs within both the private and commercial sectors, with a focus on the various standards that underpin these procedures. The stages of the process, from preparation to verification, will be examined through the presentation of relevant examples and a thorough analysis of the challenges encountered. The objective is twofold: to inform and to underscore the significance of rigorous testing in fostering the sustainable development of the industry. In summary, a synthesis of the results and a discussion of prospects for the future will be offered, when UAVs are integrated into everyday life as fully autonomous systems.

10.2 Aerodynamic Performance Testing

Testing the aerodynamic characteristics of UAVs is not merely a stage of development; rather, it is a fundamental process that determines efficiency, reliability [5], and safety. While the underlying physical principles remain consistent across various drone models, the testing methodologies exhibit significant variation, contingent on the intended application of the drone, whether for personal use or the execution of commercial tasks of a serious nature.

The testing of drones for compliance with the declared aerodynamic characteristics is conducted through a series of complex procedures in multiple stages. The tests are performed in accordance with international and national aviation standards, including those established by the EU—EASA, the USA—ASTM F3298, and China GB/T 38997-2020 [3, 4].

10.2.1 General Approach to Drone Aerodynamics Testing

The aerodynamic characteristics of drones are tested in several stages, regardless of their type.

10.2.1.1 Mathematical Modeling

Prior to the initiation of manufacturing and physical testing, numerical modeling is conducted. This involves the construction of a digital 3D model of the drone and the determination of key aerodynamic characteristics, including lift, drag, moments, angles of attack, vortex ring state, ground effect, etc. Specialized software packages such as ANSYS and SolidWorks Flow Simulation are employed for this purpose. The standard atmosphere parameters are calculated and utilized, as outlined, for example, in the USA as U.S. Standard Atmosphere 1976, in the EU as ISA ICAO, and in China as GB 1920–89 [6].

10.2.1.2 Wind Tunnel Testing

The physical model of the drone is subjected to a series of wind tunnel purges. A comprehensive set of metrics is employed to assess the aerodynamic properties of the system under studying [6]. This includes the measurement of lift, drag, moments at various angles of attack, and complex vortex structures. Additionally, flow rates and the positions of airframe elements are determined. Reactions to wind changes are meticulously recorded to ensure a comprehensive understanding of the system's behavior under varied conditions. A series of tests are recorded at varying speeds and configurations, and the results are then compared with calculations and simulations.

10.2.1.3 Flight Tests

In real conditions, the UAV's behavior is evaluated for compliance with the established characteristics, including stability in flight, controllability, maximum speed, flight duration, maneuver efficiency, and the effects of external influences such as wind, rain, and turbulence. To ensure precise measurements, anemometers, inertial sensors, GPS, and specialized instruments for recording flight parameters are employed.

For a variety of industrial models, tests are conducted in accordance with established standards. In the United States, for instance, these tests are performed in accordance with the FAA standards, ASTM F3201, and F3178 [3, 4]. In Europe, the EASA and the EUROCAE have established the standards known as ED-270 and ED-280 (MOPS) [3, 4], respectively. In China, the CAAC has promulgated the CCAR-92 series CTSOA [3], which serves as the national standard for China. These standards delineate the acceptable ranges of aerodynamic coefficients for commercial drones.

All stages are carried out in strict compliance with the set of airworthiness and safety requirements described in the regulations of the ICAO, ASTM, EASA, and CAAC. A comprehensive documentation of all stages is maintained, and a unified protocol is formulated, encompassing a compendium of tests conducted, conditions observed, outcomes achieved, and any discrepancies from the stipulated values. In the event that the results obtained align with the established tolerances, a certificate of conformity or an operational permit is issued.

Following the completion of rigorous testing, the drone is then affixed with a compliance badge, which serves as a formal certification that the drone has successfully met the requisite standards. This compliance badge, which serves as a unique identifier, includes specific designations such as the FCC ID (Federal Communications Commission Identifier) for the United States, CE (Conformité Européenne) for Europe, and the SRRC Mark (State Radio Regulation Committee) for China [3, 4]. These designations serve to confirm that the drone has passed all necessary tests and regulatory checks, thereby ensuring its safety, reliability, and compliance with global standards.

10.2.2 Fixed-Wing Drone Aerodynamics Testing

The aerodynamics of fixed-wing aircraft are based on well-studied principles. However, the application of these principles to drones is subject to certain nuances.

Conventionally, it is imperative to maintain equilibrium between lift and drag [6]. The primary objective is to engineer a wing that generates maximum lift with minimal drag. This directly determines the key parameter of lift-to-drag ratio. The primary factor determining the endurance and range of a flight is the aerodynamic perfection of the fixed-wing drone.

Any flying vehicles are predicated on two fundamental components: stability and controllability [6]. The drone must demonstrate the capacity to maintain a direct trajectory, withstand gusts of wind, and execute aerobatic maneuvers. In contrast to a manned aircraft, there is no pilot who can intuitively fend off a stall; therefore, stability must be an inherent feature of the design itself.

For this particular class of UAV, the takeoff and landing phases of flight present the greatest challenges [7, 8]. Fixed-wing drones employ a variety of launch mechanisms, including manual hand launch, ejection launch, and landing gear takeoff. Non-traditional methods, such as VTOL drones, which combine the features of a multicopter and an airplane, are also employed. During the takeoff and landing phases, the aerodynamics at low speeds are of paramount importance.

The effects of mass and compactness are of particular importance for any drone. UAVs frequently exhibit more compact and robust structural designs compared to manned aircraft. This results in a reduced wingspan, consequently leading to a diminished lift-to-drag ratio [6]. The designer's objective is to identify the optimal compromise between strength, weight, and aerodynamic efficiency.

In accordance with the aforementioned steps delineated in 10.2.1, the following tasks can be identified, which are unique to this type of UAV.

The subsequent tasks are addressed during the computer modeling stage (CFD):

- The wing airfoil optimization [6]. The simulation enables rapid iteration through a multitude of airfoils, facilitating the selection of the optimal airfoil for a specific range of speeds and loads. A compromise is being sought between two conflicting objectives: maximum lift, which is necessary for takeoff and landing, and minimum drag, which is necessary for cruising.
- The analysis of the entire layout [8]. The model encompasses not only the wing but also the interaction of all components, including the fuselage, tail, nacelle, and payload pylons. CFD is a useful tool for identifying problems such as interference drag.
- The calculation of longitudinal stability [9]. The center of pressure and the center of mass of the vehicle are simulated. CFD is instrumental in predicting the stability of the drone during flight, as well as the necessity of constant autopilot adjustments, which can result in energy overruns.

At the stage of bench testing, the primary instrument employed is a wind tunnel. The results of these tests serve as the primary means of validating calculations and simulations.

- The polar graph of wings [6]. The dependencies of the lift coefficient and the drag coefficient on the angle of attack are eliminated. These graphs are instrumental in determining the lift-to-drag ratio and the critical angle of attack, beyond which flow stall occurs.
- The moments of pitch, yaw, and roll [6]. It is imperative to determine the structural stability and the requisite control measures.
- The visualization of the flow is imperative [9]. Utilizing smoke or wool tufts, engineers employ visual observation to assess the aerodynamic characteristics surrounding the airframe. The presence of a turbulent wake signifies the existence of a flow stall zone.

Additionally, for drones that utilize a specific launch method, such as catapults or lunch devices, designated tests are conducted to assess their performance and reliability. To this end, specialized benches have been developed to replicate the operation of a catapult or to assess the efficacy of the landing gear extension system.

The flight test stage is mandatory for all flying devices. Flight tests for fixed-wing drones are centered on assessing their behavior in real-world atmospheric conditions.

- The aircraft performance characteristics (APC) determination. During flight, the UAV is equipped with sensors that measure the real polar characteristics, the minimum and maximum velocity, the rate of climb, and the ceiling.
- The stability tests. The drone's behavior is currently being assessed following the transition to a straight-line flight mode. A stable vehicle is more likely to return to its original mode after being exposed to turbulence.
- The stall tests. This stage is characterized by its high level of risk. At a safe altitude, the drone is deliberately maneuvered to a critical angle of attack to assess its behavior in a stall and its ability to recover. This is of particular significance in the context of autonomous systems, where the safety of programming algorithms is of paramount importance.
- The autopilot system tests. The objective of these tests is to evaluate the autopilot's ability to effectively counteract gusts of wind and ensure precision during a landing approach.

10.2.3 Multi-Copter Drone Aerodynamics Testing

In the CFD stage, engineers meticulously craft a precise three-dimensional model of the drone and position it within a virtual wind tunnel. CFD analysis facilitates the following capabilities [10]:

- The flows visualization. It is imperative to observe the airflow dynamics surrounding each component of the airframe, as well as the interaction between the vortices generated by the propellers.
- The determination of the aerodynamic coefficients. It is imperative to obtain the coefficients of lift, drag, and torque for a range of angles of attack and velocities.
- The conduction of the rapid iterations. The configuration of the airframe components, including length, angles, and diameters of its arms, along with the dimensions of the propellers and their installation angle, can be modified. This adjusting allows for immediate evaluation of the resulting effects.

At the stage of bench tests, the initial verification of the simulation results is carried out [11]. The drone, along with its constituent components, is affixed to a designated bench. The following types of test benches are possible for this type of drone:

- The test bench for the motor and propeller. The motor, in conjunction with the propeller, is mounted on a dynamometer, which meticulously measures the thrust generated, current consumption, torque, and heat generation at various speeds with the utmost precision. This facilitates the construction of thrust-power graphs and the identification of the optimal motor-propeller-battery configuration.

- The Wind tunnel. A full-size drone is placed in a wind tunnel, where a controlled airflow is created. The measurement of all six components of the aerodynamic forces and moments acting on the drone is conducted at this location. Contemporary systems employ Laser Doppler Anemometry (LDA) or Particle Image Velocimetry (PIV) to visualize and measure velocity fields surrounding the drone, enabling the observation of intricate vortex structures.

While bench tests provide the most accurate quantitative data under controlled conditions, they are unable to fully simulate all the nuances of a real flight.

At the stage of flight testing, the final verification of all the characteristics of the drone that have been obtained is performed. A drone equipped with sensors is lunched into free flight. The following metrics are measured in real time:

- The velocities and accelerations.
- The power consumption.
- The vibrations. The detection of these phenomena is facilitated by the utilization of accelerometers. Vibrations are an indicator of aerodynamic instability.

The flight test data is then compared with the results of CFD simulation and bench tests. This approach enables the validation of mathematical models and the identification of unforeseen phenomena that could not be adequately simulated in a laboratory condition.

10.2.4 Diverging Approaches: Personal Vs. Commercial Drones

The distinction between a drone intended for personal use and one designed for commercial application becomes evident during the testing phase.

10.2.4.1 Drones for Personal Use

The objective of testing is to ensure that the flight qualities meet the "acceptable" criteria and to verify the base flight stability, the certain flight duration, and the safety of the user [3, 4, 12]. The primary criteria for the selection are cost and ease of assembly.

CFD modeling is performed on simplified models, often using standard simulators. Designs are frequently derived from time-tested models or formulated through simplified calculations. The extent of the analysis is negligible.

The scope of bench tests is constrained to rudimentary motor-propeller thrust tests utilizing economical dynamometer benches. The process is frequently automated and extensively utilized throughout the production chain to identify defects.

Wind tunnel testing is an uncommon practice, with the only exceptions being major manufacturers of ready-to-fly models.

Flight tests are qualitative in nature, as opposed to quantitative. Engineers are currently assessing the drone's flight performance. The primary emphasis is on hover stability for multicopters, the responsiveness of controls, and the behavior in standard modes. Adjustments and improvements are empirically determined, including alterations to the weight distribution or the angle of the wing set. Stall tests for fixed-wing drones are typically conducted using simplified methods. It is evident that complex scenarios, such as flight in high winds or engine failure, are frequently not subjected to rigorous testing or are tested in a selective manner.

The primary focus of this particular drone model is to minimize production costs. The aerodynamics have been meticulously calibrated to ensure the drone's capacity to execute its fundamental operations. While phenomena such as vortex ring state may be mentioned in the manual, they are not specifically investigated. The lift-to-drag ratio of the vehicle is not a primary concern.

10.2.4.2 Drones for Commercial Use

The objective of testing is to ensure reliability, predictability of mission performance, and effectiveness across a range of conditions [3, 4, 12]. The primary criteria for evaluation include flight range and duration, payload capacity, and safety.

CFD modeling is executed with the utmost degree of precision, encompassing every element from the selection of the wing (propeller) airfoil to the optimization of the flow of antennas and sensors. The scope of the simulation encompasses not only standard operating modes but also critical cases, such as engine failure, icing, and flight in a turbulent atmosphere. The utilization of cluster computing is a common practice in the processing of complex models.

Bench testing is a mandatory and exhaustive stage. These experiments are conducted in specialized wind tunnels, which are often unique in design. The drone is subjected to rigorous testing that encompasses a range of angles of attack, wind speeds, temperatures, and humidity levels. Vibration tests are conducted to assess the aeroelasticity of the structure.

Flight tests are a large-scale and expensive complex of mandatory measures and procedures. UAVs are equipped with numerous sensors. The experimental tests are conducted in authentic conditions, including alpine environments, maritime settings, and urban areas (to assess the impact of "urban canyons" on aerodynamics). A comprehensive testing strategy is being implemented, encompassing all potential cases, including those of an extreme nature. Tests are conducted to ascertain the center-of-gravity position with different location of the center of mass, thereby identifying the optimal center-of-gravity position that ensures optimal stability and controllability. Real aerodynamic graphs are drawn-out, and the exact fuel/electricity consumption is determined. The testing phase is currently underway for landing systems designed for vehicles equipped with automatic landing options. Each flight is meticulously documented, and the data is subjected to statistical analysis.

The primary focus of this type of drone is to demonstrate reliability and compliance with stringent aviation standards. For a commercial drone delivering cargo or conducting aerial photography for the building of a billion-dollar facility, a 5% increase in flight time or a 10% reduction in energy consumption due to aerodynamic optimization justifies multimillion-dollar research expenditures.

The assessment of the aerodynamic characteristics of drones is a multifaceted process that integrates the capabilities of computational fluid dynamics, the precision of bench measurements, and the rigors of actual flight operations. The discrepancy in methodologies for assessing the suitability of drones for personal and commercial applications is a direct reflection of the distinct purposes, economic considerations, and reliability requirements of these two domains.

10.3 Strength Analysis

Drone strength testing is a set of measures designed to test a design to the limit and beyond its capabilities, ensuring that it does not fail in real conditions [6]. The aerodynamics of multicopters and fixed-wing vehicles differ significantly, and as a result, their approaches to structural integrity verification differ fundamentally. Additionally, the requirements for drones for personal and commercial use differ. Furthermore, regulatory frameworks in the United States (FAA), Europe (EASA), and China (CAAC) establish stringent requirements tailored to these discrepancies [3, 4].

In this section, an analysis of approaches to strength testing for fixed-wing and multirotor UAVs will be conducted, with reference to key regulatory documents. Subsequent to this introduction, a comparative analysis of the aforementioned techniques will be conducted, with a focus on the distinctions between drones utilized for personal and commercial applications.

10.3.1 Regulatory Frameworks for Drone Structural Integrity

The underlying philosophy of strength testing can be expressed as follows: the rationale behind breaking a vehicle that is designed to fly is not readily apparent.

The objective of these strength tests is not solely to ensure the integrity of the drone's structural integrity during flight. This process addresses several fundamental tasks [6]:

- The validation of the calculations. The objective of this investigation is to provide confirmation of the data obtained by the FEA on real physical samples.
- The determination of safety margin. It is imperative to ascertain the extent to which the structure exhibits enhanced strength in comparison to conventional operating loads.

- The modeling of critical situations. The aim to simulate a variety of emergency scenarios, including but not limited to a hard landing, a collision, or an engine failure.
- The evaluation of fatigue life. It is important to ascertain the capacity of the structure to withstand recurrent cyclic loads (vibrations) throughout its designated service life.

10.3.1.1 Fixed-Wing Drones

Fixed-wing drones are engineered for long-range flights and necessitate meticulous evaluation of aerodynamic and design loads [6]. The primary evaluative procedures encompass static tests (compression, tension, and bending), dynamic tests (vibration and flutter), and fatigue testing (detection of material degradation over time). These procedures ensure that the structure can withstand the limit loads—the maximum expected in operation—and the maximum loads with a margin of safety (ultimate loads), which are usually 1.5 times higher.

In the United States, the FAA establishes regulations pertaining to these aspects within 14 Code of Federal Regulations (CFR) Part 29 Subpart C, entitled "Strength Requirements" [3, 4]. This regulatory framework stipulates that strength requirements are expressed in terms of limit loads and ultimate loads, multiplied by a safety factor. For UAVs in the experimental category, the FAA Order 8130.34D, "Airworthiness Certification of Unmanned Aircraft Systems," is applicable [3, 4]. This order prescribes structural tests to demonstrate compliance. Advisory Circular AC 25.307-1, entitled "Proof of Structure," delineates a range of methodologies, encompassing Ground Vibration Testing (GVT) to assess vibration modes and full-scale static tests on prototypes [3, 4]. For instance, fixed-wing drones with a mass exceeding 25 kg necessitate a load strain test that does not exceed 50% of the maximum limit.

In Europe, the EASA has incorporated these requirements into the "Easy Access Rules for Unmanned Aircraft Systems" (EAR-UAS). Annex E to AMC1 Article 11 addresses maintenance and continued airworthiness, encompassing mechanical strength tests [3, 4]. For fixed-wing UAVs of the C2–C3 class (weighing up to 4–25 kg), the primary emphasis is on the strength of the propellers and the structural integrity of the vehicle. Shock load and environmental resistance tests are also conducted. The EASA mandates conformity inspections that align with the FAA model, while emphasizing harmonization with the ICAO. According to the EASA prototype rules of 2016, fixed-wing drones must undergo flutter clearance tests to avert resonant vibrations at elevated speeds.

The CAAC adheres to the standards outlined in CCAR-37, "China Civil Aviation Technical Standard Order," particularly Sect. 10.2.2, which stipulates the technical requirements for UAVs, encompassing flight control and structural testing [3]. For fixed-wing vehicle certification encompasses a comprehensive evaluation of the primary materials' performance, along with rigorous structural strength and crashworthiness assessments. CAAC's Minimum Performance Requirements (2024) [13] incorporate operational identification, yet the foundation remains rigorous conformity flights and ground tests for ultimate loads. The Chinese government has announced its commitment to enhancing the safety standards for passenger drones, with a particular emphasis on flame resistance and gas toxicity. This commitment is reflected in the integration of these safety measures into the country's airworthiness certification process.

10.3.1.2 Multicopters

Multicopters, characterized by their multiple rotors, encounter distinctive challenges, including high vibration from the propellers, uneven loads on the frame, and the necessity for redundancy to ensure safety [3, 4]. The tests are designed to evaluate critical components of the vehicle, including the integrity of the rotor blades, the effectiveness of vibration isolation systems, and the durability of the vehicle in the event of a crash. Static tests are conducted to evaluate the frame's resistance to torsion, while dynamic tests assess its response to harmonic vibrations. Furthermore, fatigue tests are performed to ascertain the frame's capacity to withstand cyclic loads associated with hovering mode.

The FAA adapts AC 25.571-1D, "Damage-Tolerance and Fatigue Assessment" [3, 4], for multirotor applications, necessitating widespread fatigue damage assessments. The FAA's regulations, entitled "Part 107 for small UAS (up to 25 kg)," places significant emphasis on the inspection and testing of these vehicles, particularly with regard to rotary-wing specifications. The distinctive flight dynamics of rotary-wing vehicle necessitate a higher level of propeller containment, as compared to fixed-wing vehicle.

The EASA differentiates between multirotor and fixed-wing aircraft in the EU-UAS: classes C0-C1 (up to 900 g) necessitate propeller safety tests, encompassing blade-off scenarios. The Mechanical Strength Test (2023) [3, 4] encompasses the evaluation of drone structures under conditions of vibration and impact, in accordance with Specific Assurance and Integrity Level (SAIL) criteria for ensuring the continued airworthiness of the structures. The EASA has placed a particular emphasis on the use of tethered drones. However, the primary focus of the organization's regulatory framework is on ensuring the stability of multirotor drones during free-flying operations, particularly in the context of vertical takeoff.

According to the 2021 CAAC Guidelines for UAV Inspection [3], multirotor systems are required to meet certain specifications in order to be qualified for flight control redundancy. For instance, EHang's certification encompassed structural integrity assessments for rotor systems, along with environmental factor evaluations. The Draft National Standard/Airworthiness standards for large cargo drones (2025) [14] incorporates real-time tracking mechanisms, while prioritizing fundamental principles such as crashworthiness and material performance.

10.3.2 Drone Structural Integrity Analysis

Despite the common principles that govern both types of UAVs (i.e., limit/ultimate loads, fatigue), the disparities between them are rooted in aerodynamics. Fixed-wing tests are centered on longitudinal loads (e.g., wing bending under lift), with flutter analysis for high-speed stability. According to FAA AC 25.307-1 [3, 4], the GVT is mandatory for modes below 20 Hz.

The primary focus of multirotor design is on the generation of vertical thrust and the mitigation of vibrations. The EASA has conducted tests on rotor imbalance, a condition in which the harmonious synchronization of the rotor's turning speed with the vehicle's propulsion system is paramount to averting a resonance phenomenon known as avoidance resonance.

The CAAC has established harmonization procedures for both fixed-wing and multirotor vehicles. For fixed-wing vehicles, the CAAC mandates stricter wind tunnel simulations, while for multirotor drone, it performs drop tests with a maximum altitude of 10 m [3].

10.3.2.1 Fixed-Wing Drones

In the case of these vehicles, the primary focus is on distributed aerodynamic forces and the behavior of the wing in its entirety as a single structure.

The following key types of tests can be distinguished [3]:

- The static bending tests of the wing. The wing is fixed, and a distributed load is applied to it, simulating aerodynamic lift. In certain instances, the load is deliberately destroyed in order to ascertain the true safety margin. The deflection of the wing is measured, and it is imperative that the result falls within the permissible values.
- The torsion resistance tests of the wing. A methodical approach is warranted to assess the wing's capacity to resist twisting moments, a scenario that arises, for instance, when the ailerons deflect.
- The flutter tests. This is a particularly challenging dynamic test. The wing or the entire drone model is placed in a wind tunnel, where flows are generated at critical speeds. The objective of this investigation is to empirically ascertain the threshold for the manifestation of flutter and ensure that it lies beyond the operational speed range of the vehicle.
- The tests of the fuselage and tail unit. The fuselage undergoes testing for impact load (simulated landing with vertical velocity) and torsion. The tail unit is subject to rigorous inspection to ascertain the loads exerted by the elevators and rudders.
- The tests on the landing gear release and payload ejection systems. For commercial models, these mechanisms undergo multiple cycles of testing to ensure reliable operation.

10.3.2.2 Multicopters

In the context of multicopter testing, this phenomenon can be likened to a "battle" involving vibrations and shock loads [15].

The structural design of a multicopter comprises a frame with multiple arms that are subjected to constant dynamic loads from the engines (motors) and the external environment.

In this case, the following types of tests can be distinguished:

- The static arm bending and torsion tests. The objective is to ascertain the point of fracture and ensure that the deformation remains elastic up to this point, i.e., that the arm reverts to its original shape. The arm of the frame is fixed, and a gradually increasing load is applied to its tip. This simulates engine (motor) thrust and impacts during landing.
- The vibration resistance tests. This is a critical component of the multicopter evaluation process. The drone is mounted on a vibration bench, which replicates the vibrations generated by the engines (motors) and propellers in various flight modes. Sensors (accelerometers) are utilized to record the resonant frequencies of the entire structure and its individual components. Engineers endeavor to shift the resonant frequencies of the frame beyond the operational speeds of the engines (motors) to avert imminent destruction.
- Drop tests. The drone is released from a specified altitude onto various surfaces (concrete, ground, etc.) in multiple positions—on the arms, on the body, at an angle. These cases are designed to simulate the emergency landing of a vehicle. The evaluation encompasses not only the structural integrity of the frame but also the mounting of the electronic components and the protection of the battery.
- The centrifugal load tests of propellers. The propeller is accelerated to speeds significantly exceeding the maximum operational limits on a stand to evaluate its rupture under the influence of centrifugal forces.

Fixed-wing UAV have the advantage of longer endurance, with some models capable of operating for up to 1–2 h, in contrast to the 20–30-min range of multirotor UAV. However, fixed-wing drone require fatigue cycles for 10^5 [3, 4] flights, whereas multirotor aircraft offer redundancy due to their ability to continue operating in the event of a motor failure. The financial implications of the two options are as follows: fixed-wing tests incur greater expenses due to the necessity of a full airframe, whereas multirotor tests are more economical due to their modular nature.

Consequently, fixed-wing vehicles are employed for coverage, multirotor drones are utilized for precision, and hybrid VTOL UAVs are utilized for overlapping operations.

10.3.3 *Diverging Approaches: Personal Vs. Commercial Drones*

The discrepancy in methodologies for strength testing between these two categories of drones is analogous to the disparity observed in aerodynamics tests.

10.3.3.1 Drones for Personal Use

In such cases, the objective is to ensure fundamental safety and survivability in minor collisions [7]. The primary criterion is cost.

The evaluation procedures are meticulously designed to be both selective and streamlined [3, 4]. A manufacturer may conduct testing on a single batch of several thousand drones.

Static tests are conducted on the most loaded units (for example, wing or arms) to ascertain the minimum safety margin.

The drop tests are conducted from a low altitude (1–2 m) onto a soft surface.

Vibration tests are frequently absent or reduced to a qualitative evaluation of the vibration's intensity on the bench.

The primary method of assessment is field testing by users. A significant number of "childhood diseases" of the design are identified after the vehicle is released into the market.

The main focus is directed for adherence to established safety standards is imperative. The margin of safety is minimal. To illustrate, the FAA has established a regulatory framework for drones, encompassing basic safety requirements and registration obligations. The regulatory framework is applicable to drones with a mass exceeding 250 g, though it does not encompass airworthiness considerations. According to the EASA, drones belonging to the C0 category are not subject to certification requirements and are instead subject to a self-declaration process [3, 4]. The CAAC has established minimal requirements for hobby drones, emphasizing the identification of the drone in question [4].

10.3.3.2 Drones for Commercial Use

The objective of strength tests in this instance is to ensure reliability and fault tolerance in the most severe conditions throughout the entire service life [7]. The primary criterion is mission performance and reliability.

The evaluation process is characterized by its ongoing nature and systematic regulation [3, 4]. Each new type of vehicle is meticulously evaluated through rigorous testing.

A full cycle of static tests is conducted prior to destruction in order to ascertain the actual safety margin, which can be 2–3 times higher than the maximum operating load.

Comprehensive vibration tests are conducted to ascertain the construction of vibration patterns and to eliminate all identified resonant frequencies within the operational range.

Hardest drop tests are mandatory for altitudes that exceed the standard on hard surfaces, such as asphalt and concrete.

The fatigue tests are performed. Structures are subjected to thousands of load cycles in long-term tests. These tests simulate the entire life of the vehicle.

Climate tests are mandatory too. Concurrently, the operation is subjected to rigorous testing in conditions of extreme temperatures, humidity, and icing.

The primary objective is to adhere to stringent industry standards, such as ASTM F3201 in the USA, EUROCAE ED-280 in Europe and type certificates CAAC structural tests in China [3, 4]. The safety margin is maximized.

The strength testing of drones constitutes an integral component of their life cycle, thereby serving as a conduit between virtual calculations and the harsh reality of actual usage.

Following a series of simple tests, an personal drone is typically deemed operational and its durability is often validated under real-world conditions. The development of a commercial drone, whether it is a multicopter designed for the delivery of donated organs or a fixed-wing drone intended for the monitoring of gas pipelines in the Arctic, is preceded by a rigorous and exhaustive testing process. This discrepancy in methodologies is a direct reflection of the respective costs and responsibilities: in the first case, the experience of a single user is at stake, in the second, the success of a multimillion-dollar project and, often, a human life. As drones become more deeply integrated into critical infrastructure, the strength testing standards currently applied to commercial vehicles will inevitably become the norm for the entire market, raising the bar for reliability for the entire industry.

10.4 Avionics Testing

Contemporary aviation, encompassing both manned airliners and UAVs, epitomizes the pinnacle of sophisticated electronic systems. Radio electronics encompasses a wide array of components, including antennas, transmitters, receivers, and communication systems such as Wi-Fi, Bluetooth, or specialized protocols like CNPC (Control and Non-Payload Communications) [7]. Navigation in UAVs is facilitated by a combination of GPS, inertial measurement units (IMUs), barometers, and lidar, enabling the real-time determination of the vehicle's position, velocity, and trajectory. The reliability and safety of these devices in a saturated electromagnetic environment is not merely a consequence of technological sophistication, but rather the result of rigorous and comprehensive testing.

According to the FAA, in 2024, there were more than 100 incidents involving drones that resulted in loss of communication or navigation failures [16]. In Europe, the EASA has documented comparable issues, underscoring the necessity for meticulous testing [2]. In China, the CAAC is implementing regulatory measures to enhance security in densely populated areas [3]. This initiative follows China's dominant position in the global drone production market, where it accounts for approximately 70% of the global market share.

The testing of electronic equipment, navigation systems, and avionics constitutes a critical step in the development of a prototype, distinguishing it from a certified product that is ready to enter the market and fulfill its intended functions. These processes are subject to stringent international regulation, with the aviation policies of the United States, Europe, and China exhibiting both commonalities and notable distinctions.

10.4.1 Regulatory Frameworks for Drone Avionics

The primary objective of testing remains consistent across all jurisdictions, namely, to ensure electromagnetic compatibility (EMC), functional safety, and reliability. However, the methods employed to achieve this objective vary.

In the United States, the primary document is FAR Part 25, "Standards of Airworthiness of Transport Aircraft," and for the domain of avionics, it is RTCA/DO-160, "Environmental Conditions and Test Procedures for Airborne Equipment" [3, 4]. RTCA/DO-160 is not a law, but rather a standard developed by the RTCA committee that the FAA acknowledges as a method of adhering to its regulations. The document comprises meticulous test procedures for a range of parameters, including vibration resistance, temperature fluctuations, humidity, impact resistance, and notably, noise immunity and interference emission. It is noteworthy that the text dedicates sections to the subjects of conducted and radiated susceptibility/emission. The DO-160 tests are widely recognized as the global standard for assessing the quality of avionics. The FAA type certification process is characterized by its stringent requirements. The equipment must undergo rigorous testing in accredited laboratories, and the results must be meticulously analyzed and approved by FAA engineers.

Within the EU, the EASA fulfills the role of the regulatory authority. The regulatory framework of the EU is largely harmonized with that of the US. However, it possesses its own distinct structure. CS-25 is an analog of FAR Part 25. The requirements for airworthiness and equipment are analogous to those in effect in the US. EUROCAE has developed standards that are almost identical to RTCA/DO-160. The European version of DO-160 is designated EUROCAE/ED-14 [3, 4]. The EASA has incorporated these documents into its regulatory framework. The fundamental distinction between European and US standards pertains to the extent of centralization. Despite the technical similarities, the EASA certification process is more centralized and systematic at the EU level, while in the US, the FAA retains direct and total control of the process.

In recent years, the Chinese aviation industry has demonstrated a marked commitment to the development of its own regulatory framework. CTSOs (China Technical Standard Orders) are analogous to American TSO. The CAAC has established its own technical standards for such equipment [3]. The primary requirements are stipulated in CCAR-25 and CCAR-37. The People's Republic of China administers tests in local laboratories that have been accredited by the relevant authorities and issues certifications in accordance with its own standards. Consequently, the HB 6167 "Environmental Conditions and Test Procedures for Airborne Equipment" standard [3] is a direct and complete analogue of DO-160/ED-14 in China.

10.4.2 Test Methods and Procedures

Tests of drone avionics and navigation systems are carried out in a variety of conditions, including laboratory, field, and simulation environments.

Simulations such as X-Plane or MATLAB/Simulink are employed to model scenarios of system failure, including jamming or GPS denial. These simulations test the efficacy of failover mechanisms, such as switching to visual navigation, to ensure the continued operation of the system in the face of these challenges.

Laboratory tests concentrate on the component level, measuring signal strength, noise factor, and spectral characteristics using vector analyzers. For the domain of avionics, the RTCA/DO-362 standard [3, 4] delineates minimum performance characteristics, encompassing resistance to interference in the C-band. Navigation testing encompasses the calibration of the IMU with gyroscopic tables and the verification of GNSS accuracy employing differential base stations.

Field tests are conducted to simulate real-world conditions, including flights in areas with high signal density (e.g., urban areas), as well as in inclement weather conditions such as rain or wind. The system has been developed to estimate the communication range (up to 10 km for commercial drones) and the recovery time after signal loss (less than 1 s).

The following procedures are of the utmost importance:

- The functional testing. The aim is to verify communication protocols (UDP/TCP) and Kalman filter algorithms for merging navigation data.
- The environmental tests. According to RTCA/DO-160, the parameters of concern include vibration, temperature (−40 °C to +70 °C), and EMC.
- The security audits. A vulnerability analysis is conducted in accordance with the National Institute of Standards and Technology (NIST) SP800–53 in the US. This involves the implementation of penetration testing for radio systems. In the European Union, the standard is NIS2 Directive, in China, it is GB/T 22239.

To illustrate this point, consider the electromagnetic interference (EMI) resistance test, which is a critical test for avionics, including drones. This test is also known as the Radiated Susceptibility (RS) test.

This test is designed to evaluate the drone's capacity to navigate in a real-world environment characterized by significant radio interference. Such interference can originate from mobile towers, radars, other drones, and powerful industrial installations.

The test is conducted in accordance with the RS test standard, as outlined in the RTCA/DO-160G standard, specifically in Sect. 20.

The objective of the evaluation is to assess the functionality of critical drone systems, including but not limited to the flight controller, GPS receiver, and radio control module. It is imperative that these components maintain their functionality when subjected to a strong electromagnetic field. Failure has the potential to result in a loss of control, the loss of the GPS signal, and a fall.

The drone in its entirety, or the critical components thereof (namely, the autopilot with connected GPS receivers, sensors, and radio modem), functions as an Equipment Under Test (EUT). The drone is mounted on a non-conductive rack within an anechoic chamber, which serves to absorb external and internal reflections. The motors and propellers of the model can be started at various speeds to simulate a real load. All systems are powered by onboard batteries.

The test bench is an anechoic chamber, a specialized room whose walls are covered with a material that absorbs radio waves. This process engenders a "clean" electromagnetic environment devoid of extraneous interference.

The transmission process entails the employment of a transmitting antenna, which generates a specific electromagnetic field, in conjunction with a signal generator and amplifier. The device is capable of generating test signals and amplifying them to the required power level.

The utilization of computers and systems as monitoring equipment facilitates the acquisition of real-time data from the drone. This data encompasses telemetry, gyroscopes and accelerometers, GPS signal quality, and the stability of the radio control channel.

The streamlined testing protocol is outlined as follows [3, 4]:

1. The initial phase of the process entails meticulous preparation and precise calibration. The drone is then positioned in proximity to the camera. The vehicle is activated and transitions to flight mode (for instance, by utilizing the "hold position" function of the GPS). In the absence of interference, the quality of the "reference" signal is measured and recorded (e.g., GPS signal strength, communication stability).
2. Consideration must be given to interference effects. The generator and amplifier have been activated. A specific frequency is designated for operation (e.g., 1.5 GHz is a range that is proximate to GPS and civil aviation frequencies).
3. The field power increases in a consistent and gradual manner, ultimately attaining the level stipulated by the established standard. For commercial drones classified under the EASA "specific" category, the permissible voltage range is 20 volts per meter, spanning from 400 MHz to 6 GHz.

Concurrently, the antenna scans the entire frequency range (e.g., from 80 MHz to 6 GHz) and modifies the polarization. During exposure at each respective frequency and power level, the operator and the automated system closely monitor the behavior of the drone.

The success/failure criteria are used to evaluate the test results:

- Success (Class A). The drone continues to function normally. It is important to note that none of the deviations listed below were recorded. All systems are currently functioning within normal parameters.
- Conditional Success (Class B). Temporary performance impairments manifest as, for instance, a temporary increase in positional "drift" in positioning mode, which dissipate once the interference ceases. The drone does not lose control.
- Failure (Class C). Significant deviations from normal operating parameters are observed, necessitating operator intervention (e.g., switching to manual mode) or system restart following the cessation of exposure.
- Catastrophic Failure (Class D). In the event of a system failure, the drone may become uncontrollable, resulting in engine shutdown, loss of communication, or physical damage. This outcome is considered the most hazardous.

The following cases may be regarded as failures, such as:

- "GPS Lost." The GPS receiver's ability to maintain a connection with satellites is compromised, prompting the drone to transition to ATTI (Attitude) mode. This results in the loss of the drone's capacity to maintain its position.
- "RC Failsafe." In the event of communication failure with the control panel, the drone is programmed to initiate the "Return-to-Home" (RTH) procedure. However, under certain circumstances, the drone may malfunction, resulting in a complete system freeze or even a crash.
- "Flight Controller Failure." The autopilot system is subject to a restart, which results in a complete loss of control for a duration of several seconds.
- "Fouling of Sensors." The interference has been observed to impact the gyroscopes and accelerometers of the drone, resulting in a tilt or rotation of the drone in the absence of pilot-operator input.
- "Motor failure." Electronic speed controllers (ESC) have been found to receive erroneous signals, resulting in the deactivation of one or more motors.

For instance, to certify a commercial drone that will fly near critical infrastructure or over people, successful completion of this test at, say, 20–30 volts per meter is mandatory. The onus falls upon the manufacturer to demonstrate to the regulatory authority (e.g., EASA, FAA, or CAAC) that the product under consideration will not descend to the city from a certain altitude due to interference from a proximate police radio station or airport radar.

This example clearly demonstrates the rigorous and systematic approach to testing avionics for commercial drones, which is on par with the reliability requirements of manned aviation systems.

10.4.3 Diverging Approaches: Personal Vs. Commercial Drones

The approach to testing and certification varies significantly depending on the intended purpose of the UAV.

10.4.3.1 Drones for Personal Use

A simplified approach is used for such drones. Regulatory authorities such as the FAA and the EASA categorize them as low-risk consumer products [3, 4].

The certification process involves an evaluation to ascertain adherence to specific standards. The drone must comply with the general EMC directives for radio equipment (for example, the EU Directive 2014/53/EU "Radio Equipment Directive"). This procedure is intended to ensure that the drone does not induce unacceptable interference to other equipment and is itself resistant to interference from household appliances.

The primary concern pertains to the absence of comprehensive testing. It is not necessary to complete a full cycle of DO-160 type tests. The manufacturer is responsible for conducting basic functionality and safety tests. The primary emphasis is placed on the enhancement of ease of use and the incorporation of built-in security systems, such as geofencing.

10.4.3.2 Drones for Commercial Use

The testing of such UAVs is characterized by strict regulation [3, 4]. These vehicles are treated with an approach similar to that used for manned aircraft, albeit adapted to the risk class.

Certification is categorized. The FAA, through Part 107 for small UAVs, and the EU EASA, via Implementing Regulation 2019/947, have both established category systems ("open," "specific," "certified") that are contingent upon mass, flight location, and risk.

Concurrently, extensive testing procedures are being conducted. It is imperative that the aforementioned element be provided and proven as a result of the tests.

- The reliability and safety. Drones operating above people or in populated areas (the "certified" category) must demonstrate a level of safety comparable to that of manned aircraft. This encompasses full-scale fault tolerance tests, EMC in accordance with aviation standards, and verification of duplication systems, among other components.
- Navigation and communication. It is imperative to substantiate the resilience of GPS/Galileo/BeiDou/GLONASS systems against interference and their capacity to securely terminate flights in the event of system failure. It is imperative that

the control channels, namely radio communications, be safeguarded from any form of unauthorized interference or disruption.
- Avionics. Sophisticated instrument-guided UAVs necessitate certified avionics that have undergone rigorous testing in accordance with DO-160/ED-14 standards.

10.5 Conclusions

Testing of navigation and avionics equipment is a complex, multi-level process that is under the close attention of global regulators. Notwithstanding the aspiration to standardize (DO-160, ED-14), administrative and political complexities in the USA, Europe, and China present distinctive challenges for manufacturers. The expanding market for unmanned aircraft has further complicated this scenario by introducing a gradation of requirements depending on the level of risk. If a drone for personal use is, in fact, a "flying smartphone" with minimal regulation, then a commercial drone, especially for high-risk operations, is already a "full-fledged" aircraft, which has almost the same reliability requirements as a manned aircraft. The ongoing expansion of the industry, particularly in the domain of urban aviation and autonomous flight, is expected to further refine these standards, thereby underscoring the significance of testing and certification procedures as pivotal components of technological sovereignty and aviation safety in the forthcoming era.

References

1. Statista. Estimated global UAV sales between 2016 and 2025, by region. New York, NY: Statista Inc.; 2025. [Internet]. https://www.statista.com/statistics/748678/estimated-global-uav-sales-by-region/. Accessed 19 Sept 2025
2. EASA. Annual safety review 2024. EASA: istock. 2024. [Internet]. https://www.easa.europa.eu/en/document-library/general-publications/annual-safety-review-2024. Accessed 19 Sept 2025
3. Scott BI. The law of unmanned aircraft systems. 2nd ed. Wolters Kluwer; 2022.
4. Tarr A, Tarr J-A, Thompson M, Ellis J. Drone law and policy global development, risks, regulation and insurance. New York, NY: Routledge; 2022.
5. Sun Y, Li L, Tiniakov D. Reliability engineering. Beijing: Springer/Science Press; 2024. https://doi.org/10.1007/978-981-99-5978-5.
6. Tiniakov D, Li L. Airworthiness engineering basis. Datian Aviation Beijing Beixue Publishing House; 2020.
7. Marshall D, Barnhart R, Shappee E, Most M. Introduction to unmanned aircraft systems. New York, NY: CRC Press; 2016.
8. Li L, Tiniakov D. Aeronautics—characteristics and emerging technologies. In: Chapter: UAVs and hydrogen technologies are the future of civil aviation [Internet]. IntechOpen; 2025. https://doi.org/10.5772/intechopen.1008352.
9. Panagiotou P, Kaparos P, Salpingidou C, Yakinthos K. Aerodynamic design of a MALE UAV. Aerosp Sci Technol. 2016;50:127–38. https://doi.org/10.1016/J.AST.2015.12.033.

10. Patel Y, Gaurav A, Srinivas K, Singh Y. A review on design and analysis of the propeller used in UAV. Int J Adv Product Ind Eng. 2017;605:20–3.
11. Anekar N. Quadcopter: design, construction and testing. Int J Res Eng Appl Manag. 2019; https://doi.org/10.18231/2454-9150.2018.1386.
12. Subramaniam K, Salim W. A review of experimental approaches for investigating the aerodynamic performance of drones and multicopters. J Adv Res Exp Fluid Mech Heat Transf. 2024;14(1):1–24. https://doi.org/10.37934/arefmht.14.1.124.
13. CAAC. CAAC, editor. Minimum performance requirements. China. 2024. [Internet]. http://www.caac.gov.cn/XXGK/XXGK/ZCJD/202402/t20240227_223035.html. Accessed 19 Sept 2025
14. CAAC. Airworthiness standards for large cargo drones. China: CAAC; 2025. [Internet]. http://www.caac.gov.cn/HDJL/YJZJ/201908/t20190807_197964.html. Accessed 19 Sept 2025
15. Devi V. Design of multicopter test bench. Int J Model Optim. 3:251–5. https://doi.org/10.7763/IJMO.2013.V3.276.
16. McNabb M. Protecting the future of drones: why the industry needs to step up on safety DRONELIFE exclusive guest post. 2025. [Internet]. https://dronelife.com/2025/02/11/drone-incidents-involving-aircraft-should-be-industry-wakeup-call/. Accessed 19 Sept 2025

Chapter 11
UAV Airworthiness Compliance Validation

11.1 Introduction

As has been noted in previous chapters of this book, UAVs have undergone a transformation from a purely military apparatus and hobby vehicles into powerful commercial, industrial, and public tools. These technologies are causing a paradigm shift in a variety of sectors, including logistics, agriculture, cartography, infrastructure monitoring, cinematography, public order monitoring, and more. Nevertheless, this rapid technological progress has engendered a formidable challenge for regulators worldwide. Ensuring the safety and harmonized integration of these new flying vehicles into the existing common airspace, which has historically been adapted for manned aviation, is of the utmost importance [1–3].

The primary instrument for resolving this task is the verification of airworthiness compliance [1, 2]. This process should be systematic and documented, with the objective of demonstrating that a particular UAV is designed, manufactured, and operated in accordance with established safety requirements. In contrast to the long-standing establishment of Airworthiness standards in the domain of manned aviation, the realm of UAVs is undergoing a phase of accelerated regulatory evolution.

One of the salient issues in this regard pertains to the integration of drones within various domains of public airspace. Presently, drones are frequently operated in segregated airspaces [1, 2, 4]. However, the creation of an "ideal" segregated airspace is not feasible in most cases. Consequently, drones may enter the segregated airspace (see Fig. 11.1). Consequently, it is essential to analyze the flight risk assessments in an unsegregated airspace and their rationing.

This chapter explores the contemporary status of Airworthiness standards for UAVs in pivotal aviation hubs worldwide, namely the United States, Europe, and China. The authors carefully analyze the structure of regulatory documents and methodically describe compliance confirmation procedures for various categories of unmanned aerial vehicles.

L. Li, D. Tiniakov, *Unmanned Aircraft Design*,
https://doi.org/10.1007/978-981-95-7167-3_11

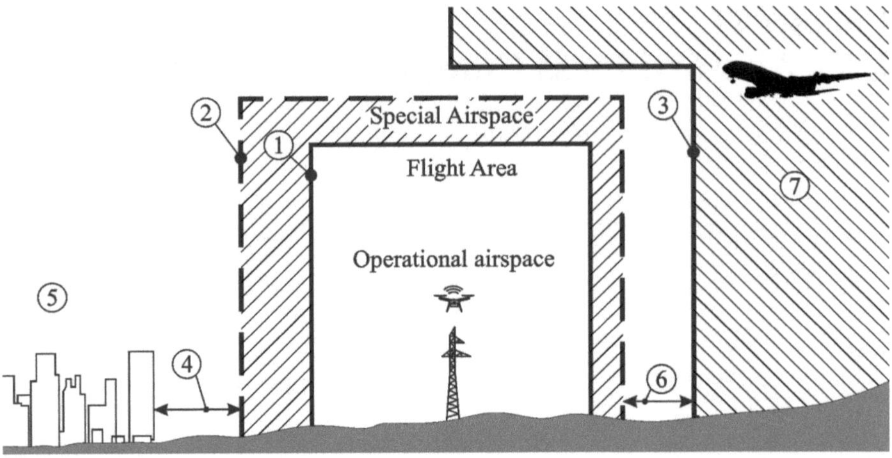

Fig. 11.1 Airspace segregation: 1—Special procedures; 2—Emergency procedures; 3—Emergency procedures for an air traffic management; 4—Ground risk buffer; 5—Adjacent zone; 6—Non-compulsory risk buffer; 7—Adjacent airspace

11.2 UAV Airworthiness Requirements

In order to comprehend the process of confirming the Airworthiness of UAVs, it is necessary to understand the general logic of building a regulatory framework for this process. The principle of a risk-based approach is employed, whereby the rigor of the requirements is directly proportional to the potential risk that the UAV poses to people, property, and other aircraft.

11.2.1 The General Structure of Airworthiness Regulations

As previously mentioned, a well-defined structure of regulatory documents and the relationships between them has been established for manned aviation throughout its entire operational life. The typical hierarchy of documents is as follows [1, 2]:

1. The Air Code represents the highest legislative act in this regard, establishing the fundamental principles and allocating powers (for instance, the U.S. Code, Title 49, Subtitle VII in the United States, the Regulation (EU) 2018/1139 in the European Union, the Civil Aviation Law of the People's Republic of China).
2. The General Rules (Regulations) delineate the requirements of the Air Code. These documents contain regulations that are legally enforceable (for example, 14 CFR Part 107 in the USA, Enforcement Regulation (EU) 2019/947 in Europe).
3. Certification Specifications (CS) are detailed technical requirements for the design, manufacture, and operation of aircraft. In the case of UAVs, they are

frequently derived from manned standards (e.g., CS-23, CS-25) or are being developed as special ones (e.g., the future CS-23 for light UAVs in the EU).
4. Advisory Circulars (AC), Acceptable Means of Compliance (AMC), and Guidance Materials (GM) are documents that provide detailed instructions on how to comply with the requirements of various rules and regulations. While adherence to these guidelines is not obligatory, it is strongly recommended that applicants adhere to them, as doing so will greatly increase the probability of demonstrating compliance.
5. Standards are defined as technical standards that are frequently developed by the industry (e.g., RTCA, EUROCAE). These standards can be referenced by regulators as acceptable methods of compliance.

This structure is adaptable and accommodates the particularities of UAVs, as evidenced by a comparison of three jurisdictions: the USA, the EU, and China.

11.2.2 The Status of Airworthiness Regulations in the United States, European Union, and China

11.2.2.1 The USA

The FAA has been a leader in establishing regulations for civilian UAVs [1, 2]. Its approach is characterized by pragmatism and a gradual tightening of rules as technology evolves.

The following key regulatory documents are available:

- The 14 CFR Part 107 "Small Unmanned Aircraft Systems" [5]. This document serves as the primary regulatory framework governing the commercial utilization of small UAVs, with a maximum weight of 25 kg. According to the provisions stipulated in Part 107, there is no prerequisite certification requirement for the drone itself, in the conventional sense of the term. Instead, the focus is on operational limitations and operator certification. In practice, the airworthiness of UAVs is validated through their registration and adherence to operational regulations, including but not limited to flying in line of sight, height restrictions, and time of day limitations.
- The 14 CFR Part 21 [6], with cross-references to Parts 23, 25, 27, 29, and so forth. This document implements a type certification process analogous to that of manned aircraft for UAVs, extending beyond the scope of Part 107. It encompasses operations such as flights beyond visual line of sight (BVLOS), the transportation of dangerous goods, and passenger transportation. The drone must comply with the airworthiness standards established for aircraft or helicopters of the appropriate class (CS-23, CS-25, FAR 27/29), with certain exceptions and adaptations.
- The flying vehicle with a special class rating. For unique UAVs that do not fall under existing regulations (e.g., multi-rotor configurations with electric

propulsion), the FAA utilizes the "Special Class" procedure according to Part 21.17(b) [6]. In such cases, the standards of airworthiness are established on a case-by-case basis, contingent upon a comprehensive risk assessment.

The FAA is currently engaged in the development of specialized standards for UAVs. The aforementioned process is initiated through the publication of notices of proposed rulemaking (NPRM) and entails close collaboration with industry entities, such as the Radio Technical Commission for Aeronautics (RTCA), which are involved in the development of standards (e.g., the DO-XXX series for UAVs) [7]. The primary objective is the integration of BVLOS operations into the national airspace system (NAS), a process that necessitates the streamlining of procedures for airworthiness compliance. This is particularly crucial in regard to system reliability, cyberattack protection, and detection and avoidance (DAA) capabilities.

11.2.2.2 The EU

The EASA is widely regarded as one of the most structured and progressive aviation regulatory bodies in the world [1, 2]. The regulation is predicated on the principle of "risk-based regulation," a concept explicitly delineated in the EU-wide rules.

The EASA has developed the following key regulatory documents:

- The Basic Regulation (EU) 2018/1139 [8]. This legislation establishes the overarching framework for aviation safety, encompassing both traditional aircraft and UAVs.
- The Execution Regulation (EU) 2019/947 [9]. This document establishes the operational framework for UAVs. It is renowned for its seminal categorization of operations, which has had a profound impact on the field of management science:

  "Open"—The risk is minimal. Prior authorization from the national authorities is not a prerequisite for this procedure. The airworthiness of a drone is confirmed by the presence of CE markings for the drone class (C0-C4) or, for Legacy drones, compliance with weight and technical restrictions.

  "Specific"—The average risk level is evaluated. Operational approval from a national authority is contingent upon a risk assessment (SORA). In this instance, the procedures for airworthiness compliance adopt a more formal posture, necessitating the presentation of evidence attesting to the safety of UAV systems.

  "Certified"—The high risks are considered. For example, the transportation of people and dangerous goods, as well as the flight over crowds of people. The regulation necessitates mandatory certification of UAVs, their systems, production, and the organization performing the operation, akin to the certification process for manned aircraft.

- The Execution Regulation (EU) 2019/945 [9]. It delineates the requirements for the design and production of UAVs for the "Open" category, introducing a class system (C0-C4) and mandatory CE marking. This development signifies a groundbreaking milestone in the realm of UAV certification, as it is the inaugural instance of type certification being granted to small drones.

The EASA is currently engaged in the development of CS for UAVs within the "Certified" category. This initiative involves the adaptation of existing standards, specifically CS-23 and CS-27, to address the unique requirements of UAVs. The agency's promotion of the concept of "faith" in technology signifies a paradigm shift in the prevailing security paradigms, whereby security is ensured not by operational limitations but rather by the inherent (built-in) reliability of the vehicle itself. The European approach endeavors to establish a thoroughly harmonized market, characterized by clear and predictable regulatory frameworks.

11.2.2.3 The PRC

China, a global leader in the production of UAVs (for instance, DJI company), is also developing a robust regulatory system. The CAAC approach aligns with global trends, yet it is distinguished by its high degree of centralization and its prioritization of state interests.

In China, there exist several key regulatory documents:

- The Chinese Civil Aviation Regulations (CCAR) [10]. This is analogous to the American CFR.
- CCAR-92 [11]. In the year 2023, a set of regulations specifically addressing the operation of unmanned aircraft took effect. The categorization system they introduce is analogous to the European system. Miniature, Ordinary, and Transport UAVs. Each class is subject to its own set of regulations pertaining to registration, certification, and operational procedures.
- Certification system. The CAAC is implementing a mandatory type certification program for UAVs utilized for commercial purposes, particularly in operations within controlled airspace [11, 12]. The assessment encompasses a meticulous evaluation of the design, control systems, navigation, communications, and data protection.

China is allocating substantial resources to the development of its own standards and technologies. The primary focus of these measures is on enhancing national security and providing support to domestic producers. The CAAC has a close coordination with the Ministry of Industry and Informatization (MIIT), reflecting an integrated approach to regulating both aviation and technological aspects [10]. The procedures for confirming airworthiness compliance in China are becoming more transparent, yet they remain under significant government control.

11.2.3 Risk Management Methodology for Integrating Drones into the Airspace (SORA)

EASA has proposed its own approach to integrating UAVs into airspace. It is called a Specific Operations Risk Assessment (SORA) [13–15]. SORA is not just a set of rules; it is a flexible, structured, and logical process that proves a planned drone operation can be performed with an acceptable level of safety. It is now the cornerstone of EASA's regulatory philosophy, which is based on risk assessment rather than strict, one-size-fits-all regulations.

11.2.3.1 The Concept of SORA

Traditional aviation is based on the principle of "what is explicitly allowed first." SORA is changing this approach for drones by introducing the principle of "proven safety" [14, 15]. Rather than attempting to create separate rules for each possible case, which is impossible due to the variety of drone operations, EASA provides the tool. SORA enables one to undertake autonomous analysis of the risks associated with a particular drone operation and to formulate proposals for their mitigation. Thus, rather than saying "you cannot fly near the city," the regulator says, "If you want to fly near the city, demonstrate with the help of SORA how you will do so safely."

11.2.3.2 The Structure of SORA

The SORA procedure can be divided into two main stages [13–15]:

1. Ground-based risk assessment;
2. Air risk assessment.

This dual nature is what makes the methodology comprehensive.

The ground-based risk assessment answers the question, "What happens if the drone falls?" This assessment focuses on the consequences for people on the ground. First, the characteristics of the drone are determined (e.g., its kinetic energy). Next, the flight zone, or the area over which the drone is flying, is analyzed (Ground Risk Buffer) (see Fig. 11.1). The key concept here is the initial Ground Risk Class (GRC). This class is determined based on the type of terrain (e.g., rural, urban, or industrial) and the expected number of people in the impact area (see Fig. 11.2). The higher the population density, the higher the initial risk level.

The air risk assessment answers the question: "What is the probability of a drone colliding with another aircraft?" It estimates the probability that a drone will end up in the same airspace as a manned aircraft, helicopter, or other drone by accident (See Fig. 11.1). The initial Air Risk Class (ARC) is determined by the type of

SAIL Level	Operation characteristics (approximately)		Lightweight UAV (<900 KJ) without DAA and parachute	Lightweight UAV (<900 KJ) with DAA and parachute	Heavy UAV (>4000 KJ) with DAA
SAIL I	Low risk rural areas, away from people		Easy Basic features are sufficient	Easy Redundant features	Easy Redundant features
SAIL II	Moderate-low risk suburban areas, small groups of people		Implemented with conditions Ground-based mitigation is required	Easy	Implemented with conditions GRC analysis
SAIL III	Moderate risk urban outskirts, crowd of people		Critically difficult There are not enough mitigations	Easy The standard scenario	Implemented with conditions Advanced procedures for GRC
SAIL IV	High risk urban areas, crowd of people		Unrealizable There are not enough mitigations	Implemented with conditions Reliable parachute Advanced procedures	Implemented with conditions Exceptional ground procedures
SAIL V	Very high risk controlled airspace, (near airports)		Unrealizable There are not enough mitigations	Implemented with conditions Highly efficient DAA coordination with ATC	Implemented with conditions Efficient DAA, Redundancy coordination with ATC
SAIL VI	Extreme risk heavy air traffic, densely populated areas		Unrealizable	Critically difficult/Unrealizable Experimental technologies direct regulatory approval	Critically difficult/Unrealizable Full integration with ATC Single cases

Fig. 11.2 Example of a UAV Matching Matrix and SAIL Levels [13–15]

airspace in which the operation takes place (controlled or uncontrolled airspace, proximity to an airfield, etc.).

11.2.3.3 Ways to Reduce Risks

The most powerful aspect of SORA is the concept of mitigation tools [13–15]. Once the initial GRCs and ARCs are identified, mitigation measures can be applied to reduce the final risk level.

For ground risk, these measures can be technical solutions, such as a parachute system or a "return home" function; operational measures, such as flights only over fenced areas or the presence of observers to clear the area under the drone; or strategic measures, such as flights at night when there are fewer people on the streets.

The most effective mitigation for air risk is the Tactical Mitigation Performance Requirement (TMPR), also known as DAA. If an operator can demonstrate that their system (e.g., a sky-watching pilot or an onboard radar) can reliably detect an approaching aircraft and prevent a collision, the air risk is significantly reduced.

The combination of all applied mitigations determines the final Specific Assurance and Integrity Level (SAIL), ranging from I (low risk) to VI (very high risk) (see Fig. 11.2) [13–15]. Depending on the SAIL level, the operation may require only a notification to the authorities, or it may require the development of a comprehensive Operation-Specific manual (Concept of Operations—ConOps) and approval from the national regulator.

EASA's SORA is not a panacea, but rather a sophisticated and thoughtful tool. It requires operators to have a deep understanding of their drones and aviation security in general. Implementing it marks the transition to a mature, responsible, and integrated ecosystem of unmanned aircraft. Though the process may seem bureaucratic, it is a necessary bridge between rapid innovation in the drone sector and traditional aviation's unshakable principle: safety first. SORA demonstrates that flexibility and safety are not mutually exclusive, paving the way for complex, commercially significant operations that will soon become an integral part of our skies.

11.3 UAV Airworthiness Compliance Methods

The procedure for confirming airworthiness compliance varies considerably depending on the risk category of the operation.

11.3.1 UAV of "Open"/Miniature/Low Risk Category

In this particular instance, the primary method employed is a declaration of conformity.

The following is a detailed description of the procedure [5–15]:

1. The drone's manufacturer is responsible for ensuring that the drone meets the established requirements, including but not limited to weight, maximum speed, marking, and remote identification function. For instance, within the EU, UAVs are required to bear the CE marking certification, which attests to their conformity with the designated class (C0–C4) [9]. This compliance is substantiated through a rigorous evaluation conducted by an authorized certification body.
2. The operator is obligated to register himself and his drone in the national registry (if it is needed). He must also undergo minimal online training and comply with operational restrictions.

Indeed, the airworthiness of a drone is confirmed by the purchase of a certified product accompanied by adherence to the stipulated operating rules.

11.3.2 UAV of "Specific"/Ordinary/Average Risk Category

The primary approach employed by this category of drones is risk assessment (SORA) and operational approval.

The following is a detailed description of the procedure [5–15]:

1. The operator conducts a risk assessment using the SORA methodology, which includes the following elements:
   (a) The initial step in the process is to ascertain the characteristics of the operation. These characteristics include the flight area and the presence of people on board.
   (b) The identification of risks both on the ground and in the air is imperative.
   (c) Determination of Integrity Levels (IL) for systems on which the mitigation of these risks depends (e.g., a collision avoidance system must have a certain level of reliability) [9].
2. The operator/manufacturer is responsible for providing evidence that the UAV and its systems meet the required requirements. These may include test results,

failure mode and effects analysis reports (FMEA) [16], and ground and flight test reports.
3. The National Authority is responsible for the review of submitted evidence and the issuance of an Operational Approval (OA).

The airworthiness compliance confirmation procedure is evolving into an engineering process that necessitates the collection of extensive evidence to substantiate the claimed level of safety.

11.3.3 UAV of "Certified"/Transport/High Risk Category

In this case, the primary method is full-scale type certification, analogous to that of manned aircraft.

The following is a detailed description of the procedure (basing on the example of EASA and FAA type certification) [5–15]:

1. The manufacturer must submit an application for type certification to the relevant regulatory authority.
2. The consideration of airworthiness standards. A set of applicable requirements has been stipulated for the "Special class" UAV (frequently based on CS-23/27/29 with adaptations) [5, 9].
3. The planning of certification. A Type Certification Plan (TCP) is currently being developed, the purpose of which is to determine the amount of evidence required.
4. The engineering and design analysis. A thorough evaluation of the project documentation, calculations, and models is conducted to ensure comprehensive understanding and analysis.
5. The tests. Extensive bench, ground, and flight tests of the systems and the entire UAV are being conducted.
6. The Production Organization Approval (POA). The feasibility of implementing serial production of identical and secure samples is currently under inspection.
7. Type Certificate Issuance. The regulatory authority issues a document confirming that the type of UAV complies with all applicable airworthiness standards.
8. The continuing of airworthiness. The manufacturer is obligated to meticulously monitor the operation and disseminate service bulletins to rectify the identified defects.

In this case, the airworthiness compliance process is a complex, multi-stage, and extremely resource-intensive process involving hundreds of specialists and lasting several years.

11.4 Cases Analysis

As previously mentioned, the airworthiness verification processes depend on the UAV category. Now, let's analyze these cases.

11.4.1 UAV of "Open" Category

The certification process for the "Open" category is structured yet flexible, combining manufacturer self-assessment and external verification.

For example, consider DJI's "Open" category drones.

As the market leader with a share of over 70% in the EU [17], DJI has become a pioneer in compliance with EASA. In August 2022, the Mavic 3 became the first drone to receive a C1 tag from TÜV Rheinland, enabling A1 flights over people weighing less than 900 g [18]. This opened the door to urban filming without permits. Similarly, the Mini 3 Pro and Mini 2 are classified as C0 (under 250 g) [19], making them ideal for A1/A3 flights without training, although an A1/A3 course is recommended.

The whole process takes 3–6 months for new models, but it takes weeks for well-known DJI models, which have a focus on minimizing downtime.

11.4.2 UAV of "Specific" Category

The "Specific" category includes drones that surpass the "Open" category but do not reach the level of passenger flights [8, 9]. This includes flights BVLOS, flights over populated areas, and flights with drones weighing more than 25 kg.

As an example, consider the certification of Zipline's "Specific" category drones.

For companies like Zipline [20], a leader in autonomous medical delivery, the "Specific" category is ideal for scaling BVLOS deliveries in urban areas. Zipline's fixed-wing drones, Zip and Sparrow, have already accumulated millions of flight hours in Africa and the United States [20]. However, they need EASA compliance to operate in the European market.

With experience in over one million deliveries of medicines and blood, Zipline is adapting its P2 platform [21] (an autonomous fixed-wing aircraft with a speed of 100 km/h and a load capacity of 1.8 kg) to the EASA "Specific" for European testing. Although Zipline's main certification is FAA Part 135 (BVLOS in the USA starting in 2023) [21], they applied to EASA in 2024 for deliveries to the EU (Germany and Switzerland) [21]. The process began with SORA: GRC II (a parachute drop minimizes damage), ARC (DAA with radars and cameras reduces risk), and SAIL III. The Dynamic Volume Reservation (DVR) for the Zip/Sparrow included tests for over 400 flight hours, redundancy (dual batteries and GPS/INS), and noise levels below 65 dB.

In 2025, Zipline received preliminary Pre-Defined Risk Assessment (PDRA) approval for Standard Scenario STS-02 (BVLOS in rural areas), enabling them to conduct operations declaratively in Switzerland in partnership with Swiss Post [21]. Full authorization for urban BVLOS is expected by July 2025, with a focus on medical delivery. This will enable Zipline to expand its operations in the EU, delivering over 1000 packages per day with no incidents.

The specific certification process is iterative and risk-based, involving the operator, manufacturer, and National Airworthiness Agency/EASA. The process takes 6–18 months.

11.4.3 UAV of "Certified" Category

The "Certified" category is designed for the riskiest scenarios, which are similar to those of traditional manned aviation. These scenarios include passenger transportation (air taxi), transporting dangerous goods, and fully autonomous flights in controlled airspace (Instrument Flight Rules (IFR), classes A–C). This category requires a full type certificate, an air operator certificate, and a certificate of airworthiness, applying the rules of Part-21 and Part-CAO [5–15].

As an example, consider the certification of Wingcopter's "Certified" category drones.

For companies like Wingcopter [22], a German manufacturer of eVTOL delivery drones, the "Certified" category paves the way for commercial BVLOS operations in populated areas, including medical logistics and humanitarian missions. The Wingcopter W198, a hybrid VTOL drone with a tilt-rotor system, can carry up to 6 kg of cargo over 100+ km at speeds of up to 150 km/h. It has already passed certification stages with the FAA (USA) [23] and Japan Civil Aviation Bureau (JCAB) [24] and is actively moving towards EASA type certification for the European market by 2025.

Founded in 2017, Wingcopter positions the W198 as the flagship of its "Certified" eVTOL series [25]. The W198 features vertical takeoff (multicopter mode) and horizontal flight (fixed-wing mode), as well as dual batteries and GPS/INS for redundancy. The primary certification is FAA Special Class, with airworthiness criteria approved in 2022 and the type certificate expected by 2025 [23]. However, Wingcopter uses the bilateral Aviation Safety Agreement (BASA) for mutual recognition with EASA. In 2023, the company began the process with a preliminary application for type 1 (cargo IFR), focusing on medical deliveries of up to 6 kg over 100 km in BVLOS conditions over cities [26].

There are key stages for EASA. The ConOps has identified a maximum altitude of 120 m and a speed of 130 km per hour for operations over people without ice certification [8, 9]. In 2024, Wingcopter underwent an initial design review, integrating DAA (radars and cameras) and low-noise propellers (less than 65 dB), and conducting tests in Germany (DLR Braunschweig) [25]. A hazard analysis revealed a Single Point of Failure (SPF) of no more than 10^{-5} per flight, which is equivalent to CS-UAS [5–15]. An issue of the type certificate for EU cargo operations is expected by November 2025 [26], similar to the FAA stage in 2022. Partnerships include ITOCHU in Japan for Level 4 BVLOS [24]. Challenges include adapting the tilt-rotor to Environmental Sustainability Level (ESL) (EASA requires 1500 h of testing), mitigating cyber threats, and ensuring U-space compliance [8, 9]. Success at the FAA and JCAB will accelerate progress at EASA through data sharing and enable expansion in the EU (Germany and Switzerland) to 1000+ shipments per day with zero incidents in trials.

11.4.4 Examples of SORA in Practice

Let's consider three different scenarios [13–15].

Case 1. Inspection of Power Lines in Rural Areas (Low Risk)
For the purposes of this operation, the following conditions are assumed: an energy company operator uses a 2-kg drone to visually inspect power lines in sparsely populated rural areas (see Fig. 11.1).

How will SORA be applied in this case?

1. Ground risk. The initial GRC will be low since the area is rural and access by people is limited along the power line where the flight will take place. Mitigation can be provided by the following procedure: "Stop flying when any person appears in the crash area."
2. Air risk. The airspace is most likely Class G (uncontrolled) and is located away from airfields. The initial ARC will be low. One tactical mitigation is visual observation of the sky by the pilot and observer.

In conclusion, after applying the mitigations, SAIL is likely to be at level I or II (see Fig. 11.2). This operation will likely be classified as "open" or require minimal approval.

Case 2. Delivery of Medicines in a Busy Urban Area (Medium/High Risk)
The expected operation plan is as follows: urgent blood tests will be delivered between two hospitals in the city using a 15-kg specialized cargo drone.

How will SORA be applied in this case?

1. Ground risk. Due to the urban environment, the initial GRC will be very high. Mitigation measures should include a certified parachute, a carefully planned route (over roads and parks, for example), flying at an altitude that reduces the risk to people, and possibly a temporary closure of the takeoff and landing zone.
2. Air risk. The airspace in the city can be challenging. The initial ARC is high. A mandatory mitigation measure should be an advanced DAA system that uses

DAGC (Direct Air-Ground Communication) to coordinate with the airfield control service.

In conclusion, the SAIL of such an operation will be high (IV–V) (see Fig. 11.2). This will require developing a detailed ConOps, conducting flight tests to confirm the effectiveness of the mitigation measures, and obtaining special permission from the national aviation regulator.

Case 3. Aerial Videography for Film Production in a Controlled Environment (Very High Risk)
Operation content: the film company is planning to conduct aerial videography near a major international airport using a drone in order to capture stunning footage of airplanes taking off.

How will SORA be applied in this case?

1. Ground risk. The initial GRS is high, as the operation may affect territories adjacent to the airport.
2. Air risk. The initial ARC will be extremely high because the flight will take place in close proximity to the takeoff and landing zones of manned airliners.

Standard mitigation measures will not suffice. The following will be required: coordination with the airport's Air Traffic Control (ATC); allocation of a strictly defined corridor and time for the flight; use of ground-based radars to track the drone; availability of a dedicated frequency for communication with the dispatcher; and possibly an escort helicopter for visual control.

In conclusion, the SAIL of such an operation will be the highest VI (see Fig. 11.2). Such an operation is exceptional. Approval will depend on direct communication between the operator, the regulator, and the ATC and will involve unprecedented safety measures.

11.5 Conclusions

The confirmation of a UAV's airworthiness is in a state of dynamic evolution. Initially based on operational constraints (FAA Part 107), the world is transitioning to complex, risk-based systems that incorporate pre-certification of equipment (EASA model) and in-depth risk assessment for specific operations (SORA).

Despite their differences in approach, the United States, Europe, and China show a common trend: the higher the potential risk from UAV operations, the more the procedures for confirming airworthiness compliance resemble classical aviation standards. This is inevitable since integrating drones into common airspace requires ensuring they behave predictably and safely, at least as well as manned aircraft.

The global task in the near future is to harmonize these national systems. Disparate regulations are becoming a barrier to international trade and transactions. The work of international organizations, such as ICAO, to create a global framework for UAVs is critically important. Success in this area will determine our ability to fully realize the enormous potential of unmanned technologies while developing a safe, efficient, and unified global airspace for all users.

References

1. De Florio F. Introduction to airworthiness. Butterworth-Heinemann; 2016.
2. Scott BI. The law of unmanned aircraft systems. 2nd ed. Wolters Kluwer; 2022.
3. Tiniakov D, Li L. Airworthiness engineering basis. Datian Aviation Beijing Beixue Publishing House; 2020.
4. Marshall D, Barnhart R, Shappee E, Most M. Introduction to unmanned aircraft systems. New York: CRC Press; 2016.
5. FAA. 14 CFR part 107—small unmanned aircraft systems. Washington, DC: FAA; 2025. [Internet]. https://www.ecfr.gov/current/title-14/part-107. Accessed 19 Sept 2025
6. FAA. 14 CFR part 21—certification procedures for products and articles. Washington, DC: FAA; 2025. [Internet]. https://www.ecfr.gov/current/title-14/part-21. Accessed 19 Sept 2025
7. FAA. Title 14. Aeronautics and space. Washington, DC: FAA; 2025. [Internet]. https://www.ecfr.gov/current/title-14. Accessed 19 Sept 2025
8. EASA. Regulation (EU) 2018/1139 of the European Parliament and of the Council. Official Journal of the European Union. 2018. [Internet]. https://www.easa.europa.eu/en/document-library/regulations/regulation-eu-20181139-european-parliament-and-council. Accessed 19 Sept 2025
9. EASA. Easy Access Rules for Unmanned Aircraft Systems (Regulations (EU) 2019/947 and 2019/945). Official Journal of the European Union. 2024. [Internet]. https://www.easa.europa.eu/en/document-library/easy-access-rules/easy-access-rules-unmanned-aircraft-systems-regulations-eu. Accessed 19 Sept 2025
10. CAAC. Chinese Civil Aviation Regulations. CAAC. 2023. [Internet]. https://www.caac.gov.cn/English/Regulations/CCAR/202307/t20230731_220835.html. Accessed 19 Sept 2025
11. CAAC. Airworthiness standards for large cargo drones. China: CAAC; 2025. [Internet]. http://www.caac.gov.cn/HDJL/YJZJ/201908/t20190807_197964.html. Accessed 19 Sept 2025
12. ICAO. Asia-Pacific Reference Materials for Regulators to Facilitate Advanced Air Mobility Operations. ICAO. A42-WP/172 Information paper. 2025
13. EASA. Specific Operations Risk Assessment (SORA). EASA. 2025. [Internet]. https://www.easa.europa.eu/en/domains/drones-air-mobility/operating-drone/specific-category-civil-drones/specific-operations-risk-assessment-sora. Accessed 19 Sept 2025
14. JARUS. JARUS guidelines on Specific Operations Risk Assessment (SORA). JARUS. 2024. [Internet]. http://jarus-rpas.org. Accessed 19 Sept 2025
15. Schnüriger P, Schreiber J, Widmer K, Lenhart PM. SORA tool—a specific operation risk assessment tool for civilian drone operations. J Drone Syst Appl. 2025;13 https://doi.org/10.1139/dsa-2024-0060.
16. Sun Y, Li L, Tiniakov D. Reliability engineering. Beijing: Springer/Science Press; 2024. https://doi.org/10.1007/978-981-99-5978-5.
17. Frąckiewicz M. Global Drone Industry: 2025 Market Report. Warsaw: TechStock2. [Internet]. https://ts2.tech/en/global-drone-industry-2025-market-report/. Accessed 19 Sept 2025
18. DJI. DJI granted world's first C1 drone certificate. DJI; 2022. [Internet]. https://www.dji.com/newsroom/news/dji-granted-worlds-first-c1-certificate-for-mavic-3-series. Accessed 19 Sept 2025
19. DJI. The DJI Mini Series and the 120m Altitude Limit in the EU. DJI; 2023. [Internet]. https://viewpoints.dji.com/blog/dji-mini-series-altitude-limit-eu. Accessed 1 Aug 2025
20. Zipline. Building the first logistics system that serves all people equally. Zipline. 2025. [Internet]. https://www.zipline.com/about. Accessed 19 Sept 2025

21. Francis S. Drone delivery: Navigating the path from high-flying hype to last-mile reality. London: Monsoon Media, Robotics and Automation News; 2025. [Internet]. https://roboticsandautomationnews.com/2025/06/12/drone-delivery-navigating-the-path-from-high-flying-hype-to-last-mile-reality/91765/. Accessed 19 Sept 2025
22. Wingcopter. Delivering the future. Wingcopter. 2025. [Internet]. https://wingcopter.com/. Accessed 19 Sept 2025
23. Crumley B. Wingcoper 198 delivery drone passes FAA special class airworthiness milestone. DroneDJ. 2022. [Internet]. https://dronedj.com/2022/05/11/wingcoper-198-delivery-drone-passes-faa-special-class-airworthiness-milestone/. Accessed 19 Sept 2025
24. ITOCHU. First type certification application for 1st class fixed-wing drones in Japan. London: UAS Vision; 2024. [Internet]. https://www.uasvision.com/2024/04/04/first-type-certification-application-for-1st-class-fixed-wing-drones-in-japan/. Accessed 19 Sept 2025
25. Wingcopter. Wingcopter 198—Wingcopter. Wingcopter. 2025. [Internet]. https://wingcopter.com/wingcopter-198. Accessed 19 Sept 2025
26. Wingcopter. Wingcopter and Lufthansa Technik collaborate on UAV type certification. Wingcopter. 2023. [Internet]. https://wingcopter.com/2023/06/20/wingcopter-and-lufthansa-technik-join-forces-to-achieve-european-type-certification-for-wingcopter-cargo-drone/. Accessed 1 Aug 2025

MIX
Papier aus verantwortungsvollen Quellen
Paper from responsible sources
FSC® C105338

If you have any concerns about our products,
you can contact us on
ProductSafety@springernature.com

In case Publisher is established outside the EU,
the EU authorized representative is:
Springer Nature Customer Service Center GmbH
Europaplatz 3, 69115 Heidelberg, Germany

Printed by Libri Plureos GmbH
in Hamburg, Germany